Canadian
Professional
Engineering
Practice
and Ethics

Canadian Professional Engineering Practice and Ethics

SECOND EDITION

GORDON C. ANDREWS
University of Waterloo

JOHN D. KEMPER
Professor Emeritus
University of California, Davis

NELSON

™

THOMSON LEARNING

Australia • Canada • Mexico • Singapore • Spain • United Kingdom • United States

For more information contact
Nelson Thomson Learning,
1120 Birchmount Road,
Scarborough, Ontario, M1K 5G4.
Or you can visit our Internet site at http://www.nelson.com

Canadian Cataloguing in Publication Data

Andrews, G.C. (Gordon Clifford), 1937–
    Canadian professional engineering practice and ethics

2nd ed.
Includes bibliographical references and index.
ISBN 0-7747-3501-5

1. Engineering — Vocational guidance — Canada. 2. Engineering ethics.
I. Kemper, John Dustin. II. Title.

TA157.A68 1999   620'.00971   C97-932405-X

Senior Acquisitions Editor: Ken Nauss
Developmental Editor: Su Mei Ku
Supervising Editor: Semareh Al-Hillal
Senior Production Co-ordinator: Sue-Ann Becker
Assistant Production Co-ordinator: Shalini Babbar

Copy Editor: Claudia Kutchukian
Cover and Interior Design: Sonya V. Thursby/Opus House Incorporated
Typesetting and Assembly: Carolyn Hutchings

Cover Art: Kell Brothers, lithographers, *Putting Up Side Plates and Top of Tube*, coloured lithograph (960 × 280.23). Photograph courtesy of the Royal Ontario Museum. © ROM.

Frontispiece: NASA photo courtesy of Spar Aerospace Limited.

This book was printed in Canada.

2 3 4 5 6 7 WC 04 03 02 01

About the Cover and Frontispiece

The cover and frontispiece illustrate the evolution of engineering in Canada over the past 140 years. Our cover features a coloured lithograph depicting the construction of the Victoria Bridge. The bridge spans the St. Lawrence River and was erected in 1859 as part of the Grand Trunk Railway (GTR) — the longest railway in the world, at 1760 km, when it was first completed in 1853. The GTR itself was built to provide a main trunk line across the "Province of Canada," and the first leg of track was laid from Montreal to Toronto. The Victoria Bridge was a key component of the trunk line and is remembered as one of the greatest engineering feats of the nineteenth century.

The Victoria Bridge is made of tubular wrought iron and rests on two abutments and 24 piers that were designed to resist the crushing ice of the St. Lawrence. The original spans were built from wrought iron rectangular boxes. In the 1890s, however, the GTR began a massive betterment program on its property, and the iron boxes of the Victoria Bridge were replaced by steel trusses. Civil engineer Joseph Hobson also replaced the superstructure with new materials. Amazingly, he accomplished this without interrupting traffic on the bridge.

The frontispiece shows the Canadarm, which was developed for the U.S. space shuttle by Canadian engineers from many disciplines. The Canadarm is Canada's contribution to the shuttle program and is used to deploy and retrieve satellites from the shuttle cargo bay. Hand controls inside the shuttle and television cameras at the Canadarm's "wrist" and "elbow" permit the astronauts to manipulate the cargo with ease, under weightless conditions.

To operate in the rigours of space, the Canadarm required the development of optically commutated brushless DC motors driven by a special servo-power amplifier, high-reduction gearboxes with zero backlash, and the use of high-strength, lightweight materials such as titanium and graphite epoxy. The harsh environment of space flight created severe lubrication and thermal expansion problems for the designers.

The arm was installed and first flew in 1981 on the second space shuttle flight. Its performance exceeded design specifications. When the space platform planned for the twenty-first century is built, suitably adapted prototypes of the Canadarm will provide the manipulative capability to assemble the components in earth orbit.

Preface

The goals of this text are to acquaint readers with the structure, practice, and ethics of the engineering profession and to encourage engineers to apply ethical concepts in their professional practice. The text is intended for senior undergraduate engineering students, recent university graduates who are beginning their careers in engineering, practising engineers, and foreign engineers who have immigrated to Canada and wish to practise engineering. The text is directed to engineers in every branch of the profession, practising in any province or territory of Canada, and should be of particular value to people preparing to write the Professional Practice Examination.

Engineering is a creative, enjoyable, and rewarding profession with a long history and a bright future. The engineering profession played a key role in establishing Canada as a nation and will be even more important in the new millennium. Canada's resources and political stability should permit the entrepreneurial spirit to flourish, particularly in the next few years as we tap the unknown potential of computer applications in communications, expert systems, and artificial intelligence. Engineers will be among the key participants in generating future prosperity.

Engineers will also be essential in addressing the challenges of international competition and the problems of pollution and waste management that are already serious threats to the Canadian way of life. As we look further into the twenty-first century, the challenges facing the world in the areas of overpopulation and providing adequate housing, food, energy, mass transit, and mass education will be felt in Canada. Many of these social problems may have engineering solutions, and all of them will have ethical implications.

ORGANIZATION

This text is organized into five parts; the first four parts each contain several chapters, and Part 5 is a single chapter, intended to assist readers who are preparing to write the Professional Practice Exam.

PART 1: INTRODUCTION TO ENGINEERING

The first two chapters introduce the reader to engineering and give an overview of the profession, its history, and the various provincial and territorial Acts that regulate engineering in Canada. A brief statistical summary shows the distribution of engineers in Canada and their job functions. These chapters discuss other basic information that will be of interest to someone entering the engineering profession: standards for admission, the significance of codes of ethics, and the definition of professional misconduct. This part also includes descriptions of the engineer's seal, the iron ring, and the engineering oath. (American practice for regulating the engineering profession is described briefly in Appendix D for comparison with Canadian practice.)

PART 2: PROFESSIONAL PRACTICE

Part 2 consists of three chapters that describe the working environment of the engineer in industry, management, and private practice. The text discusses the requirements for entering each sphere of activity and the typical problems that are encountered. Suggestions are given to help avoid some of these problems and to develop a professional approach to engineering practice.

PART 3: PROFESSIONAL ETHICS

Part 3 deals mainly with professional ethics and consists of nine chapters that outline the basic derivation of ethical theory and codes of ethics; their application to typical ethical problems in industry, management, private practice, and government; and advice on how engineers can best deal with these problems. This part also describes some ethical problems associated with the environment and with product safety, and the engineer's duty to report unethical behaviour, also called whistleblowing. The topic of whistleblowing is particularly relevant, since the provincial Associations of Professional Engineers play a key role in the process of reporting and disciplining unethical behaviour. This part closes with a chapter on professional misconduct and disciplinary powers. Twenty case studies are described in detail, and an ethical solution for each is proposed by the authors. Also, eleven case histories that ended in tragedy are narrated, since they teach important ethical lessons that were learned at great cost.

PART 4: MAINTAINING PROFESSIONALISM

The fourth part of the text consists of two chapters describing the problem of maintaining continued competence in a rapidly changing world. The authors suggest several methods of approaching this challenge. Engineers contemplating graduate studies will find the advice very useful, and all readers will recognize the vital importance of engineering societies in communicating new ideas. This part concludes with suggestions and data for choosing and enrolling in a suitable engineering society.

PART 5: EXAM PREPARATION

As a service to readers who are writing the Professional Practice Exam, the final chapter describes the examination and explains a new technique, called the EGAD strategy, for writing essay-type examinations. Sixty typical examination questions from several provinces are included, for which some answers are suggested by the authors.

FEATURES

Readers will find this a comprehensive yet readable text that follows a logical sequence in the study of professional engineering practice and ethics. The text is appropriate for individual study, for classroom use, or as a reference for practising engineers. The goals of the text are achieved through the following features:

- a logical, readable style
- comprehensive coverage of the topic, from basic to advanced concepts, suitable for every province and territory in Canada, including comments on some aspects of American practice
- twenty realistic case studies dealing with ethical problems that ask the reader to suggest the appropriate course of action, followed by the authors' recommended resolution for each case
- eleven case histories in which engineering practices that deviated from ethical standards led to disaster or personal tragedy
- sixty brief, typical examination questions, taken from the Professional Practice Exam in several provinces, to assist those readers preparing for the examination
- discussion of professional practice from three perspectives, including the engineer as employee, the engineer as manager, and the engineer in private practice
- advice for young engineers to help guide their engineering careers
- appropriate excerpts from all of the provincial and territorial Acts that regulate engineering (in appendices), including the codes of ethics and definitions of professional misconduct for all provinces and territories

- extensive reference material (in appendices) concerning guidelines for professional employees, addresses of provincial and territorial Associations, and American registration procedures
- topics for further study and discussion, in a standard format, at the end of each chapter

ACKNOWLEDGEMENTS

Professor Kemper and I have received immense help and guidance from many people during the preparation, editing, and revision of this manuscript. The task of assembling the joint work into a final manuscript with a Canadian perspective has been my responsibility, and I would like to thank all those people who contributed to making this text a reality through advice, provision of materials, or agreement to publish copyrighted articles.

FIRST EDITION
Special thanks are due to Scott Duncan of HBJ-Holt Canada, who conceived the project and pushed it through the early stages, and to Sarah Duncan and Semareh Al-Hillal, who co-ordinated the first edition. Grant Boundy, former Deputy Registrar of PEO, and Stephen Jack, Director of Communications for PEO, graciously co-operated by providing advice and materials, for which I am indebted. Georges Lozano and Wendy Ryan-Bacon of CCPE were very courteous and helpful regarding my inquiries. Colleagues Norman Ball, Dick van Heeswijk, David Burns, and the late Alan Hale provided encouragement and written material. The contribution of all those who reviewed the manuscript in its various stages is gratefully acknowledged. The reviewers for the first edition were G.A. Bernard of APEGGA; Wendy Ryan-Bacon of CCPE; Grant Boundy of APEO; E.R. Corneil of Queen's University; Dennis Brooks of APEGGA; John Gartner of Gartner/Lee Limited; C. Peter Jones, formerly of the University of British Columbia; Harold Macklin of LINMAC Inc.; and Gordon Slemon of the University of Toronto. Computer facilities provided by the Manufacturing Research Corporation of Ontario (MRCO) and the National Sciences and Engineering Research Council (NSERC) were invaluable in the preparation of the manuscript. A special thanks is due to my secretary, Kathy Roenspiess, who typed the manuscript and contributed many suggestions for improving it.

SECOND EDITION
Many of the people who assisted in preparing the first edition were also very helpful in developing the second edition, and a heartfelt thanks is extended to them all. In addition, many people who were new to the project deserve recognition. In particular, I would like to thank Har-

court Brace developmental editor Su Mei Ku, whose advice, assistance, tactful support, and insight into the topic were very beneficial and very much appreciated. Advice and assistance in revising the text were also provided by Al Schuld, Deputy Registrar of APEGGA; Kent Fletcher of Fletcher Associates; Elvis Rioux of APEGBC; Richard Furst, Anita Direnfeld, and Larry Gill of PEO; Wendy Ryan-Bacon, Laurie Macdonald, and Chris Lyon of CCPE; Jerry M. Whiting, formerly of the University of Alberta; Beth Weckman, Roydon Fraser, and Dwight Aplevich of the University of Waterloo; and Ken Nauss of Harcourt Brace. A special thanks is due to Monique Frize, NSERC/Nortel Joint Chair in Women in Science and Engineering in Ontario, Faculty of Engineering, University of Ottawa and Carleton University, for contributing Chapter 13 to this text. Dr. Frize was formerly the NSERC/Nortel Chair at the University of New Brunswick.

Finally, as anyone who has ever written, revised, or edited a book will know, the attention to detail requires long hours of concentration. I would like to thank my wife Isobelle, my children Christopher and Gail Stephanie, and my graduate students for tolerating my reclusive behaviour for many months during the preparation of both editions.

G.C. Andrews
Waterloo, Ontario

A NOTE FROM THE PUBLISHER

Thank you for selecting *Canadian Professional Engineering Practice and Ethics*, Second Edition, by Gordon C. Andrews and John D. Kemper. The authors and publisher have devoted considerable time and care to the development of this book. We appreciate your recognition of this effort and accomplishment.

We want to hear what you think about *Canadian Professional Engineering Practice and Ethics*. Please take a few minutes to fill in the stamped reader reply card at the back of the book. Your comments and suggestions will be valuable to us as we prepare new editions and other books.

Contents

Part One

Introduction to
Engineering

CHAPTER ONE

Introduction to the Engineering Profession

Welcome to engineering — a challenging, creative, and rewarding profession! Engineering has had a profound influence on society and on our daily lives. Although we may not realize it, engineered products surround us to aid, guide, and protect us. For example, engineering excellence may be evident in the sleek lines of a new car or the graceful structure of a bridge, but it is also hidden in more commonplace engineering achievements, such as the smooth, safe operation of an elevator, the ample supply of pure water from our faucets, and the digital accuracy of the communication and entertainment equipment in our homes. These examples illustrate the capability, ingenuity, and diverse interests of the people who enter the engineering profession. In fact, the terms "engineer" and "ingenuity" come from the same Latin root, *ingenium*, which means talent, genius, cleverness, or native ability.

A BRIEF HISTORICAL NOTE

Engineers have had an important and widespread influence on the development of modern civilization. Ancient achievements, from the building of the pyramids to the discovery of iron, were the beginnings of what we would call engineering today. The first group to be clearly identified as engineers were military engineers. Roman armies marched with a complement of engineers who devised and operated weapons and built the fortifications and roads needed to wage war. However, over the centuries, the nonmilitary, or "civil," engineer emerged to design and supervise the construction of roads, bridges, canals, and irrigation systems necessary for a productive society.

The Renaissance gave us Leonardo da Vinci (1452–1519), who was an outstanding artist, scientist, and engineer. His notebooks contain designs for devices such as helicopters, submarines, and gear trains — devices that were not actually invented until centuries later. The advent of printing from movable type in the fifteenth century resulted in wider dissemination of knowledge, and the scientific and mathematical discoveries that followed were guided by familiar, eminent names, such as René Descartes (1596–1650), Sir Isaac Newton (1642–1727), and many others. Descartes established the "scientific method," in which truth is discovered by logic and objective reasoning and is independent of any deference to authority. The scientific method involves stating a hypothesis and predicting the outcome of an experiment. The results of the experiment are then compared with the prediction to test the truth of the hypothesis. The scientific method is as important today as it was in Descartes' time. Newton has been described as the most important scientist of all time. His work includes the invention of calculus (an honour he shares with Leibnitz) and treatises on optics, dynamics, and gravity, which are the foundations of many engineering courses studied today. As a result of his application of dynamics to the motion of celestial bodies, Newton was one of the first to postulate convincingly the feasibility of space travel.

The scientific and mathematical revolutions of the sixteenth and seventeenth centuries led, in turn, to the Industrial Revolution of the eighteenth and nineteenth centuries. The wealth created by standardized, mass-produced products encouraged the exponential growth in engineering that we have observed in the twentieth century. In fact, while this past century has brought us much conflict, including two devastating world wars, it has also brought the more-favoured nations of the world many benefits: the widespread distribution of electricity; the invention of methods of mass communication such as telephone, radio, television, and facsimile (fax); and the start of the computer revolution that presently surrounds us.

The computer revolution holds much promise for the future. Although calculating devices such as the abacus, the slide rule, and the adding machine had long existed, the first truly electronic computer was not constructed until the end of World War II in 1945. The first computers were expensive, slow behemoths by today's standards, filling rooms with electronic machinery yet capable of only primitive calculations. The invention of the transistor and large-scale circuit integration (LSI) permitted the miniaturization of electronic devices, and the first commercial personal computer (the Apple II) was released in 1977. The desktop workstation evolved over the next two decades, providing immense and convenient computing power. As a result, entirely new fields of engineering are now emerging and evolving, such as hardware and software engineering; computer-aided analysis, design, and manufacturing; computational fluid dynamics; finite-element analysis; and system and process simulation, to name only a few.

The computer revolution is profoundly changing all of our society, and while the engineering profession is a major contributor to this rapid change, it is also simultaneously changing and evolving. Recent graduates may not recognize the significance of this massive computer revolution. Conversely, older engineers have had to learn that the only constant in today's world is change. One of the purposes of this book is to assist the engineering profession to cope with this rapid change by explaining and defining the organization of the engineering profession in Canada and showing that there are some principles that do not change, and therefore provide stability.

ENGINEERING IN CANADA

Engineering has evolved into a profession over several centuries, and the profession continues to evolve as this textbook is written. The need for formal training became evident during the Industrial Revolution. In Britain, teenage engineering apprentices typically paid for the privilege of a five-year apprenticeship in the office of a practising engineer. Standards are much higher today: in Canada, an engineer-in-training typically requires four years of engineering experience under the guidance of a professional engineer, after completing a four-year university degree, before a licence is awarded. Similarly, the duties and areas of responsibility of the engineer have changed over the centuries. The narrow engineering disciplines of the past have evolved into almost 100 modern branches of engineering.

An engineer is typically defined, in colloquial terms, as "a person who uses science, mathematics and technology, in a creative way, to satisfy human needs."[1] This definition emphasizes that engineering is the process of putting technical ideas and scientific principles to practical use for the common good. An engineer must be a combination of scientist and mathematician and must be a creative person who is capable of making decisions and solving problems. But as a professional, the engineer must also be an ethical person who is willing to put public welfare ahead of narrow personal gains.

In law, an engineer is defined even more specifically: in Canada, the title "Professional Engineer" is restricted, by law, to those persons who have demonstrated their competence and have been licensed in a provincial or territorial Association of Professional Engineers[2] or, in Quebec, the Ordre des Ingénieurs du Québec. This legal restriction on the title of Professional Engineer is closely monitored by the provincial Associations and also applies to any abbreviation or adaptation of the term "Engineer" that might lead to the conclusion that the person using the title is licensed. In 1995, there were approximately 161 000 licensed professional engineers in Canada (including geologists, geophysicists, and geoscientists, in provinces where they are regulated under the Act).[3] The distribution of engineers across Canada is shown graphically in Figure 1.1. The regulation of the engineering profession and the licensing of engineers are discussed in more detail in Chapter 2.

THE ENGINEER AND THE TECHNICAL TEAM

Engineers are primarily concerned with *design and development*, which is the creative process of converting theoretical concepts into useful applications. Engineers may occasionally work alone, particularly on small projects, but complex engineering projects usually require a team

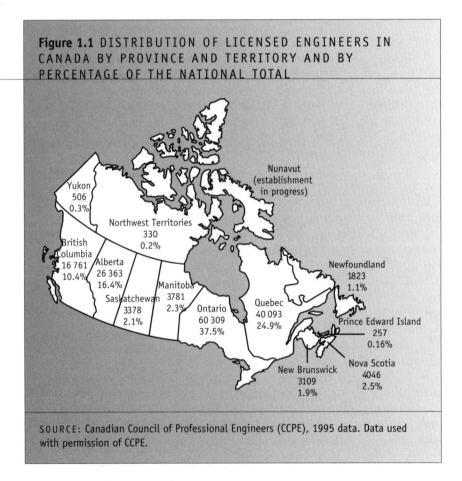

Figure 1.1 DISTRIBUTION OF LICENSED ENGINEERS IN CANADA BY PROVINCE AND TERRITORY AND BY PERCENTAGE OF THE NATIONAL TOTAL

Yukon
506
0.3%

Nunavut
(establishment
in progress)

Northwest Territories
330
0.2%

British
Columbia
16 761
10.4%

Alberta
26 363
16.4%

Saskatchewan
3378
2.1%

Manitoba 3781
2.3%

Ontario
60 309
37.5%

Quebec
40 093
24.9%

Newfoundland
1823
1.1%

Prince Edward Island
257
0.16%

Nova Scotia
4046
2.5%

New Brunswick
3109
1.9%

SOURCE: Canadian Council of Professional Engineers (CCPE), 1995 data. Data used with permission of CCPE.

approach involving people with widely different abilities, interests, and education who co-operate by contributing a particular expertise to advance the project. Engineers are only one group in the technical spectrum, although they constitute the vital link between theory and application. The full list would include research scientists, engineers, technologists, technicians, and skilled workers. The following capsule descriptions give a rough idea of the typical tasks performed by the different groups in a technical team. (It should be stressed that this is a rough categorization, and exceptions are common.)

RESEARCH SCIENTIST

The scientist usually develops ideas that expand the frontiers of knowledge but may not have practical applications for many years. The doctorate is usually the basic educational requirement, although a master's degree is occasionally acceptable. The scientist is rarely required to supervise other technical personnel except research assistants, and will usually be a member of several learned societies in the particular field of interest.

Typically, the basic task of the scientist is to perform *research* (creating new knowledge), while the basic task of the engineer is to perform *design and development* (creating new things). The roles of the scientist and the engineer overlap to some degree, and in some projects the boundary may be invisible. It is sometimes only the goal of the work, not the actual duties, that differentiates the two.

A scientist may be primarily a theoretician (that is, may specialize in creating new theoretical explanations for unexplained phenomena) or primarily an experimentalist. In either event, the final output of the scientist's work is usually a report or paper published in a scholarly journal. The scientist usually is concerned with basic research and publishing new knowledge or theories, and it is the engineer's task to carry the theory through to a useful application.

ENGINEER

The engineer usually provides the key link between theory and practical applications. The engineer must have a combination of extensive theoretical knowledge, the ability to think creatively, and the knack of obtaining practical results. The bachelor's degree is the basic educational requirement, although the master's degree may be preferred by some employers. In Canada, the engineer is required, by provincial or territorial law, to be a member of a provincial or territorial Association of Professional Engineers in order to practise engineering.

Engineers are usually concerned with creating devices, systems, and structures for human use. However, many engineers are engaged in managing engineering companies or the discovery and utilization of resources, and also are frequently employed in liaison or consulting capacities in construction, testing, and manufacturing and as agents for governmental bodies. These engineers are not directly engaged in design, but design is still at the core of their activities.

TECHNOLOGIST
The technologist usually works under the direction of engineers in applying engineering principles and methods to fairly complex engineering problems. The basic educational requirement is usually graduation from a technology program at a community college, CEGEP (Collège d'enseignement général et professionel), or CAAT (college of applied arts and technology), although occasionally a technologist may have a bachelor's degree (usually in science, mathematics, technology, or related subjects). The technologist often supervises the work of others and is encouraged to have qualifications that are recognized by a technical society. In some provinces, an association of certified engineering technicians and technologists may confer the designation Certified Engineering Technologist (CET). These associations are voluntary organizations, and the title is a beneficial but not an essential requirement for working as a technologist.

TECHNICIAN
The technician usually works under the supervision of an engineer or technologist in the practical aspects of engineering tests and maintenance of equipment. The basic educational requirement is usually graduation from a program at a community college, CEGEP, or CAAT. In some provinces, the title Certified Engineering Technician (C.Tech.) may be awarded to qualified technicians, although the title is not essential to work as a technician. The designation of technologist or technician is awarded, based on appropriate requirements, by provincial associations of technicians and technologists. By rough estimate, there were approximately 50 000 certified technicians and technologists in Canada in 1995. However, since certification is voluntary, there were probably many more people actually practising as technicians or technologists. The total number is estimated to be approximately equal to the number of professional engineers practising in Canada (161 000).

SKILLED WORKER
The skilled worker usually applies highly developed manual skills to carry out the designs and plans of others. The skilled worker's ability can be learned only at the side of a master artisan; it is the quality of this apprenticeship, not the formal education, that is important. Each type of trade worker (electrician, plumber, carpenter, welder, patternmaker, machinist, etc.) comes under a different certification procedure, which varies from province to province.

The categories described above are not, in reality, quite as clearly defined as these descriptions might indicate; the boundaries are not (and should not) be rigid barriers. Transition from one group to another is always possible, although it may not always be easy. Each group in this technical spectrum has a different task, and there are great differences in the skills, knowledge, and training required. In a large project, all five types of technical expertise will be required, although at different times. For a major project to be completed successfully, co-operation is essential between these five groups, and mutual respect for the particular skills and knowledge of each individual creates a positive and productive working environment. Although the task of the engineer is only one phase of a continuous spectrum of technical knowledge, the engineer is usually the key link between theory and practice.

BRANCHES OF ENGINEERING

Over the years, many different branches or disciplines have been established in engineering. The most general branches are civil, electrical, mechanical, industrial, and chemical engineering. However, since university engineering programs are accredited in Canada on an individual basis, many more-specialized programs, such as computer engineering and systems design, are accredited at specific institutions. In fact, as shown in Table 1.1, over 60 different engineering programs are presently accredited at Canadian universities. (The program titles in French are offered only by Quebec universities.)

The choice of a branch is important and is usually made on entry to university or during the first year of university. The applicant's branch must, of course, be specified when applying to the provincial Association for registration as a professional engineer and award of a licence to practise. Changes from branch to branch are possible, although not common. Since readers of this text are likely to be senior university undergraduates or recent engineering graduates, most will already have chosen a branch, so a detailed discussion of branches is unnecessary.

JOB FUNCTIONS OF PRACTISING ENGINEERS

Although the primary task of the engineer is typically design and development, other tasks, jobs, or activities occupy engineers and are included in the broad spectrum of engineering practice. Table 1.2 lists the percentages of professional engineers working in various job functions in Canada in 1991. The job functions in Table 1.2 are not rigidly defined; some overlap exists from job to job. Typical descriptions for most of these job functions follow.

Table 1.1 LIST OF ENGINEERING PROGRAMS ACCREDITED AT CANADIAN UNIVERSITIES

aerospace engineering	génie des mines
agricultural engineering	génie des mines et de la minéralurgie
biological engineering	génie et gestion de la construction
bio-resource engineering	génie métallurgique
building engineering	génie unifié
ceramic engineering	geological and mineral engineering
ceramic engineering and management	geological engineering
chemical and biochemical engineering	geological engineering (geophysics)
chemical and materials engineering	geomatics engineering
chemical engineering	industrial engineering
chemical engineering and management	industrial systems engineering
civil engineering	manufacturing engineering
civil engineering and computer systems	materials and metallurgical engineering
civil engineering and management	materials engineering
computer engineering	materials engineering and management
computer engineering and management	mathematics and engineering
computer systems engineering	mechanical engineering
electrical engineering	mechanical engineering and management
electrical engineering and management	metallurgical engineering
electronic systems engineering	metallurgical engineering and management
engineering chemistry	metallurgical engineering and materials science
engineering management	metals and materials engineering
engineering materials	mining and mineral process engineering
engineering physics	mining engineering
engineering physics and management	naval architectural engineering
engineering science	petroleum engineering
engineering systems and computing	regional environmental systems engineering
environmental engineering	surveying engineering
extractive metallurgy engineering	systems design engineering
food engineering	water resources engineering
forest engineering	
génie de la construction	
génie de la production automatisée	
génie des matériaux et de la métallurgie	

SOURCE: Canadian Council of Professional Engineers (CCPE), *CEAB Annual Report* (Ottawa: CCPE, 1995). Reprinted with permission of CCPE.

ENGINEERING DESIGN

The key importance of design in the practice of engineering was discussed earlier. However, *design* is sometimes confused with a related activity: *development* (discussed in the next paragraph). In practice, the term *development* is likely to refer to the early stages of a project, when various alternatives are analyzed, compared, and tested. *Design* usually

Table 1.2 PERCENTAGE OF ENGINEERS BY JOB FUNCTION

Job Function	Men	Women	Total	Percentage of Engineers
Engineering				
Engineering Design	17 813	794	18 607	16.4
Project Management	15 003	487	15 490	13.6
Operations and Production	10 036	424	10 460	9.2
Research and Development	5 473	267	5 740	5.1
Project Planning	3 258	156	3 414	3.0
Exploration	3 061	162	3 223	2.8
Construction	2 565	56	2 621	2.3
Computer Services	2 295	142	2 437	2.1
Quality Control	1 699	129	1 828	1.6
Miscellaneous	19 149	973	20 122	17.7
Engineering Related				
Business Management	12 558	178	12 736	11.2
Marketing and Sales	4 072	136	4 208	3.7
Administration	3 702	81	3 783	3.3
Teaching	2 995	114	3 109	2.7
Legal Services and Contracts	2 080	130	2 210	1.9
Corporate Planning	1 021	35	1 056	0.9
Non-Engineering	2 406	133	2 539	2.2
Unknown	16 653	936	17 589	Not counted in percentages
Total	125 839	5 333	131 072	99.7

Note: Engineers whose job function was unknown were not included in the percentages. The percentages do not total 100 percent because of rounding.

SOURCE: Canadian Council of Professional Engineers (CCPE), 1991 data. Reprinted by permission of CCPE.

refers to the later phases of a project, when the basic method has been decided and it is then necessary to establish the exact shapes and relationships of the various parts. These definitions are often mixed up, because design and development are very similar.

RESEARCH AND DEVELOPMENT

Companies usually refer to the whole spectrum of their engineering activities as *research and development,* usually abbreviated as R&D. However, as mentioned earlier, research, development, and design are stages in the engineering process, and in most companies very little research is done (although there are notable exceptions). For the most part, the activities carried out under the heading of R&D are actually those of development, particularly in Canada, where more research could (and should) be done.

By contrast, *research* is the process of learning about nature and codifying this knowledge into usable theories. Many scientific fields that were once considered the realm of physicists or chemists have become primarily the domain of engineers in the last 30 years. These fields, designated the *engineering sciences* by a committee of the American Society for Engineering Education (ASEE), are mechanics of solids, mechanics of fluids, transfer and rate processes, thermodynamics, electrical sciences, and nature and properties of materials. In Canada, the Canadian Engineering Accreditation Board (CEAB) recently added computer science to this list, thus resolving a debate over whether it is an engineering science or a branch of mathematics. Engineering science seeks new knowledge for the specific purpose of design and development, whereas science seeks knowledge without regard to application.

Sometimes the terms *basic research* and *applied research* are used. The former means the search for knowledge for its own sake, and the latter implies that there is a known use for the knowledge being sought. Engineers rarely engage in basic research; if they do, they are likely considered to be scientists. But engineers often engage in applied research. During the design of new devices or systems, critical scientific data or information may often be lacking, and a special research program may be needed to get the information. This is applied research. It is an important aspect of engineering, and engineers involved in these projects often use the title "Research Engineer."

TESTING AND QUALITY CONTROL

Engineers are frequently employed in testing or in *quality control*, which can be defined as setting standards for products, processes, or materials, and evaluating performance to see that the standards are being achieved. Some organizations have special departments for these functions that are organizationally separate from the design departments. The separation permits test departments to be more objective in their testing procedures than designers would be if they tested their own creations. Test departments usually evaluate prototypes of new designs. They may conduct tests of new parts or materials or qualification tests of products furnished by others, or they may maintain constant monitoring and inspection of a process or project.

OPERATIONS AND PRODUCTION (MANUFACTURING)

Before a product can reach the public, it must be manufactured. There are so many engineers employed in manufacturing that the field has given rise to a special society: the Society of Manufacturing Engineers (SME). The role of the manufacturing engineer can be quite diverse. Manufacturing engineers are usually responsible for solving the problems that arise during the manufacturing process. That is, they are concerned with the processes, the manufacturing equipment, and the production personnel. They are also concerned with developing and improving the production processes, including the tools and machines. Some may be in charge of the inspection process, often called quality control, as defined in the previous section.

Closely related to the manufacturing engineer is the *plant engineer*. Whereas the former is concerned with the product and the means of manufacture, the latter is concerned with the buildings and utilities that support the manufacturing process.

CONSTRUCTION

Construction engineers participate in building various engineering structures: bridges, roads, buildings, drainage systems, water supply systems, and so on. These projects are usually organized in two-pronged arrangements. That is, the owner or client engages an engineering consultant to design the structure and a contractor to build it. Both the consultant and the contractor may hire construction engineers. Construction engineers working for the consultant are concerned with quality and conformance to the plans. They monitor and examine the construction and may be resident on large projects or may make periodic site inspections. Construction engineers working for the contractor typically supervise the construction personnel and solve the day-to-day construction problems. They may be required to be resident on the site so that they can deal with crises immediately. Needless to say, construction engineers must go where the construction is located, so they travel a lot.

MARKETING AND SALES

Very few engineers enter engineering studies with the expectation of working in marketing or sales activities. However, many employers recruit engineering or technology graduates specifically for this purpose. *Sales engineering* is a field between sales and engineering that occasionally involves engineering design. Such opportunities normally arise in enterprises that manufacture and sell custom-designed systems. In a typical case, a fully operating system put together from off-the-shelf components is modified to suit the customer's unique requirements. This may require special components that have not yet been designed. The sales engineer works with the customer and makes the sale, but also designs the system to meet the customer's needs and, when necessary, works with the engineering office to develop hitherto nonexistent components. Sales engineering is therefore involved as much with technical as with financial aspects, in most cases.

TEACHING

Some engineers become teachers. If one wishes to teach at the community college level, a master's degree and a teaching credential are the usual prerequisites. Professional experience is also highly desirable. However, a master's degree is usually not adequate in colleges and universities where faculty members are required to maintain active research programs in addition to their teaching. These *research universities* require their faculty members to hold a doctorate degree, and the teaching loads are lighter than in other schools to allow time for research. Research and graduate studies are usually closely coupled in these institutions, because the very best training for graduate students takes place in a research environment.

ADMINISTRATION AND PUBLIC SERVICE

Many engineers are involved in administering engineering activities for government agencies at municipal, regional, provincial, or federal levels. These positions normally require a sound knowledge of the agency's engineering tasks, such as highway construction, electrical power generation, or telecommunications, but the engineer's job may also involve planning, funding, scheduling, regulations, or similar policy-related tasks. The government is, in fact, a major employer of engineers. Engineers working in administration and public service may face special ethical problems that result from political pressure and from decisions concerning the distribution of public resources.

MANAGEMENT

Statistics show that, sooner or later, many engineers go into management. Some engineering graduates are actively interested in management careers, but many are sceptical. Nevertheless, management is a major career destination for engineers, whether they plan for it or not. Chapters 4 and 8 of this text specifically address the challenges that face engineers in management.

CONSULTING

Consulting engineering is the most independent form of engineering, but only a very small percentage of engineers are engaged in consulting.

Some consulting engineers work as individuals offering services to the public for a fee. This is particularly true among those who are just starting out as consultants. However, most work as employees of consulting engineering corporations that hire engineers, architects, accountants, designers, CAD operators, clerks, and people of similar skills. Some consulting organizations are very large and hire hundreds of engineers of all kinds: chemical, civil, electrical, mechanical, and computer, among others. Chapters 5 and 9 of this text specifically address the challenges that face engineers in private practice.

EMPLOYMENT OF ENGINEERS IN CANADA

The number of professional engineers in Canada has almost doubled in the last ten years and is still growing. However, as we pass the millennium, there is some doubt as to whether there will be an adequate supply of engineers for the needs of the next decade. Some statistics and projections for engineering employment in Canada are given in the next few paragraphs.

Table 1.3 shows the same information that appears in Figure 1.1, divided into categories by province and gender. As we would expect, the distribution of professional engineers across Canada is not uniform. Most of them are clustered in the industrially developed areas of Ontario and Quebec, with the next largest number in the resource-rich area of Alberta. British Columbia has almost the same percentage of engineers as the four Atlantic provinces combined. Saskatchewan and Manitoba together have even fewer engineers than the combined number in Nova Scotia, New Brunswick, Prince Edward Island, and Newfoundland. The Yukon and Northwest Territories have just over 0.5 percent of the nation's engineers.

Employment prospects for engineers have always been very good, and they are predicted to become better as we pass the year 2000. To quote from *The Future of Engineering*, a report by the Canadian Council of Professional Engineers:

> Demand appears to be strong for engineers into the next century. There are two reasons for this: the first is that engineers are crucial to enable Canada to compete in an increasingly global market; and the second is that, despite the above, the number of young people entering engineering is expected to increase only moderately.[4]

This prediction should be viewed in light of the historical employment record for engineers. The unemployment rate for engineers was typically around 1 percent in the years prior to 1982, indicating a very small

Table 1.3 SUMMARY OF REGISTERED ENGINEERS IN CANADA, ORGANIZED BY PROVINCE AND BY GENDER

Province	Male	Female	Total	% of Total Registered in Canada
Alberta	24 754	1 619	26 363	16.40
British Columbia	16 157	604	16 761	10.43
Manitoba	3 661	120	3 781	2.35
New Brunswick	2 947	162	3 109	1.93
Newfoundland	1 745	78	1 823	1.13
Northwest Territories	316	14	330	0.21
Nova Scotia	3 862	184	4 046	2.52
Ontario	57 827	2 482	60 309	37.51
Prince Edward Island	247	10	257	0.16
Quebec	37 278	2 815	40 093	24.94
Saskatchewan	3 289	89	3 378	2.10
Yukon Territory	497	9	506	0.31
Total	152 580	8 186	160 756	100

SOURCE: Canadian Council of Professional Engineers (CCPE), 1995 data. Reprinted with permission of CCPE.

turnover of professional engineers. In the recession of 1982, unemployment reached a peak of 7000 engineers, or about 6 percent of the registered professional engineers. It has gradually declined and recently stabilized at about 2 percent, which is a fraction of the unemployment rate for all Canadian workers.[5] It appears that it will return to even lower values in the coming decade. Moreover, although the employment opportunities for engineers are expanding, the downsizing of university engineering faculties in the late 1990s has limited the capacity of universities to increase the numbers of engineering graduates. The employment outlook therefore appears to be very good for engineering graduates.

ENGINEERING AS A PROFESSION

The status of the engineering profession occasionally leads to differences of opinion, particularly among undergraduates. What is this status?

During the debates in the Ontario legislature that led to the passage of the Professional Engineers Act (1968–69), a *profession* was defined as "a self-selected, self-disciplined group of individuals who hold themselves out to the public as possessing a special skill derived from training and education and are prepared to exercise that skill in the interests of others."[6]

This definition is certainly satisfied by the professions of medicine, law, and theology, and it appears to be equally well satisfied when applied to engineering. The professional engineer has skill and knowledge obtained from lengthy education and practical experience and not possessed by the general public, whose members usually would not be competent to judge accurately the quality of service rendered by the engineer. Moreover, professional engineers must, by law, subscribe to a code of ethics established by the engineers for the protection of the public. Consequently, each province and territory recognizes engineering as a profession and permits engineers to regulate themselves through a provincial or territorial Association of Professional Engineers (or Ordre des Ingénieurs du Québec).

However, it might be appropriate to seek a more general meaning of the word "profession" as interpreted by non-engineers. Consider the following definition from *Webster's Third New International® Dictionary*:

> **Profession** ... A calling requiring *specialized knowledge* and often long and *intensive preparation* including instruction in skills and methods as well as in the scientific, historical, or scholarly principles underlying such skills and methods, maintaining by force of *organization* or concerned opinion *high standards* of achievement and conduct, and committing its members to *continued study* and to a kind of work which has for its prime purpose the rendering of a *public service*.[7] [Italics added.]

We can see by looking at the italicized items how well engineering fits the definition of a profession. Engineering certainly requires *specialized knowledge* and *intensive preparation*, but the degree of preparation is not quite as great as that in medicine or law, two professions with which engineers frequently compare their own. For many decades, engineering has required four years of university study and two years of professional practice, or a total of six years of preparation beyond high school for entrance into the profession, whereas medicine and law have required from seven to nine years, depending on the program. However, in almost every province, the experience requirement has increased (or is in the process of increasing) to four years. Therefore, the total engineering education and experience requirement is now eight years (in almost every province), very comparable to the requirements in the other professions.

There is no doubt that the engineering profession has a very strong *organizational structure* and that *high standards* are required for admission. Provincial and territorial Associations of Professional Engineers monitor the qualifications of applicants and, through the engineering accreditation board of the Canadian Council of Professional Engineers

(CCPE), evaluate and accredit educational programs at all engineering institutions at the university level in Canada. High standards of achievement have therefore been set for admission into the profession and are enforced. The requirements for *continued study, high standards* of conduct, and *public service* are addressed by the codes of ethics that guide the personal conduct of all engineers. Codes of ethics have legal status, since they are established by acts of the legislatures in all provinces. Failure to follow the code of ethics can lead to charges of professional misconduct and disciplinary action for the errant engineer.

Therefore, engineering clearly satisfies the definition of a profession. The difference between engineering and most other professions is the environment in which most engineers practise. In Canada, the majority of engineers are employees of large companies, working in teams on projects. Most other professionals are self-employed and work on a one-to-one basis with clients. The distinction is aptly expressed in the following assessment by a Canadian engineer:

> The hard fact of the matter is that people need physicians to save their lives, lawyers to save their property, and ministers to save their souls. Individuals will probably never have an acute personal need for an engineer. Thus, engineering as a profession will probably never receive the prestige of its sister professions. Although this may be an unhappy comparison, the engineer should take note that physicians and lawyers both feel that the prestige of their professions has never been lower, and they are mightily concerned; yet ... engineers are considered to be sober, competent, dedicated, conservative practitioners, without such devastating problems as embezzlement or absconding members and without the constant references to malpractice and incompetence.[8]

Engineers have a high status in the eyes of society. Surveys among the general public in both Canada and the United States consistently show engineering to be among the top few professions in terms of honesty, integrity, and attractiveness. The engineering profession, and engineers themselves, are held in high regard by the general public.

However, one of the fundamental facts of life for engineers is that they outnumber every other professional group except teachers and nurses. For example, for every physician in Canada, there are typically three engineers. For a profession that is based upon individual creativity, such large numbers may have disturbing implications. Moreover, engineers usually work in teams. It is sometimes difficult for engineers to cultivate individuality or to maintain a sense of personal responsibility under these conditions. However, important challenges are still met on an individual basis. New designs, products, and structures do not spring forth without creative effort at every level. Technical achievements require competent, professional, creative effort from every member of the engineering team. Although the team may be larger now than it was in simpler, bygone days, the projects are even larger and the individual challenge still exists.

TOPICS FOR STUDY AND DISCUSSION

1. This chapter defines the term "engineer" in a very general way. However, the term "engineer," meaning "professional engineer," has a legal definition in each province or territory (see Chapter 2 and Appendix A). How many definitions of "engineer" can you find in dictionaries or encyclopedias? How well do these definitions agree with the legal definition for your province or territory? Which definition is closest?

2. Newspaper and magazine writers occasionally refer to "scientific achievement" when things go well but use the label "engineering failures" when disasters occur. Examine some recent newspapers and magazines for examples of the use of terms such as "scientific," "engineering," "technological," and "technical." Have the terms been properly used? Do you detect any trends in erroneous usage? What rules (if any) should the media follow in using these terms? If serious errors in attributing credit or blame exist in the examples you find, write a letter to the offending newspaper or magazine to inform them.

3. Discuss the boundary between engineering and science. Does the role in the design process clearly identify the difference between the engineer and the scientist? What reasons would motivate a person to select a career in one of these disciplines over the other? What factors motivated you?

4. Compare the roles in the technical spectrum discussed in this chapter (scientist, engineer, technologist, technician, skilled worker). What rewards (work environment, prestige, personal satisfaction, pay, etc.) are generally associated with each of these occupations? What characteristics or traits would generally be helpful for people entering each of these occupations? Considering your own characteristics and expectations, which of these occupations is best for you?

5. Examine the various engineering functions discussed in this chapter (engineering design, research and development, testing and quality control, manufacturing, construction, marketing and sales, teaching, administration and public service, management, consulting). Considering your own interests, list these activities in order of desirability as lifelong careers. (This list of functions is not exhaustive, so please add to the list.) Are your present activities leading toward the career at the top of the list? If not, what should you do to ensure that your actions will lead to your desired career?

6. Examine your university course catalogue and classify the content of the courses you have taken or will take. Create a matrix and categorize each course. Each row of the matrix should correspond to one of your courses, and each column should correspond to one of the various engineering functions discussed in this chapter (engineering design, research and development, testing and quality control, manufacturing, construction, marketing and sales, teaching, administration and public service, management, consulting, etc. — add categories if necessary). Rate each course on how it contributes to preparing you for employment in the various functions that engineers perform. Does your program of courses show a broad coverage or a narrow coverage of the various functions? Can you draw any trends or conclusions from the pattern that might influence your course choices or career plans?

7. History textbooks usually describe the wars, treaties, and dynasties that controlled the wealth of past empires and governments. However, they rarely describe the achievements of the people, many of them engineers, who contributed to the generation of the wealth and the peaceful enjoyment of the fruits of labour. Write a brief essay on this topic.

8. Although the Romans were excellent organizers and rulers, and the Latin word *ingenium* has evolved into the definition of engineering, their piping systems were made of lead. In fact, the term "plumbing" comes from the Latin word for lead, *plumbum*. Investigate the allegation that lead poisoning from water supply pipes played a role in the decline and fall of the Roman Empire.

NOTES

1. Gordon C. Andrews and Herbert C. Ratz, *Introduction to Professional Engineering*, 5th ed. (Waterloo, ON: University of Waterloo, 1995), 3–4.
2. Throughout this text, the term "Association of Professional Engineers," or simply "Association," when capitalized, refers to the legal entity established by statute in the reader's province or territory to regulate the practice of professional engineering. These Associations are listed in Appendix A.
3. Similarly, the term "Professional Engineering Act," or simply "Act," when capitalized, refers to the statute itself. The provincial and territorial statutes are listed in Chapter 2, and excerpts from the statutes are included in Appendix B. Similarly, the term "engineer," when used in this text, also refers to geologists, geophysicists, and geoscientists for those provinces that include these specialties under the Act.
4. Task Force on the Future of Engineering, *The Future of Engineering* (Ottawa: Canadian Council of Professional Engineers, 1988), 29.
5. Ibid., 33.

6. Hon. H.A. MacKenzie [Opening address for the debate on the Professional Engineers Act, 1968–69], Ontario, Legislature, *Debates.*

7. *Webster's Third New International® Dictionary, Unabridged* © 1993 by Merriam-Webster, Incorporated. Reprinted with permission.

8. J. Carruthers, former Director of Communication, Professional Engineers of Ontario, personal communication, June 1977.

CHAPTER TWO

Regulation of the Engineering Profession

Most countries in the developed world impose some form of regulation, licensing, or similar government control on the professions. The purpose of this control is to protect the safety of the public, to restrict unqualified persons from practising, and to discipline unscrupulous practitioners. In Canada, the legal right to regulate the professions falls under provincial authority; similarly, in the United States, the states have this regulatory power.

THE EVOLUTION OF LICENSING LAWS IN CANADA

Canadian efforts to put engineering on the same professional footing as law and medicine began formally as early as 1887, when the Canadian Society of Civil Engineers (CSCE) held its first general meeting. The campaign to regulate the engineering profession was led by the CSCE (which became the Engineering Institute of Canada, or EIC, in 1918). In the years after Confederation, Canada followed the British model: engineers entered the profession through a period of apprenticeship, and few engineers were university graduates. However, from its beginning, the CSCE established and maintained high standards for admission to the Society, with the goal of improving professional engineering practice. Applicants were required to be at least 30 years of age and have at least ten years of experience, which could include an apprenticeship in an engineer's office or a term of instruction in a school of engineering acceptable to the CSCE Council, and also had to show "responsible charge of work" for at least five years as an engineer designing and directing engineering works.[1]

In spite of the early Canadian initiative, however, the United States was, in fact, the first country to regulate the practice of engineering. The State of Wyoming enacted a law in 1907 as a result of many instances of gross incompetence observed during a major irrigation project.[2] More than a decade passed before Canadian engineers overcame professional rivalries, business competition, class structures, and other challenges and agreed on proposals that would improve professional standards generally and the status of engineers indirectly. In August 1918, the question of seeking licensing legislation was put forward by F.H. Peters, an Alberta engineer, at a general meeting of the CSCE in Saskatoon. The basis for his proposal was that engineers had developed the resources of the nation but had received neither the remuneration nor the respect they deserved.[3] At that time, World War I was drawing to an end, and the returning soldiers, some of whom had been involved in various aspects of military engineering, dramatically increased the number of engineers, depressing salaries, increasing competition, and putting quality standards at risk.

A model act was drafted by the CSCE (which had just changed its name to the EIC) and was published in the EIC *Journal*. In September 1919, the *Journal* announced that 77 percent of EIC members had approved the model act by mail ballot. By the spring of 1920, all provinces except Ontario, Saskatchewan, and Prince Edward Island had passed licensing laws. In Ontario, a joint advisory committee redrafted the bill, and it was passed in 1922. The laws enacted in British Columbia, Manitoba, Quebec, New Brunswick, and Nova Scotia were "closed," meaning that a licence was needed to practise engineering or to use the title of Professional Engineer. In Alberta and Ontario, the laws were "open," meaning that, in effect, the title was protected but licensing was voluntary. Alberta amended its Act to close it in 1930, and Ontario closed its Act in 1937.[4]

In the years that followed, the provinces and territories of Canada and all of the United States amended or passed licensing laws to regulate the engineering profession and the title of Professional Engineer. Prince Edward Island, in 1955, was the last province to enact closed legislation.

There is a difference between the Canadian and the U.S. laws. The engineering profession is self-regulating in Canada: Associations are created under the Acts and, in each province or territory, most members of the governing council are elected by the Association members and must, by vote, approve regulations and by-laws. In the United States, the state governments establish the regulations and license the engineers directly.

In some countries, there is no licensing of engineers and the possession of a degree or membership in a technical society is used as a gauge of the person's competence. In some countries, the term "engineer" is not regulated by law. In Britain, for example, it is often used to mean "mechanic" (the sign "Engineer on Duty" is found outside many garages). Professional competence is established in Britain by gaining membership in one of the technical societies, which call their members Chartered Engineers.

PROVINCIAL AND TERRITORIAL ACTS

The engineering profession is regulated in each province or territory of Canada by Acts of the provincial legislatures (for the provinces) or legislative councils (for the territories). The names of these Acts are listed below, and relevant excerpts from these statutes are included in Appendix B of this text.

Alberta	Engineering, Geological and Geophysical Professions Act
British Columbia	Engineers and Geoscientists Act
Manitoba	Engineering and Geoscientific Professions Act
New Brunswick	Engineering Profession Act
Newfoundland	Engineers and Geoscientists Act
Northwest Territories	Engineering, Geological and Geophysical Professions Act
Nova Scotia	Engineering Profession Act
Ontario	Professional Engineers Act
Prince Edward Island	Engineering Profession Act
Quebec	Engineers Act
Saskatchewan	Engineering and Geoscience Professions Act
Yukon Territory	Engineering Profession Act

Each of the Acts contains the basic elements of a self-regulating engineering profession. Although there are variations, each Act typically includes

- the purpose of the Act
- the legal definition of engineering
- the procedure for establishing a provincial or territorial Association of Professional Engineers, and the purpose (or "objects") of the Association
- standards for admission to the Association (or granting of a licence)
- procedures for establishing specific regulations to govern the practice of engineering
- procedures for establishing by-laws to govern the Association's administration and to elect a governing council
- a code of ethics to guide personal actions of the members
- disciplinary procedures

Throughout this textbook, the term "Professional Engineering Act," "provincial Act," or simply "Act" refers to the relevant Act or ordinance for the reader's province or territory. Similarly, the term "provincial engineering Association" or simply "Association" refers to the Association of Professional Engineers (or Ordre des ingénieurs) for the reader's province or territory. The term "engineer" also includes geologists, geophysicists, and geoscientists in those provinces and territories that include those specialties under the Act.

LEGAL DEFINITION OF ENGINEERING

Each provincial and territorial Act defines the term "professional engineer" or "the practice of professional engineering." These definitions are important, because they delineate the boundaries between engineering and other professions (such as architecture and town planning) and between engineers and other personnel in the design and development spectrum (such as scientists and technologists). Since the definitions of professional engineer and the practice of professional engineering vary slightly in each province and territory, the Canadian Council of Professional Engineers (CCPE) has proposed a simple, national definition. (CCPE is a federation of the twelve provincial and territorial bodies, described in more detail later in this chapter.) CCPE defines the practice of professional engineering as

any act of planning, designing, composing, evaluating, advising, reporting, directing or supervising, or managing any of the foregoing, that requires the application of engineering principles and that concerns the safeguarding of life, health, property, economic interests, the public welfare or the environment.[5]

The CCPE definition includes an exemption for anyone who

either holds a recognized honours or higher degree in one or more of the physical, chemical, life, computer or mathematical sciences, or who possesses an equivalent combination of education, training and experience, ... from practising natural science which ... means any act (including management) requiring the application of scientific principles, competently performed.[6]

The goal of this national definition is to increase the unity of the engineering profession, to permit easier movement of engineers throughout Canada, and to simplify licensing problems. This definition was ratified by CCPE, and it is anticipated that each province and territory will examine its Professional Engineering Act in due course and eventually amend it to agree with the CCPE national definition.

The legal definitions of engineering or the practice of professional engineering for all the provinces and territories of Canada are included in Appendix B. New Brunswick has the distinction of brevity, and British Columbia, Quebec, and the Northwest Territories share the distinction of length. No two definitions are identical.

Some provinces (notably British Columbia and Quebec) include a list of types of machinery or structures (such as railways, bridges, highways, and canals) that are within the engineer's area of practice. This makes the definition very clear and specific, but also very long and difficult to read or understand. As time passes, the list will get out of date as some components, such as steam engines, disappear, and new areas, such as engineering software, must be added.

The shorter definitions are easier to understand and remember, but they may contain terms (such as "engineering principles") that are very general and need further interpretation and definition. For example, in Ontario, the definition of the practice of professional engineering is almost identical to the CCPE definition:

> any act of designing, composing, evaluating, advising, reporting, directing or supervising wherein the safeguarding of life, health, property or the public welfare is concerned and that requires the application of engineering principles, but does not include practising as a natural scientist.[7]

For comparison, we could also examine the definitions of engineer and the practice of professional engineering as defined in the United States. The following definitions are from the Model Law (1996 revision) prepared by the U.S. National Council of Examiners for Engineering and Surveying. The Model Law, like the CCPE definition, serves merely as a guide for lawmaking bodies and has no legal effect unless written into law by state legislatures.

> **Engineer.** The term "Engineer," within the intent of this Act, shall mean a person who is qualified to practice engineering by reason of special knowledge and use of the mathematical, physical, and engineering sciences and the principles and methods of engineering analysis and design, acquired by engineering education and experience.

> **Practice of Engineering.** The term "Practice of Engineering," within the intent of this Act, shall mean any service or creative work, the adequate performance of which requires engineering education, training, and experience in the application of special knowledge of the mathematical, physical, and engineering sciences to such services or creative work as consultation, investigation, expert technical testimony, evaluation, planning, design, and design coordination of engineering works and systems, planning the use of land and water, teaching of advanced engineering sub-

jects, performing engineering surveys and studies, and the review of construction for the purpose of monitoring compliance with drawings and specifications; any of which embraces such service or work, either public or private, in connection with any utilities, structures, buildings, machines, equipment, processes, work systems, projects, and industrial or consumer products or equipment of a mechanical, electrical, hydraulic, pneumatic, or thermal nature, insofar as they involve safeguarding life, health, or property, and including such other professional services as may be necessary to the planning, progress, and completion of any engineering services.[8]

The above definitions, although far from identical, show considerable similarity, so it may not be a large step to a North American definition of engineering once a national definition has been accepted. However, all of the above definitions use the term "engineering principles," which creates a circular definition unless we provide further explanation.

The difference between "engineering" principles and "scientific" or "technological" principles lies in the depth and purpose of the study. The difference can be explained by referring to the policy statement of the Canadian Engineering Accreditation Board, which is a standing committee of CCPE and has the task of accrediting Canadian engineering programs. One of the key criteria is

to identify those programs that develop an individual's ability to use appropriate knowledge and information to convert, utilize and manage resources optimally through effective analysis, interpretation and decision-making. This ability is essential to the design process that characterizes the practice of engineering.[9]

The important words in the above definition are "appropriate knowledge," "manage resources optimally," and "design process." The term "engineering principles" includes mathematics, basic science, and engineering science appropriate to the specific discipline, but these concepts must be applied to the goal of optimal use of resources in the design process. Engineering differs from science mainly in the goal of the study: engineering involves putting scientific phenomena and principles into practical application. It differs from technology mainly in the depth of study and application of the appropriate subjects.

In particular, the determination of the "factor of safety" between expected usage and ultimate capacity of the system, device, or structure is the responsibility of the engineer, and the knowledge needed to set the factor of safety is obtained through study of engineering principles.

PROVINCIAL AND TERRITORIAL ASSOCIATIONS OF PROFESSIONAL ENGINEERS

To administer the provincial or territorial Act, each province or territory has established a self-governing Association of Professional Engineers. The name of the province is included in the Association name, as can be seen in the list of addresses for the Associations in Appendix A. Quebec has an "Ordre" rather than an Association, and the Associations in Alberta, British Columbia, the Northwest Territories, and Newfoundland include geologists, geophysicists, or geoscientists. In Ontario, the Association has adopted the simpler working name of Professional Engineers Ontario (PEO), although its legal name, the Association of Professional Engineers of Ontario, remains unchanged.

Although attempts to regulate engineering in Canada were begun in the late 1800s, the first Acts were not passed until the 1920s. Early forms of the Acts were "open" in that they protected the title Professional Engineer (P.Eng.) but did not prevent nonmembers from practising engineering.

Through amendments, each Act is now "closed," and a licence is required to practise engineering in every province and territory. The Act can be found in any public library, and copies are available from the provincial or territorial Association.

Under the authority of the Act, the Association is delegated the responsibility for administering it. Regulations or by-laws and a code of ethics have been written for each province or territory. In most cases, the usage is as follows:

- *Regulations* are rules set up to implement or support the Act; they concern topics such as qualifications for admission to the Association and professional conduct.
- *By-laws* are rules set up to administer the Association itself. They concern the methods for electing members to the Association Council, financial statements, committees, and so on.
- The *code of ethics* is a set of rules of personal conduct to guide individual engineers. Every engineer must be familiar with and endeavour to follow this very important document. The code of ethics is a component of the Professional Practice Exam and therefore is discussed in detail in this text.

Since the regulations, by-laws, and code of ethics are set up under the authority of each Professional Engineering Act, they govern the profession *with the force of law*. Engineers regulate their profession by electing the majority of members to the Association council (which also contains members appointed by the provincial government) and by confirming (by ballot) the regulations and by-laws passed by the council. For this reason, engineering is called a self-regulating profession. Obviously, for this system to work, engineers should be informed when voting on changes to regulations and by-laws, and they must be willing to serve in the elected positions in their Association, particularly at the council level.

ADMISSION TO THE ENGINEERING PROFESSION

Each provincial Association admits applicants to the profession by registering them as members of the Association and granting them licences to practise. The standards for admission are similar in all provinces. An applicant is typically admitted to the profession and awarded a P.Eng. licence if he or she satisfies six conditions:

- *Citizenship:* The applicant must be a citizen of Canada or have the status of a permanent resident.
- *Age:* The applicant must be the minimum age of 18 years or the legal age of majority in most provinces except the Yukon, which has a minimum age of 23 years.
- *Education:* The applicant must prove compliance with academic requirements, as discussed below.
- *Examinations:* Examinations may be required, as discussed below.
- *Experience:* The applicant must prove compliance with experience requirements, as discussed below.
- *Character:* The applicant must be of good character, as determined mainly from references.

Although these six conditions apply to almost every jurisdiction in Canada, some differences do exist. The following paragraphs give clearer explanations of some of the requirements. Readers should check with their local Association regarding these requirements, as the admissions process is constantly under review.

ACADEMIC REQUIREMENTS

Academic requirements are usually evaluated by a board of examiners or an academic requirements committee. The most important requirement for admission to the engineering profession is the academic accomplishment of the applicant. Graduation from a recognized (CCPE-accredited) program at a Canadian university grants full exemption from the examination program, except for the Professional Practice Exam. (In Quebec, graduation from a Quebec university grants exemption from the Professional Practice Exam as well.)

Degrees in engineering from many U.S. universities accredited by the U.S. Accreditation Board for Engineering and Technology (ABET) are recognized as equivalent to Canadian degrees. However, examinations may be required, even for those holding degrees from ABET-accredited universities, depending on individual circumstances. Some degrees are not recognized, and applicants holding them are required to pass a series of examinations confirming the engineering knowledge they have. Applicants must provide documents to substantiate their claims of academic qualifications.

EXAMINATIONS

People who have not completed university-level engineering degrees may apply for admission to the engineering profession, but will usually be assigned to write examinations. There are approximately sixteen to eighteen three-hour examinations for each branch of engineering. Applicants may be required to write a subset of exams to make up deficiencies in their academic qualifications. Permission to enter the examination system varies widely. In British Columbia, it is virtually open, whereas in Quebec, it is fairly tightly controlled. In Ontario, the examination system is open only to those who hold, as a minimum, a three-year engineering technologist diploma from a college of applied arts and technology, a technologist-level certificate from the Ontario Association of Certified Engineering Technicians and Technologists (OACETT), or other acceptable education as determined by the Association's academic requirements committee.

Although the examination system provides an alternative route into the profession, the examination system is *not* an educational system. People applying for Association exams present themselves as being qualified and prepared to write and pass the exams. The Associations do not offer classes, laboratories, or correspondence courses.

The number and type of examinations assigned will depend on the applicant's academic achievements to the date of evaluation. The possession of one or more postgraduate degrees beyond the bachelor's degree may be taken into consideration when determining exam assignments. However, postgraduate degrees by themselves are not adequate, since admission requires a knowledge of engineering principles covered in accredited engineering programs at the undergraduate level.

PROFESSIONAL PRACTICE EXAM

Most applicants, regardless of academic qualifications, previous courses taken at university, or membership in other Associations, are required to pass an examination covering topics in professional practice, law, contracts, liability, and ethics. Exceptions are made for people who transfer from province to province and have written the examination in the previous five years. Quebec does not require its university graduates to sit the exam.

Figure 2.1 shows the typical admissions process for graduates of CCPE-accredited university engineering programs. The process is fairly simple for such graduates, since only the Professional Practice Exam is required. For applicants who are not graduates of accredited programs, a more extensive set of exams may be required.

EXPERIENCE REQUIREMENT

Admission to the engineering profession is not based on academic achievement alone; satisfactory experience is a major requirement for obtaining a licence. It is the responsibility of the applicant to prepare accurate and complete documentation demonstrating to the Association that the experience requirements have been met. This documentation will be assessed by a committee of the Association to ensure that the applicant can meet the generally accepted level of skill required to engage in professional practice. Standards for evaluating engineering experience, developed by CCPE, have been adopted by most provincial Associations. The applicant's experience is evaluated on its nature, duration, currency, and quality. These standards are explained below.

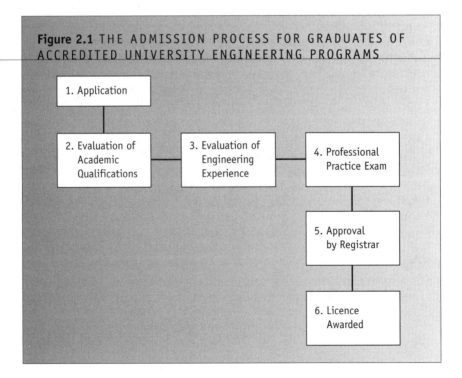

Figure 2.1 THE ADMISSION PROCESS FOR GRADUATES OF ACCREDITED UNIVERSITY ENGINEERING PROGRAMS

1. Application

2. Evaluation of Academic Qualifications

3. Evaluation of Engineering Experience

4. Professional Practice Exam

5. Approval by Registrar

6. Licence Awarded

NATURE

Engineering experience is expected to be similar to the applicant's area of academic study. However, if there is an incompatibility (for example, if a mechanical engineering graduate should be working in an electrical engineering job), the Association may ask why this incompatibility exists and, in some cases, may require additional experience or additional studies before granting a licence. Also, some jobs may have similar descriptions but, depending on the actual activities performed, may be given different credit as engineering experience. An applicant should not presume that his or her area of employment will automatically be accepted (or rejected) as valid engineering experience. In particular, if your employment falls into any of the following categories, you should consult the provincial Association for more specific advice:

- teaching (at any level)
- sales and marketing
- military service
- project management
- operations and maintenance
- computer engineering

DURATION

The length of experience required has increased over the past decade. Until recently, the general experience requirement was two years of direct engineering work after obtaining the engineering degree. However, since 1993, almost every province has implemented a four-year experience requirement or is in the process of implementing a four-year requirement.

CURRENCY

The "recentness" of the experience is important. In general, more recent experience is preferred over experience acquired in the distant past, for several reasons. The most important reason is probably that engineering procedures and standards are changing as the computer revolution sweeps through all of society. Experience obtained in pre-computer days may be obsolete.

QUALITY

Each applicant must prepare an experience résumé and explain how the experience satisfies the requirements. Experience is usually judged against five quality criteria:

- application of theory
- practical experience

- management of engineering
- communication skills
- social implications of engineering

The application of theory is a mandatory experience requirement that must be demonstrated over a substantial part (but not necessarily all) of the experience period. It must be supplemented by exposure to, or experience in, the remaining four criteria. Each province may put a slightly different emphasis on each of these criteria, so you should consult your Association's guidelines while preparing your experience documentation.

MISCELLANEOUS EXPERIENCE CREDIT
Some provinces interpret and evaluate experience slightly differently. The following miscellaneous points may or may not apply in your province:

- Many provinces (such as Ontario) require all applicants, including engineers whose experience has been gained in other countries, to obtain twelve months' experience in Canada under the direction of a professional engineer who is licensed in Canada.[10]
- Applicants can usually obtain credit for twelve months' experience for completing a postgraduate degree in engineering. In fact, in some provinces, more credit may be obtained (up to the total time spent in postgraduate studies), depending on how well the postgraduate experience satisfies the five quality criteria described above.
- In some provinces where the four-year experience requirement is in place, the Association will grant up to twelve months' credit for experience obtained prior to graduation. The experience must satisfy all of the standards mentioned above.

GRADUATES OF FOREIGN ENGINEERING SCHOOLS

An applicant who received his or her engineering education from a foreign college or university must provide the Association with originals (or certified copies) of all transcripts and diplomas. The academic requirements committee (or board of examiners, depending on the province) will assess those documents and determine the admissibility of the applicant. Each case is evaluated individually. However, since foreign universities are not accredited by the Canadian Engineering Accreditation Board, admission is not generally awarded on the sole basis of foreign educational documents; some further evidence of engineering competence is usually required.

Such evidence can be provided in several forms. For example, senior engineers with foreign education may be able to document their actual engineering achievements. Other applicants with a foreign bachelor's degree may validate their earlier education by completing a master's degree (or even a series of advanced undergraduate engineering courses) in an accredited Canadian university. In most provinces, an applicant who is not a graduate of a CEAB-accredited engineering program, but who has completed engineering studies judged by the Association to be equivalent to the *Syllabus of Examinations* published by the Canadian Council of Professional Engineers (CCPE), is normally assigned to write a set of four Confirmatory Examinations. These examinations are usually set by the Association, and they cover the advanced topics in a small portion of the full engineering program. Readers should contact their provincial or territorial Association for more information on validating credentials.

Applicants from foreign universities should understand that the request for corroborating evidence of academic ability does not imply any lack of respect for the foreign university or for the individual. The Association is required by law to assess the qualifications of applicants and, in the absence of an accreditation process for the foreign university, the Association must evaluate other evidence to justify admission.

NONRESIDENT OR TEMPORARY LICENCES

Most provinces provide two types of engineering licence: full membership for residents of the province, and nonresident or temporary licences. The procedures for obtaining a nonresident or temporary licence vary slightly from province to province. In Ontario, an applicant must provide evidence that he or she is

- a member of an Association of Professional Engineers in another province or territory with equal admission requirements,
- qualified to work on the specified project and is familiar with the applicable codes, standards, and laws relevant to the project,
- widely recognized in the field of practice of professional engineering relevant to the project, and
- collaborating with a member of the Association in completing the project specified for the temporary licence. This last requirement may be waived if the applicant is highly qualified.

LICENSING OF ENGINEERING CORPORATIONS

The licensing of corporations varies from province to province. Controversy has occasionally arisen over this licensing, since its purpose is to protect the public against incompetence, negligence, and professional misconduct, and qualities such as competence can be evaluated accurately only for human beings.

The main issues at stake concern the protection of the public: which individuals are responsible for providing the engineering services, and does the corporation carry liability insurance? A more recent question is one of quality assurance: how can the public be assured of the continuing competence of the corporation's engineers? Each province has developed slightly different methods to deal with these issues.

In every province, the Act requires (or the Association is developing regulations to require) corporations providing services to the general public to identify the individuals who are responsible for providing these services. (In some provinces, sole proprietors are exempt from the regulations.) In about half of the provinces, the Act requires corporations or other business entities offering professional engineering services to the public to hold a Certificate of Authorization. To obtain a Certificate, the business must assign one or more of its members the responsibility for and supervision of the engineering services provided. The requirements for liability insurance vary drastically from province to province. Some provinces make the insurance mandatory, while others make it voluntary.

For example, in Ontario, all Certificate of Authorization holders must carry professional liability insurance or disclose in writing to every client that they are not insured. Failure to conform to this regulation is considered to be professional misconduct.[11]

Employees of corporations that have a corporate certificate (or that do not offer engineering services to the public) do not have to apply for individual certificates of authorization. However, if an engineer "moonlights" at night or on weekends, then a certificate may be necessary. In some provinces (such as Ontario), every entity — whether an individual, a partnership, or a corporation that offers or provides professional engineering services to the public — requires a Certificate of Authorization. If you are planning to provide services to the general public, be certain to explore these points with your provincial Association.

CONSULTING ENGINEERS

The designation "Consulting Engineer" is not regulated at present in any province but Ontario. It is controlled in Ontario by the regulations under the Professional Engineers Act. To qualify as a consulting engineer, a member

- must have been continuously engaged for at least two years in private practice
- must have at least five years of satisfactory experience since becoming a member

- must pass (or be exempted from) examinations that may be prescribed by the Association Council

Since applicants for the Consulting Engineer designation must be engaged in private practice, and therefore offering their services to the public, they must also be holders of a Certificate of Authorization in Ontario or associated with a partnership or corporation that is a holder of a certificate.

THE ENGINEER'S SEAL

In every province, the Professional Engineering Act provides for each engineer to have a seal denoting that he or she is licensed. All final drawings, specifications, plans, reports, and other documents involving the practice of professional engineering, when issued in final form for action by others, should bear the signature and seal of the professional engineer who prepared and approved them. This is particularly important for services provided to the general public. The seal has important legal significance, since it implies that the documents have been competently prepared and indicates clearly the person responsible for them. The seal should not, therefore, be used casually or indiscriminately. In particular, preliminary documents should not be sealed, but marked "preliminary" or "not for construction."

The seal denotes that the documents have been *prepared* or *approved* by the person who sealed them. This implies an intimate knowledge of and control over the documents or the project to which the documents relate. An engineer who knowingly signs or seals documents that have not been prepared by himself or herself or by technical assistants under his or her direct supervision may be guilty of professional misconduct and may also be liable for fraud or negligence if the misrepresentation results in someone suffering damages.

A fairly common problem involves engineers who are asked to "check" documents, then sign and seal them. This is usually not ethical. The engineer who prepared them or supervised their preparation should seal the documents. If they were prepared by a non-engineer, then perhaps he or she should have been under the supervision of an engineer. The extent of work needed to "check" a document is not clearly defined, and many disciplinary cases have resulted when engineers "checked" and sealed documents that later proved to have serious flaws.

Alberta recently reviewed its policies and revised its standards of practice to permit engineers, geologists, and geophysicists to stamp the work of others, provided that the stamping is accompanied by a thorough review of the work and a full acceptance of responsibility for the work. The Practice Standards Committee of the Association of Professional Engineers, Geologists and Geophysicists of Alberta (APEGGA) examined the legislation of other associations, and, while the wording of the legislation was similar to the Alberta Act, seven associations permitted or encouraged the interpretation that stamping the work of others is acceptable when accompanied by a thorough review and acceptance of responsibility. Therefore, the problem is reduced to defining what "checking" means. If you have performed sufficient review and analysis of the work that you are willing to accept full responsibility for it, then it is acceptable to stamp the work, at least in Alberta.[12]

In some projects, a proper review would require complete duplication of the analysis, and if the work is completely redone, then it would be appropriate to assume responsibility for it. However, this still leaves grey areas. Do not assume responsibility for work that you have not thoroughly and independently reviewed, and if necessary, where two or more engineers assume responsibility for different areas of a project, specify the areas of responsibility unambiguously. This topic is discussed in more detail in Chapter 9.

THE ENGINEERING CODE OF ETHICS

Each provincial Association of Professional Engineers subscribes to a code of ethics, which sets out a standard of conduct that members must follow in the practice of professional engineering. The code of ethics for each province and territory of Canada is included in Appendix B. Each code defines, in general terms, the duties of the engineer to the public, to the employer (or client), to fellow engineers, to the engineering profession, and to him- or herself. The major purpose of the code is, of course, to protect the general public from unscrupulous practitioners; however, by instilling public confidence in engineering, the code also raises the esteem of the entire profession.

In most provinces, the code of ethics is specifically mentioned in the Act and therefore has the full force of law. Discussion of the code of ethics as a guide to personal professional conduct is the main topic of the latter half of this text, which describes the evolution and application of codes in detail.

PROFESSIONAL MISCONDUCT

The main purpose of the provincial Associations of Professional Engineers is to protect the public welfare. Therefore, it is occasionally necessary to discipline errant engineers. Under the terms of each provincial Act, the Association is awarded the authority to reprimand, suspend, or expel a member who is guilty of professional misconduct, which is usually defined as negligence, incompetence, or corruption. These terms, while widely used in common speech, require formal legal definitions to be enforceable, and the Acts and regulations include many pages defining these terms and giving specific examples of what constitutes professional misconduct.

Each provincial Association, in order to enforce the Act and regulations and to prevent or discipline misconduct, has staff or council members who receive complaints, prosecute persons practising engineering under false pretences, and arrange disciplinary hearings for engineers charged with misconduct. Disciplinary decisions are made not by the staff members, however, but by the committee of engineers appointed by the council. Since the council is mainly elected by the engineers of the province, the "self-regulating" aspect of the profession is carried through to disciplinary actions as well. The results of disciplinary hearings are usually published and circulated to all members. (For more on the disciplinary process, see Chapter 14.)

The most frequently reported complaints concern violations of the code of ethics. In most provinces, the code of ethics has been specifically included in the Act and is therefore enforceable under the Act. The terms of the code of ethics are based on common sense, natural justice, and basic ethical concepts. Although everyone should read and understand the code, it is not usually necessary to memorize it; most engineers find that they follow it intuitively and never need fear charges of professional misconduct.

THE IRON RING AND THE ENGINEERING OATH

In addition to the code of ethics, which by law enjoins each engineer to act in an honest and conscientious manner, there is a much older voluntary oath, written by Rudyard Kipling and first used in 1925, called the Obligation of the Engineer. Those who have taken the oath can usually be identified by the iron ring each wears. The iron ring is awarded during a rather solemn ceremony called the Ritual of the Calling of an Engineer, which is conducted by the Corporation of the Seven Wardens. Although the corporation is not a secret society, it does not seek or require publicity.

The ceremony is generally conducted, at universities that grant engineering degrees, during April or May of each year, and is made known to the graduating students. The ceremony permits nonuniversity engineers to participate, but it is not open to the general public. The iron ring does not indicate that a degree has been awarded, but it shows that the wearer has participated in the ceremony and has voluntarily agreed to abide by the Obligation of the Engineer. The Obligation is fairly brief and directs the engineer to high standards of performance and thought.

A detailed discussion of the engineering oath and a historical note on the iron ring are included in Appendix C.

CANADIAN COUNCIL OF PROFESSIONAL ENGINEERS (CCPE)

The Canadian Council of Professional Engineers (CCPE) was established in 1936 as a federation of the provincial and territorial Associations that license engineers and oversee the profession across Canada. As a federation of Associations, CCPE does not have individual members, but every licensed engineer is indirectly a member of CCPE. The role of CCPE is to co-ordinate the engineering profession on a national scale. To achieve this goal, CCPE has three important boards or committees, described below.

CANADIAN ENGINEERING ACCREDITATION BOARD (CEAB)

In 1965, the Canadian Council of Professional Engineers established the Canadian Accreditation Board (CAB), now known as the Canadian Engineering Accreditation Board (CEAB). The concept of accreditation was implemented by CCPE to test and evaluate undergraduate engineering degree programs offered at Canadian universities and to award recognition to programs that meet the required standards. With the consent of the engineering Associations, CEAB was empowered to develop minimum criteria for undergraduate engineering degree programs and, through a process of direct investigation, to provide engineering schools with a means of having their programs formally tested against these criteria. The criteria for accreditation are formulated to ensure that graduates receive an education satisfying the academic requirements for professional engineering registration throughout Canada.[13]

An accreditation visit is undertaken at the invitation of a particular engineering school and with the concurrence of the Association for that province. A team of senior engineers is assembled, composed of specialist engineers for the subjects involved and at least one engineer from the provincial Association. Armed with documents, including a detailed questionnaire completed by the institution beforehand, the team proceeds to consult with administrators, faculty, students, and department personnel.

The team examines the academic and professional quality of faculty and the adequacy of laboratories, equipment, computer facilities, and so forth. It also evaluates the quality of the students' work on the basis of face-to-face interviews with senior students and assessment of recent examination papers, laboratory work, reports and theses, records, models or equipment constructed by students, and other evidence of the scope of their education.

Furthermore, the team performs a qualitative analysis of the curriculum content to ensure that it meets the minimum criteria. Finally, the team reports its findings to CEAB, which then makes an accreditation decision. Based on the information provided, CEAB may grant or extend accreditation of a program for a period of up to six years, or it may deny accreditation altogether. CEAB publishes an annual listing of the accreditation history of all programs that are presently or have ever been accredited.

Engineering accreditation is extremely important to the university and to the graduates of the program, because accreditation is a measure of quality. A program is accredited only by demonstrating academic and professional competence. In recognition of this quality assurance, every provincial Association grants exemption from all the technical examinations required for licensing as a professional engineer. Although many engineering students (or even graduate engineers) may not be aware of CEAB's role in assuring quality in programs of study, CEAB nevertheless has a significant influence on their lives.

CANADIAN ENGINEERING HUMAN RESOURCES BOARD (CEHRB)

Since 1972, the Canadian Engineering Human Resources Board (formerly the Canadian Engineering Manpower Board) has been a window on the engineering profession, providing detailed information about the supply of and demand for Canadian engineers in business, government, and educational institutions. CEHRB is a standing committee of the Canadian Council of Professional Engineers. Its mandate is to collect and maintain data relating to the engineering profession in Canada. It does this through surveys and studies and through its engineering human resources database.

In addition to these sources of information, CEHRB's close links to government, industry, and educational institutions afford it an excellent source of information on trends that affect the engineering labour market and the profession in general. CEHRB makes this information available to its members through its reports, a newsletter, and a specialized information service.

CANADIAN ENGINEERING QUALIFICATIONS BOARD (CEQB)

The Canadian Engineering Qualifications Board (CEQB) deals with matters concerning qualifications for entering the engineering profession. This is particularly important for evaluating the credentials of candidates who have studied in foreign countries. The CEQB also monitors engineering admission procedures, engineering practice guidelines, and continuing competence requirements in the various provinces and territories, and strives to encourage commonality while maintaining standards. The CEQB maintains a CCPE *Syllabus of Examinations* as part of this mandate. A recent major achievement of the CEQB was the signing of an interim agreement on mobility between the constituent associations of the CCPE. This interim agreement will be reviewed in about the year 2000, but paves the way for more convenient transfer of engineering registration from one province or territory to another.

CANADIAN ENGINEERING PUBLIC AWARENESS BOARD (CEPAB)

The Canadian Engineering Public Awareness Board (CEPAB) is the most recently established standing committee of the Canadian Council of Professional Engineers. Its goal is to raise public awareness of "the important role that professional engineers play in the sustainable growth and development of the Canadian economy and in the creation of wealth for all Canadians."[14] The CEPAB is charged with the responsibility of developing a leadership role for professional engineers in public policy matters that affect the engineering profession, and with suggesting to CCPE and its constituent members methods of improving the image of professional engineers and informing the public about the engineering profession's achievements. These are worthwhile tasks that have frequently been overlooked. The welfare of the profession and its members depends, to a great degree, on public awareness of its achievements.

THE CANADIAN ACADEMY OF ENGINEERING

The Canadian Academy of Engineering is an independent, self-governing, and nonprofit organization established in 1987 to serve the nation in matters of engineering concern. The Fellows of the Academy are professional engineers from all disciplines and are elected on the basis of their distinguished service and contribution to society, to the country, and to the profession. The total number of Fellows may not exceed 250 at any one time.

The mission of the Academy is to enhance, through the application and adaptation of science and engineering principles, the promotion of well-being and the creation of wealth in Canada. The Academy fulfils this mission by

- promoting increased awareness of the role of engineering in society
- recognizing excellence in engineering contributions to the Canadian economy
- advising on engineering education, research, development, and innovation
- promoting industrial competitiveness while preserving the environment in Canada and abroad
- speaking out on issues relevant to engineering in Canada and abroad
- developing and maintaining effective relations with other professional engineering organizations, academies, and learned societies in Canada and abroad

The Academy is self-financing and does not receive grants from government, although it may agree to carry out studies and surveys on a contract basis. The Fellows of the Academy can therefore bring into corporate activity, in a completely independent manner, the wide experience and expert knowledge that they have acquired as practising members within the engineering profession of Canada.[15]

The Academy provides an expert forum for the development of advice on a wide range of matters of a technological and engineering nature for the engineering profession, for the educational system, and for the governments of Canada, as these matters are of growing importance in an increasingly technology-dependent society. Among the Academy's recent studies are a comprehensive study of engineering education in Canada[16] and communiqués on the role of engineering research in Canada, to mention just two of several such publications. The Academy is located in Ottawa.

SUMMARY

Figure 2.2 illustrates how the engineer interacts with the various organizations mentioned in this chapter. The licensed engineer is a member of a provincial Association and is usually a member of at least one engineering society. All provincial Associations are federated members of CCPE. The members of the Association elect the council, which appoints the staff.

TOPICS FOR STUDY AND DISCUSSION

1. Should the "professional" person be more concerned about the welfare of the general public than the "average" person? Should persons in positions of great trust, whose actions could cause great harm to the general public, be required to obey a higher code of ethics than the average person? How does your answer apply to engineering?

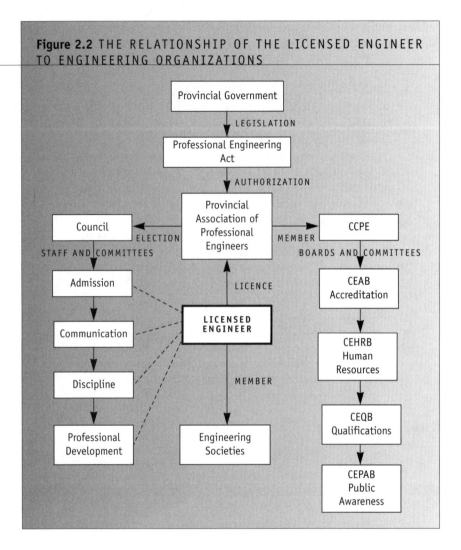

Figure 2.2 THE RELATIONSHIP OF THE LICENSED ENGINEER TO ENGINEERING ORGANIZATIONS

2. The definitions of "engineering" and "engineering practice" vary from province to province (to territory), as described in this chapter and in Appendix B. Consult Appendix B and review the definitions for your province or territory. Do they agree with the CCPE definition and the U.S. Model Law definitions given in this chapter? Can you think of any activities that are clearly engineering activities but that would *not* be covered by the definition for your province or territory? From the various definitions available in this text, select the one you consider to be most accurate, and explain why.

3. Most professions in Canada have a two-pronged structure in which one organization regulates the members of the profession, and an

independent organization works on behalf of the members by setting fees and organizing pension plans and the like. A good example is law: the Law Society regulates members, and the Bar Association works on their behalf. Similarly in medicine, the College of Physicians and Surgeons regulates the profession, and the Medical Association works on behalf of the members. In engineering, a similar two-pronged structure does not exist. In some provinces the provincial Association is empowered to act on behalf of engineers, but in other provinces the Association is not legally permitted to act on behalf of engineers as directly as the Bar Association and the Medical Association can. Discuss the advantages and disadvantages of the two-pronged structure. Should it be implemented in engineering? Please explain the reasons for your decision.

4. To regulate the engineering profession effectively, the provincial Associations need a definition of "engineering" (or "the practice of engineering") that is sufficiently specific that it does not encroach on other professions (such as science and architecture) yet sufficiently general that it can include new areas of engineering practice (such as computer engineering and software engineering) as they are established. Which of the various provincial definitions (if any) best satisfies these conflicting criteria? How do we distinguish between *computer engineering* and *computer science*? How do we distinguish between *software engineering* and *software development*?

NOTES

1. J.R. Millard, *The Master Spirit of the Age: Canadian Engineers and the Politics of Professionalism, 1887–1922* (Toronto: University of Toronto Press, 1988).
2. H.A. MacKenzie [Opening address for the debate on the Professional Engineers Act, 1968–69], Ontario, Legislature, *Debates*.
3. Millard, *The Master Spirit of the Age*.
4. Ibid.
5. "CCPE Set to Ratify National P.Eng. Definition," *Engineering Dimensions* 11, no. 5 (September/October 1990): 21.
6. W. Ryan-Bacon, Director, Educational Affairs, Canadian Council of Professional Engineers (CCPE), personal correspondence, 1995.
7. Professional Engineers Act, RSO 1990, c. P.28, s. 1.
8. National Council of Examiners for Engineering and Surveying (NCEES), *Model Law* (Clemson, SC: NCEES, 1996), 1, 2. © 1996 All Rights Reserved National Council of Examiners for Engineering and Surveying. Reprinted with permission of NCEES.
9. Canadian Engineering Accreditation Board, *1996 Accreditation Criteria and Procedures* (Ottawa: CCPE, 1996), 10.
10. Professional Engineers Ontario, *Guide to Required Experience for Licensing as a Professional Engineer in Ontario* (Toronto: PEO, 1996), 7.

11. Professional Engineers Ontario (PEO), *Guideline to Professional Practice* (Toronto: PEO, 1996), 16.

12. Association of Professional Engineers, Geologists and Geophysicists of Alberta, "Stamping Policy Undergoes Change," *The Pegg* (November 1995).

13. Canadian Engineering Qualifications Board, *1995 Annual Report* (Ottawa: CCPE, 1995); and C. Ella, "Ensuring Quality Engineering Programs: The Canadian Engineering Accreditation Board," *Engineering Dimensions* 11, no. 3 (May/June, 1990): 40–41.

14. Canadian Council of Professional Engineers, *CEPAB Policy Statement* (Ottawa: CCPE, undated), 1.

15. The Canadian Academy of Engineering is located at 130 Albert St., Suite 1414, Ottawa, Ontario, K1P 5G4.

16. Task Force of the Canadian Academy of Engineering, *Engineering Education in Canadian Universities* (Ottawa: Canadian Academy of Engineering, 1993).

Part Two

Professional
Practice

CHAPTER THREE

Entering
Engineering
Employment

Most professional engineers in Canada are employees of corporations or private companies. In fact, since four years of satisfactory engineering experience are typically required to obtain an engineering licence, almost all engineers are employees at some time in their careers. This chapter describes the typical challenges and decisions that a recent university graduate will face during his or her first employment as an engineer (or engineer-in-training) and gives some advice for meeting these challenges. The reader of this text is expected to be a recent university graduate or senior undergraduate, so this chapter should provide useful insight and advice for most readers.

MAKING THE TRANSITION

Moving from the theory of the university to the reality of the workplace is very exciting, but sometimes a little threatening. The excitement comes from moving to a new location and meeting new colleagues, and from being involved in important projects that will be implemented, constructed, or manufactured and will influence other people's lives. It is very enjoyable to see one's ideas influencing the progress of an engineering project. This creative aspect of engineering was described by a recent graduate:

> Although most companies assign a new engineer to a training program, a new engineer at Digital Equipment reported that she was merely handed an orientation manual and a set of equipment specifications. She read them, and watched her colleagues a little apprehensively for the next two

days. After this initial aloofness, she says: "The first guy I was introduced to brought a design problem over to my desk and said, 'Here, see if you can do this.' I worked out an answer, and we went over it." This pulled her directly into the design activity of a major new computer system. As time passed, she was delighted that her ideas found their way into a new product. She said: "The feeling was *Wow!* It's mine and it's on the market!" After three years her responsibilities had expanded and her salary had grown by nearly 50 percent. She also reported that her working day typically extended from 8 A.M. to 6 P.M. and that she occasionally worked 80 hours a week when a project deadline approached.[1]

Obviously, this type of work is very different from secluded study in a university library. Recent graduates may notice another difference: university courses have rather short and precise durations, but on the job, projects may take a long time, perhaps a year or more. Even when one aspect of a project is finished, the graduate engineer may be drawn into later parts of the project. For example, even when a design project is completed, the design engineer may be drawn into the production, construction, or fabrication phases, and it may take a long time (and possibly many redesigns) before the project yields a profit for the company (and recognition for those involved). This is quite different from the clearly defined time span of a typical university course.

Other surprises may be how little supervision you receive, how much responsibility is assigned to you because you are "fresh out of school and familiar with the theory," and how important your personal time management becomes when you have deadlines. The following advice may help the new engineering graduate during the first few weeks on the job:

- Begin your employment on a professional note: apply to your provincial Association for professional registration (or to be recorded as an engineer-in-training).
- Make the best use of the employer's training program. If you were unable to get a commitment for such a program during your job interview, discuss the possibility of a training program or familiarization period with your new supervisor. Once you take on the full responsibility for your job, it may be too late to discuss training.
- Begin to document your engineering experience. The provincial Associations have criteria for experience that should be examined at the beginning of your employment.
- Study your company and learn its organization and information flow.
- Develop a professional attitude. Examine your own work habits to see if there are ways to improve your efficiency, productivity, and self-confidence in the work environment.

These topics are important to newly hired engineering graduates, and are discussed in more detail in the following sections.

APPLYING FOR PROFESSIONAL REGISTRATION

It is important to apply to your provincial or territorial Association for admission to the engineering profession as soon as possible. The addresses of the twelve Associations are in Appendix A of this text, and a simple letter or phone call will get the process started.

Many recent graduates have the misunderstanding that they must wait until they satisfy all of the admission requirements, including the engineering experience requirement, before they apply. This is a mistake; applicants may apply at any time after university graduation. Almost every provincial and territorial Association has an engineer-in-training category that provides a link to the profession, and this exchange of information may simplify the licensing process. Foreign-trained applicants also have an incentive to apply as soon as possible, since the Association may be able to provide advice that speeds up the qualification process.

SURVIVING THE TRAINING PERIOD

Although almost every company claims to have a training program, the term is used loosely and may cover a wide spectrum. A recent study reveals that formal training in Canada lags behind that provided in other countries,[2] so in some cases, formal training may not be provided at all (particularly if the supervisor received no such training when he or she was hired). The new employee is typically put directly to work, and the supervisor provides instruction as the need arises.

Interestingly, many recent university graduates prefer to be put directly to work rather than undergo a formal training program. One explanation may be that the new graduate, saturated with formal classroom education, is eager to get started on practical work. However, while on-the-job training may be more popular, it is not more effective. Therefore, if a formal training program has been organized for you, make the most of it by participating fully. If, for any reason, a training program was not offered during your job interview, you should discuss it with your supervisor when you arrive on the job. It may be possible to negotiate a training period, particularly if there is no immediate crisis facing the unit and you can be spared for a short period. Even a few days or even hours of orientation are better than nothing.

To give you some insight into the goals of training programs, the following summary of *induction strategies* used by organizations to ac-

.aint new employees with their engineering duties is reproduced with permission from the work of Edgar H. Schein, professor of Industrial Management at the Massachusetts Institute of Technology. These strategies are arranged roughly in order from least desirable to most desirable. An enlightened employer would rarely follow either of the first two "strategies," since they are somewhat inefficient and unlikely to produce a committed or enthusiastic employee.

1. Sink or Swim The new graduate is simply given a project and is judged by the outcome. No information or guidance is provided, and the newcomer is partially judged by how well he or she defines the goal and organizes the assignment. This requires the new employee to take vaguely stated objectives and translate them into specific tasks that can be dealt with one by one. However, with no guidance, the project may be an insurmountable hurdle, and the results may be inconclusive or disheartening.

2. The "Upending" Experience The intent of this type of strategy is to jar the new employee loose from the presumed impracticalities acquired in college and to confront him or her with the "realities" of industrial life. In one approach of this nature reported by Schein, each new engineer is asked to analyze a special electric circuit that violates several theoretical assumptions. When the new engineer reports that the circuit will not work, it is demonstrated that it not only does work but has been in commercial use by the company for several years. Chastened, the engineer is then asked to find out *why* it works. When this proves to be impossible, the newcomer develops an appreciation for the new colleagues and is then considered ready to tackle a real assignment. Conversely, in some cases, the individual may become demoralized over the fact that the first encounter with industry has been so negative.

3. Training While Working This is the typical on-the-job type of induction program. The new person is given an assignment commensurate with his or her experience level and carries it out under the close guidance of a supervisor.

4. Working While Training The new person is considered to belong to a formal training program but is given small projects involving real work. Rotation through several different departments may occur during the course of the program. It is sometimes difficult to decide whether programs of this type should be classified as *on-the-job* or *formal* training.

5. Full-Time Training These programs clearly belong in the formal category. They usually involve class work and rotational assignments that call for the trainee to observe the work being done by others; direct participation is minimal. Schein observes that trainees occasionally criticize such observational activities as mostly meaningless or "Mickey Mouse." The future application of the knowledge should be made clear to the new employees.

6. Integrative Strategies In an approach of this type, an attempt is made to adjust to the different needs of different trainees. In one such program, new employees are given regular job assignments for a year and then are sent to a summer-long, full-time university training program. Ideally, the supervisors for these initial assignments are selected and trained to be sensitive to the problems of new employees. Some programs lead to advanced degrees and are clearly much sought after.[3]

Of the six strategies recorded by Schein, the first two seem unlikely to achieve the goal of any inductive strategy, which is to turn a new engineer into a productive employee as soon as possible. (It should be mentioned that Schein did not endorse these two strategies; he merely recorded them.) Initial assignments should give the new employee as much responsibility as possible, for the sake of the individual and the organization. A training program, however brief, will shorten the time it takes the employee to start working at full potential, and this is important for the employee's self-esteem as well as company profitability. There is much to gain by using people at their highest potential and much to lose by using them at their lowest.

OBTAINING ENGINEERING EXPERIENCE

As you begin your new employment, you should give some thought to documenting your experience and judging how well it will satisfy the experience requirements for your licence. As explained in the previous chapter, you will need four years of engineering experience to fulfil the licensing requirements (in every province except Quebec, which, as this text goes to press, remains at two years of experience). The engineering experience must be documented in the form of a personal résumé for evaluation by the provincial Association. The Association will typically require that the experience satisfy the following quality criteria:[4]

- **Application of Theory** This is the best form of experience, and includes analysis, design and synthesis, testing methods, and implementation of projects. Most professional associations would expect to see a sizable portion (typically 20 percent) of the experience submitted for licensing to be in this category.

ctical Experience Most technical experience that does not fall ıder the other headings would likely qualify as practical experi-
ınce.

Management of Engineering This experience is difficult for a new graduate to obtain, but as time passes, the new graduate will typically become involved in various aspects of management, such as planning, scheduling, budgeting, supervision, project control, and risk assessment.

- Communication Skills Engineering is critically dependent on accurate and effective communication: preparing written work, such as reports, and specifications, making oral reports or presentations, and even making presentations to the public.
- Social Implications of Engineering This is a difficult area to define clearly, but it typically includes any experience that increases the awareness of an engineer's professional responsibility to guard against conditions that are dangerous to life, limb, property, or the environment, and to call such conditions to the attention of those responsible.

LEARNING THE COMPANY ORGANIZATION

An important task of a new employee is to learn how the company is organized. With this knowledge, the employee is better able to anticipate problems, to recommend improvements, and thereby assist in achieving the company's goals. In addition to the company's formal organizational structure, there is a less formal (perhaps invisible) information flow within the company. The formal structure is usually explained in the company policy manual or in annual reports. However, the information flow may not be obvious; to learn it, you must observe and ask questions. The next few paragraphs deal with both of these characteristics of the company. As a new employee, your goal is to see where you fit into the formal structure and how you can most efficiently communicate with your colleagues to advance the company's activities.

FORMAL ORGANIZATIONAL STRUCTURE

Consider the most typical case: a company involved primarily in design and development, possibly with some research activity also. For such a company, three basic types of organization are common: the project structure, the functional structure, or the hybrid structure.

The *project structure* (Figure 3.1) results in a very decentralized corporation. Each subgroup of the department is responsible for a com-

Figure 3.1 EXAMPLE OF PROJECT STRUCTURE

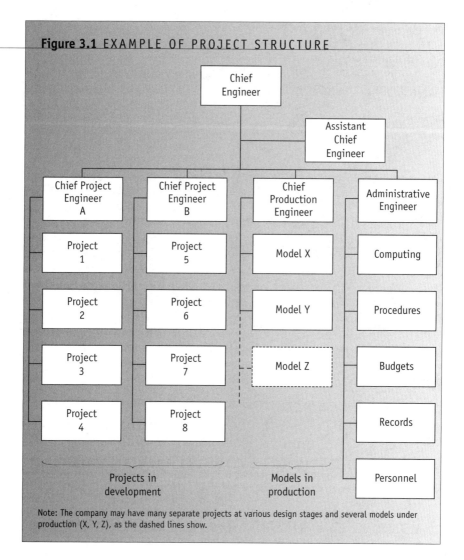

Note: The company may have many separate projects at various design stages and several models under production (X, Y, Z), as the dashed lines show.

plete project (or projects) and contains within itself all the functional specialities necessary to complete the projects. The major advantage of such a structure is that the boundaries of responsibility are crystal clear. The major disadvantage is that functions are duplicated among groups.

The *functional structure* (Figure 3.2) is highly centralized. A department is split into its functional specialities, and the functional subgroups operate on all projects passing through the department. One company might separate the functions into electronic design and mechanical engineering design. Another company might separate the functions into aerodynamics, stress analysis, weights, materials, and

Figure 3.2 EXAMPLE OF FUNCTIONAL AND HYBRID STRUCTURES

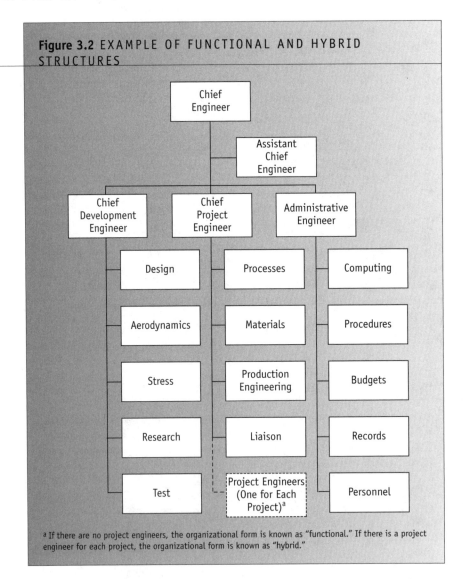

[a] If there are no project engineers, the organizational form is known as "functional." If there is a project engineer for each project, the organizational form is known as "hybrid."

test groups. Obviously, the kind of grouping depends on the type of industry. The major advantage of the functional structure is that greater technical competence can be achieved in the various engineering specialties than it can in the project structure. For example, it is more likely that the company will possess true expertise in material science if a group can focus all of its attention on this activity rather than each project group having to develop its own expertise. The principal disadvantage of a functional structure is that it is difficult to pinpoint responsibility, and certain important matters may get overlooked.

The *hybrid structure* (Figure 3.2) combines the foregoing two forms and is characteristic of large companies. Functional groups exist, but each project is under the supervision of a project manager who shepherds it through the various functional activities. Those who choose the hybrid structure generally do so in the hope that they will obtain the advantages of both the project and the functional form.

INFORMATION FLOW

To contribute in an optimum way, a new employee also needs to know how the information flows through the company. The information flow may or may not follow the formal structure described earlier. The key characteristic is whether the company's approach to information flow is "closed" or "open."

A *closed system* typically exhibits a strong adherence to the organizational structure, with directives flowing from the top down along the lines of the organization chart and accountability for results flowing in the opposite direction. Great emphasis is placed on productivity, on enforcement of budgets and schedules, and on requiring information to flow through proper channels. Even though such practices seem properly efficient and businesslike, some negative by-products can occur, such as high dependency of individuals upon superiors, low autonomy, low opportunity for interaction, and low individual influence potential.

In the *open system*, greater emphasis is placed on the autonomy of individuals and less reliance is given to achieving results by means of administrative control. It might be said that the open system emphasizes subject matter, while the closed system emphasizes the methods by which the subject matter is to be handled. In a closed system, new jobs tend to be adapted to the organizational structure, while in an open system the structure is adapted to the jobs. Not surprisingly, the open system usually results in higher individual satisfaction, less conflict, higher group performance, and greater individual opportunity.

In 1982, T.J. Peters and R.H. Waterman, Jr., published a book that gave strong support to open organizational systems: *In Search of Excellence: Lessons from America's Best-Run Companies.*[5] It has been called the best-selling management book of all time. Peters and Waterman spent three years conducting in-depth studies of 62 major American corporations that had shown long-term success as measured by such things as equity growth and return on investment. They found that the "excellent" companies had in common a number of organizational precepts, mostly unwritten. Among these were a bias for action, an emphasis on autonomy and entrepreneurship, a willingness to take risks, an obsession with the importance of people as a key to productivity, and a constant pressure to maintain simple organizational forms. These elements are exactly what is meant by an open organizational form.

Although it may be impossible for you, as a new employee, to influence the organizational structure of the company, understanding how it works will permit you to foresee problems, to help the organization succeed, and to advance along your own career trajectory.

DEVELOPING PROFESSIONAL ATTITUDES

University graduates entering their first full-time engineering jobs usually need a period of training and adjustment. The older employees may be as apprehensive as the new employee during the adjustment phase. Having a friendly attitude, speaking courteously, and making minor allowances for personality traits usually help to make the workplace friendlier and to make allies out of colleagues who might otherwise be more aloof. Developing these characteristics is a matter of practice, and they should easily be mastered by a university graduate. Here are some suggestions that might help you to develop professional habits and a professional image:

- Get a good pocket calendar, and begin to schedule and monitor your time efficiently. You probably did this in university, or you likely would not have graduated. Good scheduling is even more important in the workplace. As a student, many courses and activities were rigidly scheduled. On the job, it may be necessary to balance conflicting time commitments, and it is not always possible to focus on one assignment at a time. You may have to say no to some people and tasks, but it is important to do so tactfully.
- Remember that you are now in a professional environment. Dress appropriately.
- Select competent, energetic engineers as role models.
- Once you are acclimatized to the new environment, accept tough jobs. Challenging and unusual assignments will help you develop your skills and knowledge.
- Listen closely, and ask questions when you are unsure of something; otherwise, you might get the right answer to the wrong problem.
- Double-check your work — make sure it's right. Your work may influence important and expensive decisions, and you won't have a chance to correct and resubmit it.
- Don't resort to long or vague words or jargon when writing reports, letters, or memos. Learn to describe what you are doing in simple, concise terms. Buy a dictionary and use it.
- Be tactful and positive. Don't be afraid to disagree, but do so constructively.

MOVING UP THE LADDER

Getting a job and adjusting to the new workplace both are major hurdles. However, as time passes, young engineers will eventually want to "move up the ladder," and new problems may be encountered. Fortunately, some serious work has been done to identify and analyze the kinds of problems that are commonly experienced by young professionals. In his book *Characteristics of Engineers and Scientists*, Lee Danielson reports the results of interviews with 367 U.S. engineers and scientists conducted by the Bureau of Industrial Relations at the University of Michigan in the 1960s.[6] Although this is a rather old study, these characteristics are "people problems" that have changed little in recent years. Danielson found the following to be the most frequently mentioned problem areas of young professionals, in order of their frequency of mention:

1. adjusting to company practices
2. advancement slow or uncertain
3. accepting routine jobs
4. learning what is expected
5. finding one's own niche
6. unrealistic ambitions
7. lack of initiative
8. gaining social acceptance
9. lack of specialized courses
10. lack of recognition

Obviously, the first problem listed above could be alleviated (or even eliminated) by company training programs. Problems 4, 5, and 8 will diminish with the passage of time as the new employee adjusts to the company (and vice versa). The remaining items are probably the most important ones on the list and are also the most difficult to deal with constructively.

The essence of some of these points is that young engineers are often simultaneously criticized for not being ambitious enough (problem 7) and for being too ambitious (problem 6). Great things are expected of them because of their education; at the same time, they are told that they do not know enough to handle anything important and are given routine tasks. To achieve the desired amount of humility and yet avoid apathy is a considerable challenge for the young engineer. One writer warns against "producing the apathetic, uncreative, passive kind of employee whom most organizations seem to welcome at the outset but regret being saddled with at a later time."[7]

Enthusiasm is one of the most valuable assets a young engineer possesses. Yet it is this very enthusiasm that might sometimes be interpreted as overaggressiveness or unrealistic expectations. If consistently re-buffed, enthusiasm can easily degenerate to a permanently low level. This would be a tragedy, because enthusiasm is too precious to be smothered — and lost — in this fashion. One of the best ways to move out of routine assignments is to handle each of them with as much skill, dispatch, and enthusiasm as one possesses. There are few, if any, better ways to come to the attention of management and move into more challenging tasks.

The problem of slow promotion is something else. Very few people — perhaps no one — are ever promoted as fast as they hope. Moreover, most young people have heard the adage that "the squeaky wheel gets the grease" and test its truth in practice. But management is generally unimpressed by any agitation for promotion unless it is coupled with proven ability and achievement. The best advice for a new engineer who is in a position with low promotion potential is to devote time to improving technical knowledge, developing management and people skills, and letting one's ability speak for itself. Intelligent employers ignore the "squeaky wheels" and recognize the value of promoting capable people.

CONSTRUCTIVE ACTION BY EMPLOYERS

Once the newly graduated engineer has adjusted to industry and demonstrated professional attitudes of enthusiasm, alertness, and thoroughness, the employer has a duty to see that these qualities are rewarded. The engineer's responsibility is to do a good job of engineering; management's responsibility is to do everything possible to help the engineer do that job.

One constructive point upon which everyone appears to agree, at least in principle, is that good communication must be maintained between engineering and management. This means that engineers must be supplied with information concerning company objectives, particularly those that affect engineering projects. There is a natural limit to the degree to which such a program can be pursued, however. Much information about company objectives may also be precisely the kind of information that will aid and comfort the competition if it should come into their hands. Management can hardly be blamed if it tends to hold back on such sensitive information, particularly to new employees, since it knows full well that engineers do quit the organization from time to time and sometimes join competitors. Hence, the goal of complete communication is never achieved, although with honest effort it can be approached.

Another important area is that of salaries, reviews, and promotions. It is generally agreed that salaries should reflect the contributions made by the engineers. This is by no means an easy task, and many companies do a less than adequate job in this area. In one survey involving 350 U.S. engineers and engineering managers, 88 percent of the engineers thought it was imperative for their companies to establish salary progressions that reflect engineers' contributions; when asked if their companies realistically followed such a policy, only 23 percent answered in the affirmative.[8]

Good facilities and support personnel are also emphasized in virtually every list of recommendations. This includes such things as secretaries and clerical support; provision of technicians, equipment, computers, and telephones; and reasonably private quarters. The day of the giant "bullpens," with engineers stacked at desks ranged row on row, is nearly a thing of the past. Most companies have gone to considerable trouble and expense to provide semi-private quarters for engineers, with perhaps two or three people per office, privacy screens, and sound-deadening carpeting.

A survey of engineers and engineering managers in 1987 sought to find out just what motivates engineers and what is most important to them.[9] The respondents were asked to make choices between contrasting pairs; the following list shows the percentage of respondents making that choice.

- *Interesting work* was more important than *prestige and recognition* (100 percent).
- *Technical tasks* were more important than *managerial tasks* (87 percent).
- *Diversified work* was more important than *concentrating on a single project* (80 percent).
- *Work on one's own ideas* was more important than *knowing what is expected* (76 percent).
- *Sense of individual accomplishment* was more important than *good supervisor and associates* (68 percent).
- *Working in a group* was more important than *working alone* (67 percent).
- *Good salary* was more important than *recognition for work well done* (64 percent).
- *Opportunity for advancement* was more important than *job security* (61 percent).

PROFESSIONAL EMPLOYMENT GUIDELINES

The NSPE *Guidelines to Professional Employment for Engineers and Scientists* is an American document that is equally relevant to Canadian engineers.[10] The *Guidelines* are discussed in more detail in Chapter 7 and are printed in their entirety in Appendix F. The emergence of these *Guidelines* was a significant event in the history of the U.S. engineering profession, since they recognized that most engineers are employees, and the usual codes of ethics, which apply primarily to engineers in private practice, sometimes are not very helpful to engineers in industry, government, and education. The *Guidelines* set out to remedy the problem, and as a result, they contain far more injunctions for employers than for employees.

In brief, the *Guidelines* state that the employee is expected to be loyal to the employer's objectives, to safeguard the public welfare, to avoid conflicts of interest, and to pursue professional development programs. The employer is expected to keep professional employees informed of the organization's objectives and policies, to establish equitable compensation plans, to minimize new hirings during layoffs, to provide for early vesting of pension rights, to assist in professional development programs, to provide timely notice in the event of termination, and to assist in relocation efforts following termination. The *Guidelines* are not legally binding, but they have had a constructive impact on employee–employer relationships.

The *Guidelines* were first published in 1973. In 1981, the American Association of Engineering Societies (AAES) conducted a survey to see how well the *Guidelines* had been incorporated into industry employment practices. The survey showed that the *Guidelines* had been effective in improving the conditions of engineers in the United States. They are reproduced in Appendix F as ideals in employment standards. Readers may wish to review and compare them with their own employment conditions, and, if necessary, see how their conditions might more closely approach the ideals.

TOPICS FOR STUDY AND DISCUSSION

1. Consider the company for which you are presently working and a company for which you might like to work, and compare them using the information in this chapter. For example, which form of company organization most clearly represents each company's structure? What form of training program is usually offered to recent engineering graduates? How do the companies compare in terms of the characteristics listed in the section "Constructive Action by Employers?" Does the comparison show one company to be a more attractive employer than the other?

2. Make a list of those factors, in priority order, that you believe are important to you in professional employment, such as salary, location, potential for advancement, and type of work. Then assuming you are an employer, make a similar list, putting down those things

that a new employee is likely to be seeking from you. Compare the two lists, looking for inconsistencies in the two sets of expectations.

3. The codes of ethics for provincial Associations (in Appendix B) all state that the professional engineer should be loyal to the employer and should have due regard for the health, safety, and welfare of the public. Consider some situations in which these guidelines could come into conflict, and in each, state what you believe to be the proper course of action for the professional engineer. This conflict is discussed in more detail later in this text.

4. One issue that occasionally causes controversy is the payment of overtime to engineers. Should engineers be paid overtime, or should they expect to work some additional time because of the professional nature of their work, which permits them, in most cases, to set their own priorities and work schedules? If overtime should be paid, under what circumstances should it be monitored? Should engineers punch a time clock, or should their attendance be checked by supervisors? Should a "bonus" take the place of an overtime payment? Would it make any difference if the company had a "flex-time" system, in which personnel have much more latitude in the times that they arrive at and leave work?

5. Another controversial issue in engineering employment is the formation of engineering unions. From your knowledge of unions (and from any other sources, if you wish), write a list of potential good effects and bad effects of the unionization of engineers. Using your own beliefs and values, decide for yourself whether a union would be good or bad for you. After you have come to a conclusion, read Chapter 7, "Ethical Problems of Engineers in Industry," which refers briefly to the unionization dilemma. Does this discussion support your previous conclusion?

NOTES

1. M. Everett, "What It's Like to Work at Digital," *Graduating Engineer* (December 1986): 35–37.
2. Premier's Council of Ontario, *People and Skills in the New Global Economy* (Toronto: Queen's Printer, 1990): 91–92.
3. Edgar H. Schein, "How to Break in the College Graduate," *Harvard Business Review* (November/December 1964): 68–76. Reprinted by permission of *Harvard Business Review*.
4. *Admission to the Practice of Engineering in Canada*, CCPE/CEQB publication G01-92 (Ottawa: CCPE). Available at http://www.ccpe.ca/english/boards/CEQB/admission.html.
5. T.J. Peters and R.H. Waterman, Jr., *In Search of Excellence: Lessons from America's Best-Run Companies* (New York: Warner Books, 1982).

6. L.E. Danielson, *Characteristics of Engineers and Scientists* (Ann Arbor: University of Michigan, 1960), 55–78.

7. Schein, "How to Break in the College Graduate," 68–76.

8. *Engineering Professionalism in Industry* (Washington, DC: The Professional Engineers Conference Board of Industry, 1960), 35–37.

9. E. Raudsepp, "What Motivates Today's Engineers?" *Machine Design* (21 January 1988): 84–87. Reprinted by permission of *Machine Design*, a Penton Publication.

10. National Society of Professional Engineers (NSPE), *Guidelines to Professional Employment for Engineers and Scientists*, 3rd ed. (Alexandria, VA, 31 October 1989).

CHAPTER FOUR

Engineers in Management

Engineers' attitudes toward management show a curious ambivalence. Many engineers believe that moving into management is the natural path of advancement, and most salary surveys show that management responsibility is rewarded,[1] but some engineers are not convinced. Their view is that if they wanted to be managers they would have studied business, not engineering. Fortunately, well-organized engineering companies will always provide dual career paths for good engineers — one path into management and one path into engineering specialization — because developing first-rate engineering expertise is as essential to a company's success as good management. This chapter does not try to persuade readers one way or the other. However, since so many engineers *do* eventually go into management (perhaps as many as two-thirds), we must explore the topic.

WHAT DO MANAGERS REALLY DO?

There is a simple answer to the question of what managers do: they manage money, materials, and people to meet corporate objectives. During the workday, however, the manager is seldom alone. There are many meetings as well as informal interactions with associates, because the manager, as an executive, is primarily concerned with people. The most distinctive characteristic of an executive's job is getting numbers of people moving in the same direction.

The following description of the duties of a company president was written some years ago by the chair of a board of directors (the president's boss) of a very large corporation. It is even more relevant today.

Although the company is extremely large, with six manufacturing plants and thousands of employees, it still has aggressive competitors, and the president is under pressure from many sources. As the chair explains:

> [O]ur president knows that to maintain his share of the market and to make earnings which will please his directors, he must accomplish the following very quickly: design and perfect a brand-new and more advanced line of products; tool up these products in such a way as to permit higher quality and lower costs than his competitors; purchase new machinery; arrange major additional long-term financing. At the same time his corporation's labor contract is up for negotiation, and this must be rewritten in such a way as to obtain good employee response and yet make no more concessions than do his competitors. Sales coverage of all customers has to be intensified, and sales costs reduced. Every one of these objectives must be accomplished simultaneously, and ahead of similar efforts on the part of his competitors — or the future of his company is in great danger. Every head of a corporation lives every day with the awareness that it is quite possible to go broke. At the same time he lives with the awareness that he cannot personally accomplish a single one of these vital objectives. The actual work will have to be accomplished by numerous individuals, with some actually unknown to him, most of them many layers removed from his direct influence in the organization.[2]

Another key aspect of a manager's job is decision making. Of course, all engineers are decision-makers, so they are generally well-equipped for this management role. However, management decisions and engineering decisions are slightly different, since management decisions usually concern money or people, not products. Also, while engineering decisions may be equally important and far-reaching, the results of management decisions may be more immediately apparent. A manager, especially at the executive level, may be acutely aware that a decision today to cut back factory production may mean that hundreds of people are out of work tomorrow, and the manager's decision may be headline news in the local newspaper. Making management decisions is a lonely privilege, and developing an effective decision-making process and management style (as discussed later in this chapter) is important to engineers who want to be effective managers.

LEVELS OF ENGINEERING RESPONSIBILITY

Good management is essential in every engineering organization, and while managers are usually selected because they are competent engineers, they must *want* to lead and must develop or improve their leadership skills. Conversely, for those engineers who want to specialize in engineering areas that are profitable to the company, moving into management should not be essential for promotion. Enlightened engineering companies encourage and reward excellence in engineering achievement. Salaries for engineering specialists (and especially those experts who achieve world-class ability) must keep pace with salaries of managers to ensure that the creative engineering spirit is encouraged to flourish.

The list that follows is a rough guide to the various engineering levels that might be found in a large manufacturing company. Note that the list recognizes the dual management/technical career paths, and the levels are based on responsibility, which may mean management responsibility or engineering responsibility. A word of caution in interpreting this list: there may be much overlap in these levels, and some companies may recognize more or fewer levels depending on company size. Moreover, the typical company (particularly in manufacturing) usually employs many more managers than specialists.

A — Entry Level A bachelor's degree in engineering or applied science, or its equivalent, is usually required for entry into engineering-oriented jobs. Recent university graduates, usually with little or no practical experience, receive training in the various phases of office, plant, field, or laboratory engineering work as on-the-job assignments or occasionally as classroom instruction.

B — Junior Engineer Level, Engineer-in-Training During the first two to three years of work experience after graduation, an engineering employee will receive assignments of limited scope and complexity, usually minor phases of broader assignments. This is normally regarded as a continuing portion of the engineer's training and development. The employee is usually registered with the provincial Association as an engineer-in-training, member-in-training, or geoscientist-in-training, depending on the province.

C — Professional Engineer This is typically regarded as a fully qualified professional engineering level. The engineer carries out responsible and varied engineering assignments requiring general familiarity with a broad field of engineering and knowledge of reciprocal effects of the work upon other fields. Normally, this stage requires a minimum of three to five years of related work experience after graduation. The incumbent would normally be licensed as a professional engineer.

D — First Supervisory or Specialist Level Job titles associated with this level (and for levels E and F) may have several variations, such as project engineer, lead engineer, engineering specialist, and so on. This is the first level of direct and sustained supervision of other professional engineers or the first level of full specialization. This stage requires applying mature engineering knowledge in planning and conducting projects, and the duties have scope for independent accomplishment and co-ordination of difficult and responsible assignments. Normally a minimum of five to eight years of experience in the field of specialization after graduation is required to achieve this level.

E — Middle Management or Senior Specialist Level Job titles at this level may be chief project engineer, chief industrial engineer, group head, senior specialist, and so on. This level usually requires knowledge of more than one field of engineering or requires performance as an engineering specialist in a particular field of engineering. The incumbent participates in short-range and long-range planning and makes independent decisions on work methods and procedures within an overall program. Originality and ingenuity are required for devising practical and economical solutions to problems. The engineer may supervise large groups, containing both professional and nonprofessional staff or may exercise authority over a small group of highly qualified professional personnel engaged in complex technical applications. This level normally requires a minimum of nine to twelve years of engineering and/or administrative experience after graduation.

F — Senior Management or Senior Consultant Level Job titles at this level may be chief engineer, director of engineering, plant manager, senior consultant, and so on. This level and the next may show considerable overlap depending on company size (a chief engineer in a large corporation may have essentially the same duties as the vice-president of engineering in a medium-size corporation, for example). The incumbent is usually responsible for an engineering administrative function, directing several professional and other groups engaged in interrelated engineering responsibilities or may be an engineering consultant, achieving recognition as an authority in an engineering field of major importance to the organization. The engineer independently conceives programs and problems to be investigated, participates in discussions, determines basic operating policies, and devises ways of reaching program objectives in the most economical manner and of meeting any unusual conditions affecting work progress. The job requires extensive engineering experience, including responsible administrative duties.

F+ — Senior Executive Level Job titles at this level may be president, vice-president of engineering, vice-president of manufacturing, general manager, partner (in a consulting firm), and so on. Within the framework of general policy, the incumbent conceives independent programs and problems to be investigated. He or she plans or approves projects requiring considerable amounts of human financial resources. The engineer determines basic operating policies and solves primary problems or programs to accomplish objectives in the most economical manner to meet any unusual condition. This level requires many years of authoritative engineering and administrative experience. The incumbent is expected to possess a high degree of originality, skill, and proficiency in the various broad phases of engineering application.

When an engineer first goes into management, the goal is usually to help the company by staffing a new project or by filling a void created by the departure of a colleague; very rarely does the engineer have his or her eyes on the top job. However, it is important to give some thought to one's career objectives, especially in those companies that have dual paths for promotion, since developing the leadership skills needed to deal with people may take time away from achieving excellence in engineering, analytical, and consulting skills. One should make this decision consciously rather than just drifting into it. Obtaining clear ideas of the promotion opportunities in your industry, observing the career paths of the more senior engineers, and developing your own long-term career goals will help you make better career decisions.

ATTRACTIONS OF MANAGEMENT CAREERS

Probably the first and most obvious attraction of a management career is money. Managers must be rewarded for the extra responsibility that they assume and the leadership and initiative that they must display. Most managers also take pride in meeting the challenges of management, that is, by using their initiative and creativity to find new answers to old problems. Many managers simply like to compete, and management gives them a socially acceptable opportunity to do so. These financial and psychological rewards benefit the individuals and society in general.

However, a word of caution is appropriate: a few people aspire to management because they like to exert authority or power over others. This is the wrong reason to enter management, and the careers of such people will be mercifully short, since such authoritarian motives are generally despised. In Canada's advanced democratic society, true authority arises from motivating workers to do your bidding willingly. Unless managers use their authority ethically, for the mutual benefit of the individuals and the corporation, the motivation of the workers will disappear.

SALARY EXPECTATIONS

Over one's career, salaries depend to a great degree on factors (such as economic depressions and recessions) over which there is no personal control, so historical data, even if corrected for inflation, is not necessarily a prediction of the future. However, a glimpse of what may lie ahead can be obtained by looking at today's salaries for engineers at all responsibility levels and at all stages of engineering careers. Such a salary survey is conducted annually by Professional Engineers Ontario (PEO). The "Ontario Engineers' Salaries: Survey of Members" is accompanied by a very comprehensive checklist so that each engineer can determine the appropriate responsibility level that corresponds to his or her duties, years of experience, supervision received, supervisory (or leadership) authority exercised, and so on. These responsibility levels are labelled A to F+, roughly similar to the descriptions provided in the last section. Table 4.1 shows the salary levels for 1996 (the most recent year, as this text goes to press) as a function of responsibility level.

For comparison, the median salary as a function of years since graduation is shown in Figure 4.1. This shows the increase in salary as a function of time as the engineer moves into management or specialist positions. The graphs are plotted separately for male and female engineers and show a rough parity, except for the years 1977 to 1985, where it is clear that salaries for female engineers who graduated during this period are consistently below those for male engineers. Note that these data apply only to Ontario and are a small "snapshot" of the comprehensive data in the PEO survey. For more detailed information, readers should consult the most recent edition of the survey.[3]

CHALLENGE AND CREATIVITY

When a manager speaks of the job as being challenging and creative, this is not just idle talk — the statement is usually sincere. Engineering

Table 4.1 ONTARIO ENGINEERING SALARIES BY RESPONSIBILITY LEVEL, 1996

Responsibility Level	Low Quartile	Median	High Quartile
Level A	32 500	36 000	40 000
Level B	36 500	40 100	44 600
Level C	46 800	53 000	60 000
Level D	56 000	63 074	71 000
Level E	68 480	75 000	82 400
Level F	78 000	88 000	96 000
Level F+	87 000	102 000	122 000

SOURCE: Professional Engineers Ontario (PEO), "Ontario Engineers' Salaries: Survey of Members, 1996," *Engineering Dimensions* 18, no. 2 (March/April 1997). Reprinted by permission of PEO.

Figure 4.1 MEDIAN SALARIES FOR ONTARIO ENGINEERS FOR GRADUATION YEARS 1975–1996

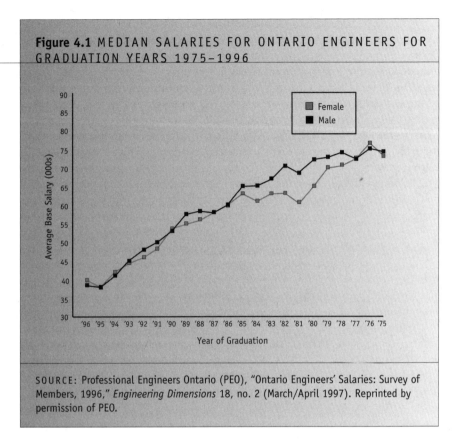

SOURCE: Professional Engineers Ontario (PEO), "Ontario Engineers' Salaries: Survey of Members, 1996," *Engineering Dimensions* 18, no. 2 (March/April 1997). Reprinted by permission of PEO.

managers, like all engineers, take great pride in seeing their plans and inventions being built, and also enjoy seeing programs that they originated take shape and prosper. Moreover, most engineering managers are motivated by a genuine desire to make the world a better place. Like all human beings, they want to feel that their existence has meaning and gives value to the rest of the world. This provides a constant sense of challenge. Many engineering managers also like the competition involved in leading the corporation in its constant race for new products and improvements. This is the positive motivation that drives our free enterprise system. If you respond well to such challenges, and if you are attracted to the competitive atmosphere of the marketplace, then engineering management may be a rewarding career path for you.

LEADERSHIP REQUIREMENTS FOR MANAGERS

Engineering organizations need managers who are leaders and not just paper-shufflers. However, very few people are born leaders, and most managers must develop leadership skills if they want to advance in management. In fact, the essential ingredient for management success is a high level of energy and drive to develop the skills needed to succeed in this leadership role. A person who does not have this high level of drive but who aspires to a management career may be making a serious mistake. However, for those who are willing and able to make the commitment, there is a certain amount of excitement and exhilaration in the challenge of leadership. In fact, the challenge itself propels most people into management careers, particularly at the higher levels.

The key aspects of leadership are explained in an article by John Farrow, reproduced here in full from *Engineering Dimensions* with permission.

The following advertisement was published in the *Wall Street Journal* by United Technologies Corporation:

> People don't want to be managed. They want to be led. Whoever heard of a world manager? World leader, yes. Educational leader. Political leader. Scout leader. Community leader. Labour leader. Business leader. They lead, they don't manage. The carrot always wins over the stick. Ask your horse. You can *lead* your horse to water, but you can't *manage* him to drink. If you want to manage somebody, manage yourself. Do that well and you'll be ready to stop managing and start leading.

People today are better educated and have different expectations from those of 20 years ago. Professionals, including engineers, are one of the groups leading this change towards a different management style. They seek involvement and anticipation at the work place. Their managers must be leaders; in fact, most working enterprises acknowledge the need for better leadership at all levels.

Leaders are not just born; they can also be created. If you can manage yourself, and develop your skills, you have the discipline needed to lead others.

PHILOSOPHY OF LEADERSHIP

For some, the word "leadership" implies that one person is a dictator who makes all the decisions and does all the work of leading others. However, in groups with more than two or three members, there are usually too many factors for one person to do everything. A good leader works as a senior partner with other members of the group to achieve the task, build the team, and meet individual needs. From moment to moment, the leadership of the group can shift from one member to another, depending on the circumstances.

The leader must also recognize that most groups have three categories of need. Leadership functions, therefore, break down into three groups: task needs, group maintenance needs, and individual needs. The leader must not only recognize these different categories of need, but must also understand that others will share the job of satisfying these needs. Leadership itself must be a team effort.

There are six key leadership requirements: *vision, planning, communicating, monitoring, organizing,* and *role modelling.*

VISION

Having a vision is the basis for all leadership. It is a big task in leading big organizations, but it is also necessary for smaller groups, because it involves bringing to life objectives that can be understood in personal terms. We often create a vision to help ourselves achieve a desired goal. Athletes trying to achieve peak performance at a prescribed moment use this technique a great deal. For Muhammad Ali, it was "to float like a butterfly, sting like a bee." My guess is that his coach coined this phrase to help Ali focus more closely on his performance in the boxing ring.

Individuals within organizations need vision as well. In the early days of Polaroid, Edwin Land inspired his team with the vision that they could "achieve the impossible." This compelling vision led to achievements for Polaroid that other organizations thought unrealizable.

However, these corporate visions are not easily created. They often require much thought, study, trial and error. Effective visions for organizations are not the product of semi-mystical revelations, but the result of hard work — by the leaders. Furthermore, a good vision is not simply offered once and then allowed to fade away. It must be repeated often so it becomes part of the organization's culture and is reinforced through the decision-making process. It must also be re-evaluated often against changing circumstances and, if necessary, updated.

A vision often starts with the leader, but must then be "owned" by the whole organization. To achieve this, a leader must explain the vision, what is involved, and why it exists. The vision must also be related to individual roles, and hence must be broken down into tasks individuals can relate to.

PLANNING

Planning involves bridging the gap between where you and the organization are today and where you and the organization want to be at some point in the future. Planning is a response to the group's need to determine how the task will be achieved. This, in turn, leads to such questions as: "Who does what?" and "When does it have to be done?" A leader without a plan is unlikely to be effective.

The development of a plan is not always as easy as it seems. Good planning is usually, but not necessarily, participatory. An effective plan development process ensures that a broad range of people have a common understanding of the objectives. It involves examining alternatives, in order to select the best solution. Finally, it involves a detailed description of tasks that will become the responsibility of small groups or individuals.

Vision is what; planning is how, who, and when.

COMMUNICATING

Good leaders communicate their visions, in a way that many can relate to. When they are communicating specific planned activities, they must conduct detailed briefings which are precise and clear. Good communication requires:

- *Preparation*. Create a structure that people can remember and into which they can fit details.
- *Clarity*. Think through potential difficulties and ensure that complexities are made sufficiently simple to be understood.
- *Energy*. Ensure that the message demands attention and is easy to remember and pass on.
- *Sincerity*. Communicate what you believe and what can, in turn, be trusted.
- *Follow-up*. Has the message been received? Follow-up can be done through question periods and/or observation. If the message hasn't been fully understood, try again.

MONITORING, EVALUATING, CONTROLLING

These three functions conform to what some used to see as "management." Each of these steps is closely linked with the others.

Monitoring involves checking progress against the plan. Groups must be sensitive to the need for this to ensure that the task is realized. A leader is often responsible for monitoring, but it is a task that is readily delegated. For example, monitoring the financial performance of many companies is often the primary responsibility of the vice president, finance. However, deciding *what* needs to be monitored is a more critical decision, and usually involves leadership.

Similarly, *evaluating* the results of the monitoring to determine what it means can be simple or complex, and again, is usually a leadership responsibility. Ideally, a group that is experienced, skilled, and able to determine the appropriate action will respond well to the majority of situations. However, for the group to demonstrate this ability, the circumstances must have been anticipated and responsive action decided upon in advance. This, in turn, requires planning and training.

Controlling usually involves some type of intervention. In most cases, teams should be self-controlling; however, one of the roles a leader may assume is that of ensuring that this self-controlling mechanism is working effectively, and of taking responsibility for intervention when it is not.

Engineers are usually familiar with these process control skills because they apply to production equipment. The process of intervening in groups is much more difficult and requires that the basic technical approaches to monitoring and controlling be supplemented by people-management skills. The type of intervention required will depend on the circumstances. Hence, from a management perspective, the key is to match the type of intervention to the needs of the group or individuals being managed.

ORGANIZING

To achieve anything, some structure must be imposed, especially if the group is large and the task is complex. The structure may be permanent, but could be temporary if the group is responsible for a single task. In organizing, three principles are especially useful:

1. Smaller work groups tend to be more effective than larger ones.
2. The capability of the leaders of the sub-groups should be considered when sizing groups.
3. Creating additional levels in the organization is costly and often slows communication.

How you decide to structure a group is dependent on both the task and the people. It is not a science, but an art; therefore, if what exists now isn't working, change it — but make the changes in consultation with those concerned with an eye to minimizing short-term disruption.

ROLE MODELLING

A good example is the most powerful individual communication and leadership tool. One key role for the leader is to represent, time and again, what he or she wants the organization to be. The larger the organization, the more important it is for the leader to play a symbolic role that can be consistently interpreted by a large number of people. However, leaders faced with the complexity of larger organizations must be prepared for greater difficulty in being consistent and meaningful models.

Trustworthiness is perhaps the most effective example. In any organization, many different individuals must learn to trust each other. Trust within a group is something that the leader can strongly influence, especially by exemplifying openness and consistency. Being a model in this regard is vital for a good leader.

CONCLUSION

Leaders are born less often than they are created. This creation comes, in part, from circumstances and from individual commitment to developing the requisite skills.[4]

ACADEMIC REQUIREMENTS FOR MANAGERS

Engineers appear to make good managers and executives; for example, a 1986 survey showed that one-third of the 50 largest corporations in the world were run by chief executive officers (CEOs) who had engineering backgrounds. In Japan, fully two-thirds had engineering backgrounds. While these executives were not necessarily promoted through engineering departments, they received their basic degrees in engineering, and their leadership set the pace for the productive decade that followed. One of the reasons for the popularity of engineering as a management background is the training engineers receive in solving problems, which is obviously a useful skill. Another reason is the growing technical complexity of industry. Engineers have an obvious advantage when it comes to managing high-tech companies. However, to progress to the highest levels of management, engineers must usually develop their leadership skills and financial expertise. These skills can be obtained through personal study and practice, but in recent years, formal education has become a faster route to personal advancement.

Many university students interested in management careers progress to the master's degree in business administration (MBA). The MBA program certainly contains the kinds of organizational and financial material that engineering managers need, but there may be a danger if one moves directly from the bachelor's degree in engineering to the MBA before acquiring some engineering experience. In fact, graduates who move directly into management may never get sufficient engineering experience to be effective as engineering managers (or even to be licensed as professional engineers). The number of people with MBAs has become quite large, and there is a risk of overrating the formal education. As *Time* magazine reported, some people with such academic credentials often offend prospective employers with their attitude. The attitude arises, according to *Time*, from their belief that they are destined to take command of the organizations they join. A growing number of corporate managers look on them as "arrogant amateurs, trained only in figures and lacking experience." MBA graduates are seen as too expensive, too aggressive, and lacking in loyalty.[5] Nevertheless, the skills acquired through an MBA program are important for management. The best advice for aspiring managers is that they do need to acquire leadership and financial skills, but not at the expense of solid engineering experience.

Another excellent preparation for engineers with management aspirations are the management-oriented master's programs in industrial engineering, management sciences, or engineering management offered by a number of universities (usually engineering faculties) in Canada and the United States. These programs combine management and advanced engineering courses. Typical courses are organizational theory, marketing management, production management, human relations, accounting, engineering law, project management, value analysis, computer science, and operations research. In addition to improving their engineering knowledge, participants in these programs acquire basic management skills. Graduates usually obtain jobs in production engineering, industrial engineering, or marketing.

Finally, the traditional engineering master's degree should not be overlooked as preparation for management. As the depth of knowledge increases and our society becomes more computer-oriented, the need for specialists increases. To advance to a management position in such high-tech industries as digital communication, aircraft design, robotics, electrical power generation, or computer manufacturing, an advanced degree in engineering is a definite asset. These degree programs are entirely technical and highly specialized, but they prepare the graduate to give the technical leadership needed to advance in a high-tech company.

Obviously, the "ideal" career path is different for each individual and depends on personal interests, abilities, and attitudes. Whether an engineer aspiring to management uses personal study or pursues an advanced degree, the person must first have the determination to acquire the necessary leadership and financial skills.

DEVELOPING LEADERSHIP SKILLS

Earlier, we described a philosophy of leadership that incorporates vision, planning, communicating, monitoring, evaluating, controlling, organizing, and role modelling. All of these skills can be developed or improved through personal study. In fact, many "self-help" books have been written to help people do just that. See Further Reading at the end of this chapter for some recommendations. There are, of course, hundreds of other titles. The good news is that a wealth of information is available to the determined engineer who wants to move into a management position but cannot afford the time or money for a master's program.

CHOOSING A PERSONAL LEADERSHIP STYLE

Leadership styles take many forms. Choosing, or more accurately, developing an appropriate style is essential for a successful career in management. The best leadership style depends on many factors, such as the personality, maturity, and competence of the manager; the type of corporation; the degree of initiative, creativity, and independence required from employees; and, to a lesser degree, the knowledge or skill of the employees and their willingness or motivation to achieve the corporate goals. The spectrum of leadership styles ranges from the collegial to the military style, as shown in Table 4.2. Most people would find that this table is arranged in order of desirability, with the collegial style most preferred, and the military style least preferred. Where does your usual management style — or what you expect your style to be someday — fit in this rough spectrum?

If we consider a typical engineering activity, such as a design office, and we assume that we have a motivated, mature, and competent manager, and professional and well-motivated personnel engaged in creative activity, then it should be possible for the engineering manager to adopt a collegial or team-oriented leadership style. However, even when dealing with the same employees, it might be necessary to be more authoritarian in some matters — for example, in specifying engineering goals such as incorporating safety features into a design — but a good manager would try to give the maximum latitude as to how the goals are achieved. Consistency is important, but flexibility is necessary to deal with exceptional issues. Leading highly skilled and well-motivated personnel is difficult, because it is the manager's task to ensure that they are sufficiently challenged and rewarded to stay that way.

DEVELOPING NEW PRODUCTS

One of the duties of an engineering manager might be to seek new business, whether that means new products, services, or contracts. In

Table 4.2 A COMPARISON OF LEADERSHIP STYLES

Leadership Style	Typical Example
Collegial	Manager treats employees as colleagues, permits them to function independently within agreed terms of reference.
Team-oriented	Manager defines goals, but asks employees to suggest solutions and make decisions.
Interactive	Manager presents problem, obtains ideas and suggestions of employees, then makes decision.
Responsive	Manager presents tentative decision to employees, subject to change after discussion.
Paternal	Manager presents decision, invites questions, and may change decision.
Authoritarian	Manager makes decision, explains it to employees.
Military	Manager makes decision, instructs employees.

the specific case of a manufacturing corporation, for example, an engineering manager might be required to take a leadership role in developing new products. Similar initiatives may be required of engineering managers in other types of corporations.

When a corporation's product line reaches its peak in sales volume, it may already have started a downhill slide in profits because competitors copy success. A continuous stream of new products is therefore essential to the long-term well-being of the corporation. The manager of the new products department is faced with the tasks of

- generating new ideas
- seeking, analyzing, and screening ideas from all other sources
- co-ordinating new product ideas with market needs
- conducting pilot tests of new products under market conditions

Generating new ideas may require the manager to initiate research projects. Most companies, in fact, rely on their R&D departments to generate new ideas for products, but other sources may be equally useful.

Seeking ideas from other sources is often more useful than doing internal research. New ideas come from many sources: the patent literature, government, competitors, and occasionally top management itself. If one wished to find a route to the largest number of creative minds in this country, the trail of issued patents would surely be the best starting place. An avalanche of new patents is issued every year, but very few of these ideas ever reach the market because their owners cannot afford commercial development. As a result, many patents are available for licence or sale. In addition, examining expired patents may generate entirely new product ideas. Supplying goods needed by the government is also big business. Many companies have offices in Ottawa and Washington whose job is to keep abreast of government needs and respond to government Requests for Proposals (RFPs). Inspecting competitors' products can also stimulate new ideas. As well, senior managers who have been involved with the company's past successful products are a relevant source of overlooked ideas.

After new product ideas have been collected, they must be carefully screened. Out of this group, only a few will actually become successful products. Ideas that survive the screening step must be analyzed on a return-on-investment basis. To do this, an estimate of the future market must be made and usually this estimate is coupled with a market survey. However, the information acquired must be examined with an experienced and sceptical eye, and predictions must be independently evaluated.

Many products fail because of insufficient field testing under actual market conditions. The customer can be relied upon to find all sorts of defects in the product that did not show up during in-plant testing. If it is efficiently conducted, field testing can show whether the customer's needs have been correctly identified *before* costly and irrevocable steps have been taken to implement mass production and distribution.

RETURN-ON-INVESTMENT ANALYSIS

Before a project goes too far, someone must take a hard look at the future and try to decide whether the proposed product has any chance of yielding a profit. In order to do this, two steps are necessary: someone must devise a reasonably clear configuration of the proposed product so that production costs can be estimated, and someone must predict the anticipated annual sales.

Table 4.3 shows a typical return-on-investment analysis for a consumer product that will sell for $300. For this product, it is estimated that $1 million will be required for development and another $1 million for plant expansion, tooling, and distribution start-up (including sales training costs, service training costs, and the expense of staffing new offices). Such a capital investment is by no means unusual; in fact, the usual tendency is to underestimate the amount of capital required.

In the analysis shown, it is assumed that the third year after introduction of the product represents what will be a steady-state condition, and that the return of 27 percent per year will continue from then on. (Return on investment is calculated by dividing the gross annual return by the total capital investment, in this case $820\,000/\$3\,080\,000 \cong 0.27$.) At this rate, the company's investment would be recovered by the fifth year. However, such an assumption might be naïve. For one thing, the competition cannot be expected to stand still during this five-year period, and there is a high likelihood that additional R&D investments will be required during the five years to keep ahead of the competition. Furthermore, if the sales estimates are optimistic, as is often the case, or the manufacturing cost estimate turns out to be too low, the whole proposition could become unattractive. The manager must make key, long-range decisions based on approximate data and speculative analyses. Good managers develop nerves of steel.

DRAWBACKS OF MANAGEMENT CAREERS

Management careers have obvious benefits, such as higher salaries, greater autonomy, and challenging tasks. However, any discussion of management careers would be incomplete without mentioning that there are also some drawbacks. If you are interested in eventually being a manager, as you read through the following descriptions give some

Table 4.3 SAMPLE RETURN-ON-INVESTMENT ANALYSIS FOR A CONSUMER PRODUCT

	First Year	Second Year	Third Year
Initial investment (R&D, plant expansion, tooling, etc.)	$2 000 000		
Number of units required for sales inventory	4 000	8 000	12 000
Dollars tied up in inventory, at factory value	($150×4 000) = $600 000	($100×8 000) = $800 000	($90×12 000) = $1 080 000
Total capital investment	$2 600 000	$2 800 000	$3 080 000
Annual sales (number of units)	3 600	7 200	12 000
Manufacturing cost per unit	$150	$100	$90
Selling price per unit	$300	$300	$300
Selling expense (at 37.5%)	$112	$112	$112
General and administrative expense (at 10%)	$30	$30	$30
Net return per unit	$8	$58	$68
Gross annual return	$28 800	$417 600	$816 000
Annual return on investment (%)	1	15	26

thought to how you feel about these drawbacks. Are they major obstacles, minor inconveniences, or not relevant to you?

LOSS OF PERSONAL FREEDOM

One infringement on a manager's freedom is the right to select social companions as one pleases. Some managers make it a rule never to socialize with other company people for fear that such an arrangement may someday prove embarrassing. Conversely, others feel compelled to socialize with company people, especially if the company is in a small town. Both conditions curtail personal freedom.

The young executive may find that much of his or her socializing is in the form of company obligations. For example, entertaining out-of-town VIPs may be required, and the question may arise as to what all this has to do with the job. Nevertheless, it must be recognized that many executives enjoy this part of their work. While some regard it as a curtailment of freedom, others look upon it as a kind of fringe benefit.

STRESS

The executive with a stomach ulcer is a popular stereotype in the modern business world. Like all stereotypes, this one is usually false; nevertheless, many managers do experience physical disorders that have their origin in emotional stress. It is pressure, of course, that causes this situation, but what causes the pressure?

In some instances, pressure is used as a conscious corporate tool to maintain an atmosphere of urgency and to make sure everyone is working at maximum output. Managers who object to the stress are almost sure to hear the admonition "If you can't stand the heat, get out of the kitchen." Hence, any adverse feelings may be forced under wraps. The result is more stress.

Many other sources of stress are associated with management, such as anxiety concerning job security, slowness of promotion, and intense competition with rivals. Probably one of the biggest ulcer-producers is the requirement that the ideal executive always present a calm, self-assured façade. Feelings of weakness and self-doubt can never be allowed to show. When people are forced into jobs they do not want, they occasionally experience great anxiety and emotional conflict, a state of mind called "promotion neurosis." Although stress is a factor in some management jobs, many techniques have been developed to make it more tolerable, and should be investigated if you are concerned about the effects of stress.

IMPACT ON FAMILY

A manager, even at the executive level, is rarely told that he or she must put the job before everything else. Nevertheless, to be a success in any profession, some additional stress or sacrifice is inevitable, and the demands on managers are such that family life often comes second. Naturally, every case is unique, and people assign different priorities to family life. Some managers see company contact as an expansion of family life, many accept or ignore the intrusion on family life, and a few feel that the demands of management put an unacceptable strain on family life. This should be considered if you are thinking about moving into management.

Also, new employees learn very quickly that corporations generally expect instant mobility from their management hopefuls. For example, if an engineer working in Calgary is offered a transfer to Ottawa (presumably a promotion), the company usually expects a prompt "yes." If the offer is declined, then the next person on the list will be chosen, and it can be assumed that the first engineer's climb up the promotion ladder will be slower.

Most companies consider mobility important. Obviously, a person's spouse will have a great deal to say about such matters. In many families, both partners hold jobs, and an instant move may not be easy to arrange. Three possibilities exist: one of the partners must commute, jobs for both have to be found at the new location, or one of them has to quit his or her job. More and more, the second choice — jobs for both at the new location — is becoming possible. Companies are beginning to accept such conditions as a part of doing business, and most enlightened companies are willing to try to find jobs for the spouses of the employees being moved. Families with children have special problems: parents may see a move as a promotion and a welcome advancement, but children may see it as the loss of familiar friends, home, and community.

Another problem with moving is the impact on family finances that results from moving from a relatively inexpensive part of the country to a major city such as Toronto or Vancouver, where housing prices are significantly higher than elsewhere in Canada. Although some companies have policies for helping employees to sell their houses, there is always the risk of serious financial loss for the employee who is promoted to a job in a big city.

ETHICAL CONFLICT

Another potential difficulty in a management career arises in the sphere of ethical conflict. Some companies demand that their managers follow a policy of seeking the good of the company to the exclusion of other considerations. This may create a conflict of interest when the good of the company conflicts with the good of society as a whole. However, it might be argued that, as a manager, the engineer is in a better position to influence company policy and to prevent such problems from arising. Ethical conflict is one of the major issues discussed in this text, and we return to it in later chapters.

TOPICS FOR STUDY AND DISCUSSION

1. Most engineers eventually have to decide whether they want to move into management or stay in purely technical work. Make a list for yourself of the pros and cons of each alternative.
2. Many engineers argue that engineering *is* management, because engineers are usually the only professionally recognized group in a

typical engineering company, their work is decision-oriented, and their decisions have a direct effect on the profitability of the company. However, some engineers disagree; they say that engineers are employees, and their bosses make the key decisions. Which argument do you believe to be closer to the truth? Are both arguments true, to some extent? Does it perhaps depend on the type of industry, type of company, or size of company? Do you believe an engineer's performance on the job might be affected by holding one view or the other? Write a brief essay explaining your position on the question of whether engineering is management.

3. In the book *Entrepreneurial Megabucks*, A.D. Silver describes a six-step entrepreneurial process:[6]

   a. Identify the problem.
   b. Create the solution.
   c. Plan the business.
   d. Select the entrepreneurial team.
   e. Test the prototype.
   f. Raise the venture capital.

   How does this process compare with the six-step problem-solving process described in Chapter 6 and applied to ethical problems? Are there any similarities? Are there any gaps in the above six steps? Is this process really several interconnected problem-solving processes? Why or why not? Each of the six steps would really be composed of many substeps. Make a list of the substeps that you envision carrying out for each major step. It is acceptable to look up Silver's book to answer these questions.

4. In Chapter 3, two kinds of organizational systems were described: closed and open. Write down what you consider to be the good and bad points of each system, and try to formulate a set of reasons why you might prefer to work under one system rather than the other. Would your answer be different if you were evaluating the organizational system from the viewpoint of management rather than as an employee?

5. Visualize yourself some twenty years in the future. You are now the chief engineer in a medium-sized manufacturing company employing about 50 to 100 engineers. Decide for yourself what kind of product your company is manufacturing, picking something with which you have some familiarity. Now, assuming you are going to organize your department into a functional structure (see Figure 3.2), decide what functions need to be represented in your engineering department. Describe how the development of new products would occur, with the follow-up activities of product design and testing, given your functional structure.

6. In the return-on-investment analysis shown in Table 4.3, what happens if the manufacturing cost per unit can never be brought

below \$110? How many years would it then take to recover the total capital investment? What would you do if sales cannot be increased beyond the level of 3600 units per year? What would you do if the analysis proceeds as shown for the first three years, but annual sales in the fourth year are 7200 and in the fifth and succeeding years are 3600 because competitors have introduced similar products?

FURTHER READING

VISION
A radical book for inspiring vision and creativity is Roger von Oech's *A Whack on the Side of the Head: How You Can Be More Creative* (New York: Warner Books, 1990). It includes several exercises on how to break out of the box of routine thinking, a key step in developing a vision for any enterprise.

PLANNING
Planning is an activity that varies from discipline to discipline, and few self-help books are available on this topic. A general approach for personal planning is provided in *The Seven Habits of Highly Effective People* by Stephen Covey (New York: Simon & Schuster, 1989).

COMMUNICATION
Communication is the backbone of any organization, and there are several good books on this topic, such as *Communication in Organizations* by K. Roberts (Fort Worth, TX: Dryden, 1988) and *Business Communication: Strategy and Skills*, Fourth Canadian Edition, by Richard Huseman et al. (Toronto: Dryden, 1996).

MONITORING, EVALUATING, AND CONTROLLING
These are the key aspects of the manager's role, and there are many books on these topics. Probably the most-read book is *A Passion for Excellence* by Tom Peters and Nancy Austin (New York: Random House, 1985), which suggests that managers should monitor, evaluate, and control productivity using a technique called "managing by wandering around." A more recent book by Peters called *Thriving on Chaos: A Handbook for a Management Revolution* (New York: Alfred A. Knopf, 1987), challenges some of his own principles, and stresses that modern corporations must first develop effective techniques for coping with change. An earlier book by K. Blanchard and S. Johnson, *The One-Minute Manager* (New York: Berkley Books, 1981), stresses the importance of time in the manager's life, and recommends many time-saving techniques for monitoring, evaluating, and controlling activities.

ORGANIZING

The role of incentives in organizing people is discussed in *Performance Criteria and Incentive Systems* by S. Globerson (Amsterdam: Elsevier, 1985). The use of performance appraisals in evaluating how well corporate goals have been achieved is discussed in *Practical Guide to Performance Appraisals* by R. Henderson (Reston, VA: Reston Publishing, 1986).

ROLE MODELLING

There are many helpful books in this area, such as *Leadership in Organizations* by G.A. Yukl (Englewood Cliffs, NJ: Prentice-Hall, 1988) and *In Search of Excellence: Lessons from America's Best-Run Companies* by T.J. Peters and R.H. Waterman, Jr. (New York: Warner Books, 1982). Several intriguing books with subversive titles are available: *What They Don't Teach You at Harvard Business School: Notes from a Street-Smart Executive* by M.H. McCormack (New York: Bantam Books, 1984), *Swim with the Sharks (without Being Eaten Alive)* by H. Mackay (New York: William Morrow, 1988), and *Leadership Secrets of Attila the Hun* by W. Roberts (New York: Warner Books, 1987).

NOTES

1. Professional Engineers Ontario (PEO), "Ontario Engineers' Salaries: Survey of Members, 1996," *Engineering Dimensions* 18, no. 2 (March/April 1997).
2. J. Irwin Miller, "The Dilemma of the Corporation Man," *Fortune* (August 1959): 103. Reprinted with permission of J. Irwin Miller.
3. PEO, "Ontario Engineers' Salaries."
4. J. Farrow, "Getting Rid of Management," *Engineering Dimensions* 10, no. 5 (September/October 1989): 22–24. Reproduced in full with permission of Professional Engineers Ontario (PEO).
5. "The Money Chase," *Time* (4 May 1981): 58–69.
6. A.D. Silver, *Entrepreneurial Megabucks: The 100 Greatest Entrepreneurs of the Last Twenty-Five Years* (New York: John Wiley & Sons, 1985).

CHAPTER FIVE

Engineers in
Private Practice

Engineers who offer their services directly to the general public are said to be in *private practice*. Such engineers are frequently called *consulting engineers* or *engineering consultants* although, as discussed below, these terms are regulated in Ontario. Working in private practice is usually more entrepreneurial than working for an employer and can be varied and interesting, particularly if new clients with challenging problems are continually being engaged. However, engineers in private practice also have more responsibilities: they are usually subject to increased licensing and insurance requirements, and, in addition to all of the tasks usually associated with engineering, they must seek new clients and compete for new projects.

In 1996, the Association of Consulting Engineers of Canada (ACEC) listed 730 consulting engineering firms (sole proprietorships, partnerships, and corporations) in its directory.[1] In that same year, the total number of licensed professional engineers in Canada was over 160 000. Of course, not all engineers in private practice are members of ACEC, but these numbers seem to indicate that only a minority of engineers are in private practice.

SERVICES PROVIDED BY ENGINEERS IN PRIVATE PRACTICE

Private practice is a good example of free enterprise in the provision of engineering services, and consulting engineers expand their capabilities to suit the demands of the engineering marketplace. The range of engineering services provided to the public is therefore very broad, and the firms listed in the ACEC directory reflect this wide range. Although civil engineering is probably the largest single specialty, the ACEC directory lists engineers who can provide advice on almost any subject and who can assist with any type of project. Some typical tasks that might be required of a consulting engineer are listed below:

- **Engineering Advice** Advice may be needed for specific projects, or on a continuing basis, in the areas of design, development, inspection, testing, quality control, management, and so on.
- **Expert Witness** An independent consultant may be needed to provide engineering opinions or advice to a court, commission, board, hearing, or similar government or judicial body.
- **Feasibility Studies** Consultants are particularly useful in the preliminary stages of a project when feasibility, financial justification, cost estimates, and completion dates are being determined. It is obviously not advisable to make commitments to hire personnel until a project has been shown to be feasible and economically affordable, so hiring a consultant to conduct a study is convenient and prudent. The results of the study could decide whether the project goes ahead or not.
- **Detail Design** Consulting engineers have the expertise to carry out detailed design, including preparing drawings, specifications, and contract documents.
- **Specialized Design** Custom design and development are available, particularly for manufacturing processes, machine design, mining, and other specialized areas. Consulting engineers may work independently or in conjunction with the client's staff. Consultants may also provide assistance in developing inventions or preparing patents.
- **Project Management** Supervision of part or all of a project is commonly carried out by consulting engineers. This could include the design, manufacturing, construction, or assembly phases of a project, or the commissioning (initial start-up) of a large plant.

In summary, engineers in private practice can be found performing any task that requires professional engineering knowledge, usually on behalf of a client who lacks the personnel or expertise to conduct the work.

LICENSING REQUIREMENTS FOR PRIVATE PRACTICE

When engineers offer their services to the public, they are generally subject to more licensing requirements, in addition to the basic P.Eng. licence, and liability insurance requirements. The main reason for this additional scrutiny and insurance is, as mentioned in Chapter 2, to protect the public. Also, most engineers in private practice are principals or employees of incorporated consulting companies. When an incorporated company offers engineering services, it is prudent to ask who are the individuals providing those services, and who is responsible for ensuring their quality. How can the public be assured of the continuing competence of the corporation's engineers? Does the corporation carry liability insurance? Each province has developed slightly different methods of answering these questions.

As shown in Table 5.1, every provincial or territorial Act (except the Yukon's) requires or will eventually require corporations providing services to the general public to identify the individuals who are responsible for providing that service. In some provinces, sole proprietors are exempt from the regulations. In about half of the provinces, the Act requires businesses offering professional engineering services to the public to hold a Certificate of Authorization or some equivalent credential. To obtain a Certificate, the business must designate one or more of its experienced professional engineers to assume responsibility for and supervise these services.

The requirements for liability insurance vary significantly from province to province. Some provinces make the insurance mandatory, while others make it voluntary. For example, in Ontario, all Certificate of Authorization holders must carry professional liability insurance or disclose in writing to every client that they are not insured. Failure to conform to this regulation is considered to be professional misconduct.

LEGAL RESPONSIBILITIES AND LIABILITY INSURANCE

Engineers in private practice typically have additional legal obligations and liabilities. While an employee engineer is usually covered by the liability insurance of his or her employer, the engineer in private practice must arrange this protection. To illustrate the type of problem that may arise in the course of carrying out a contract, consider the following examples:

Many contracts in the past have used the statement "The engineer shall have general supervision and direction of the work." The word *supervision* sometimes causes trouble. In some court decisions, it has been held that the design professionals (a term including both engineers and architects) were responsible for defective construction techniques in cases where they undertook the responsibility of supervising the construction, even though the construction methods are generally agreed to be under the control of the contractor. Claimants in liability cases invariably attempt to portray the engineer (or architect) as a supervisor, and therefore a guarantor of the contractor's work.[2]

Table 5.1 COMPARISON OF CERTIFICATE OF AUTHORIZATION AND LIABILITY INSURANCE REQUIREMENTS

Province/Territory	Certificate of Authorization (or equivalent) Required	Liability Insurance Requirements
Alberta	Yes, although sole proprietors are exempt.	[Voluntary.]
British Columbia	Planned, but not implemented (August 1997).	Voluntary, but uninsured practitioners must inform clients.
Manitoba	No, although 1997 draft of revised Engineering Act includes provisions to license groups.	Draft Act would require client notification of uninsured status.
New Brunswick	Yes, although sole proprietors are exempt.	Voluntary.
Newfoundland	Yes (includes sole proprietors).	Coverage is advised but not mandatory.
Northwest Territories	Yes, although sole proprietors are exempt.	Coverage is advised but not mandatory.
Nova Scotia	No, but firms must register and name employee Professional Engineers.	Voluntary.
Ontario	Yes (includes sole proprietors).	Voluntary, but uninsured practitioners must inform clients.
Prince Edward Island	Yes (mainly applies to out-of-province firms working in PEI).	Voluntary.
Quebec	1997 draft regulations make provision for separate licence for engineers in private practice.	Mandatory for all members.
Saskatchewan	Yes (includes sole proprietors).	Voluntary, but uninsured practitioners must inform clients.
Yukon	No.	No.

SOURCE: Adapted from Professional Engineers Ontario (PEO), *The Link* 2, no. 2 (August/September 1997): 5. Reprinted with permission of PEO.

The word *inspection* in a consultant's contract may also cause trouble, because it has sometimes been interpreted to mean exhaustive and continuous inspection of all details of the construction. Most often, this was not the function that the engineer had in mind when "inspection of the work" was included in the contract. More likely, some kind of educated spot-checking was envisioned. Hence, the word *observation* has been proposed as a substitute that more accurately describes the service intended. If actual detailed inspection is desired, then it is recommended by professional groups that the contract provide for a full-time project representative whose task it is to perform detailed and continuous inspections.

Special legal hazards are involved in using new materials or equipment. Courts have generally held that the engineer or architect is obliged to conduct tests of a new material or to have reliable information concerning the results of tests conducted by others. Sole reliance upon manufacturers' sales literature and specifications has been held to be insufficient. The question of the obligations of the design professional in using new materials remains a tricky legal matter.

Moreover, the written language of the contract may be insufficient to protect the engineer if services are provided in an area beyond the scope of the contract. Having once moved into that area, the engineer may be charged with the responsibility for all the functions involved, such as failure to exercise reasonable care in performing the services or failing to do what one experienced in the field would do in the exercise of reasonable care.

Being sued is, unfortunately, an occupational hazard for an engineer in private practice. Legal hazards such as those described above could lead to personal financial tragedy if a project fails. Because of this, every provincial Association requires or recommends that engineers maintain professional liability insurance, often known as *errors and omissions insurance.*

As recent history shows, obtaining adequate liability insurance for engineers at reasonable cost can sometimes be difficult. In the 1980s, the cost of professional liability insurance increased so rapidly in the United States that some consulting firms decided to go uninsured. In 1987, about 23 percent of the members of the American Consulting Engineers Council reported that they carried no liability coverage. For small organizations (one to five persons) the proportion was 41 percent without coverage.[3] For some kinds of work, such as hazardous waste reduction, it was claimed that liability insurance was almost impossible to obtain.[4] One spokesperson said that liability insurance is the most disruptive problem facing engineers, and another declared, "What you see today is the engineer brought in as one of 15 people in the lawsuit because his name is on the building, or on the drawings, or in the phone book.... The chance of being sued is not a function of committing sins or errors; it's a function of how much money, or how big an insurance policy, you have."[5]

In Canada, similar problems have been experienced. In Ontario, for example, when a revised Professional Engineers Act was announced in

1984, engineers providing services directly to the public were required to have a Certificate of Authorization and a minimum of $250 000 liability insurance (or to be employees of a company with both of these). However, the major insurance companies were unable to provide this insurance at reasonable cost to the large number of engineers and firms who applied. Because of the confusion over a requirement that could not reasonably be satisfied, the Association obtained permission from the Attorney General to postpone the requirement. The regulations were later amended to permit engineers in private practice to advise their clients in writing that they are practising without liability insurance, before entering into an agreement to provide professional engineering services.[6] Statistics reported by PEO in 1990 showed that, of 2389 Certificate of Authorization holders, 1027 were insured in accordance with the regulations, 186 were insured in equivalent ways, 60 were exempted as practising in a field for which insurers will not provide coverage (for example, environmental and nuclear projects), and 1116 elected to disclose to clients that they do not have professional liability insurance.[7] Consequently, whether by choice or by exemption, roughly 49 percent of the Certificate of Authorization holders were offering professional engineering advice to the Ontario public in 1990 without liability insurance.

However, the problem of liability coverage is improving as the insurance companies get a better idea of the risks involved. In recent years, several insurance companies have responded to this need, and they may be contacted through the provincial Associations.

When a project goes awry, aggrieved claimants tend to take a "shotgun" approach and sue everyone associated with the project. Coverage is required to cover two aspects of liability: the cost of defending a claim and the cost of paying a claim.[8] The first of these — defending the engineer against a claim — is frequently overlooked, but the costs associated with preparing a defence can be extremely expensive, even in a case where the engineer is an innocent bystander caught in the crossfire. To obtain the best liability coverage at a reasonable rate, the engineer in private practice must consult an insurance broker, and must document and explain the engineering company's capabilities and area of practice to define the risk as specifically as possible.

CONSULTING ENGINEERS

Although all consulting engineers are in private practice, not all engineers in private practice are consulting engineers. In the province of Ontario, the title Consulting Engineer (or any variation with the same meaning) is regulated under the Professional Engineering Act and may not be used without authorization from Professional Engineers Ontario.

Although regulation of this title is unique to Ontario, the requirement was generated by a desire to protect the general public, and it may be of interest to readers whether they are residents of Ontario or elsewhere.

To be designated a consulting engineer in Ontario, a licensed professional engineer must apply to PEO and satisfy several additional requirements:

- **Authorization** Permission must be obtained to engage in private practice, as indicated by a Certificate of Authorization, awarded by Professional Engineers Ontario.
- **Experience** Consulting engineers must have five years' experience in addition to that required for registration; at least two years of the experience must be in private practice.
- **Professional Liability Insurance** Proof of liability insurance must be filed with Professional Engineers Ontario, although permission to offer services to the public is awarded even in the absence of liability insurance (subject to certain requirements, such as notifying all clients of this fact using a specified written format).

PROVINCIAL CONSULTING ENGINEERING ASSOCIATIONS

Organizations devoted to assisting consulting engineers have been set up in every Canadian province and territory except PEI. The names of these organizations follow, and their current addresses can be obtained from the telephone directories of the cities indicated.

- Consulting Engineers of Alberta (Edmonton)
- Consulting Engineers of British Columbia (Vancouver)
- Association of Consulting Engineers of Manitoba (Winnipeg)
- Consulting Engineers of New Brunswick (Moncton)
- Consulting Engineers of Newfoundland and Labrador (St. John's)
- Consulting Engineers of Northwest Territories (Yellowknife)
- Nova Scotia Consulting Engineers Association (Halifax)
- Consulting Engineers of Ontario (Toronto)
- Association des Ingénieurs-Conseils du Québec (Montreal)
- Association of Consulting Engineers of Saskatchewan (Regina)
- Consulting Engineers of Yukon (Whitehorse)

The goal of these organizations is to promote the interests of the consulting engineers that form their memberships. There are several ways in which this is done:

- **Publications** Each organization publishes a directory of its members and the services provided by the members. The directories are available in most public libraries. Other publications, on such topics as selecting a consulting engineer and various aspects of professional practice, are also available.

- **Communication** The organizations communicate with their members on issues that affect the profession, and they provide a central source of information on consulting engineers for the public, the industry, and government.
- **Advocacy** The organizations make representations to municipal, regional, and provincial governments on behalf of their members when requested, or when an issue affects consulting engineers as a group.

ASSOCIATION OF CONSULTING ENGINEERS OF CANADA (ACEC)

Membership in one of the provincial consulting engineering organizations includes membership in the Association of Consulting Engineers of Canada (ACEC). Membership is voluntary and is limited to companies primarily engaged in providing consulting engineering services directly to the public. ACEC is a national Canadian nonprofit organization founded in 1925 with the goals of promoting satisfactory business relations between its member firms and their clients, and fostering the exchange of professional, management, and business information. When necessary, ACEC acts to safeguard the interests of consulting engineers, to improve the high professional standards in consulting, and to provide liaison with the federal government. The American counterpart of ACEC is the American Consulting Engineers Council (also known as ACEC).

INTERNATIONAL FEDERATION OF CONSULTING ENGINEERS (FIDIC)

Both ACECs (Canadian and American) are member associations of the International Federation of Consulting Engineers/Fédération Internationale des Ingénieurs-Conseils (FIDIC) founded in 1913, which is made up of the national consulting engineers associations of more than 50 countries. Member associations must comply with FIDIC's code on professional status, independence, and competence. FIDIC publishes an international directory and, like provincial and national organizations, works on behalf of consulting engineers at the international level.

THE CONTRACTING PROCEDURE

Selecting a consulting engineer is one of the most important decisions a client makes, and experience shows that selecting one based on qualifications ultimately produces the best value for money spent.

Procuring engineering services is not the same as purchasing material, where there may be little variation in quality and the goal is simply to obtain the lowest price. Although it is certainly not illegal or
unethical to select a consultant based solely on lowest bid (and provincial Associations cannot and would not discourage such a process),
it is preferable to separate the evaluation of qualifications from the
negotiation of the fee.

The fee is negotiated after the consulting engineer is selected. In
fact, the fee can be set only after the scope of the project has been fully
defined, and in many cases, the scope can be defined only after the
client and the consulting engineer are working together on the project
team. In most projects, the most significant cost savings are achieved in
the early stages of the project, and employing the best-qualified engineers will usually generate more cost savings. Penny-pinching at the
design stage may result in much higher capital, operating, or maintenance costs.

Guidelines for the contracting process are available from the provin-
cial Associations, the provincial consulting organizations, ACEC, and
FIDIC. The Quality-Based Selection (QBS) process recommended by
ACEC is reproduced here with the permission of ACEC.[9]

QUALITY-BASED SELECTION (QBS)
Many clients rely on an engineering firm with which they have a long-
term relationship (this is called "sole sourcing"). However, clients may
want to assess the merits of several engineering companies before choosing.

The global trend is for clients to select consulting engineers in accordance with the Quality-Based Selection (QBS) system. QBS means that the
client chooses a consultant according to:

• Technical competence
• Managerial ability
• Experience on similar projects
• Dedicated personnel available for the project's duration
• Proven performance
• Location and/or local knowledge
• Professional independence and integrity.

. . .

Quality-Based Selection is recognized and used around the world. Since
1972, the United States Federal Government has applied it to all federal
work. More than 30 U.S. state departments also use it. The World Bank
and the Asian Development Bank are only two of the many international
financing institutions advocating applications for QBS for their projects.
In Canada, QBS is used by the Province of Quebec and the Municipality

of Metropolitan Toronto. The system is strongly endorsed by the International Federation of Consulting Engineers (FIDIC) and by the Association of Consulting Engineers of Canada (ACEC).

ADVANTAGES FOR THE CLIENT

A good client–consultant relationship is assured from the beginning of the process. Adversarial relationships are avoided. By first agreeing on the scope of the project, the client can make clear the required emphasis on factors such as environmental impact, cost, schedules, and social implications before fees are negotiated. Fees are fairer to both client and consultant because they are negotiated after the parameters of the work are established. Provincial and territorial fee guidelines published by associations of professional engineers should be used as a basis for such negotiations.

EVALUATION AND SELECTION

The recommended steps in the selection of a consulting engineer are:

- Prepare a list of qualified consultants (long list). For small projects, ACEC recommends selecting a single candidate based on quality of earlier work rather than going through the entire process;
- Contact consulting engineers so selected, outlining the nature of the project and enquiring about the engineer's interest;
- Identify three to five candidate consultants from the long list to develop a short list;
- Request proposals [from the candidate consultants on the short list];
- Interview candidate consultants separately to examine their qualifications and discuss the project and scope of services required. Clients may also wish to check carefully with recent clients of each consulting engineering firm;
- Select the consulting engineer who appears best suited for the project;
- Negotiate fees and execute a contract with the selected consulting engineer. If negotiations are not successful, negotiate with second choice;
- Notify all those interviewed when a contract has been awarded.

FACILITATING THE SELECTION PROCESS

(ACEC recommends the following actions on the part of the client and the consulting firm)

The Client:
- Describes in general terms the needs of the proposed project and its purposes and objectives;
- Designates the various phases into which the work is to be divided;
- Sets out a desired timetable for the work;
- Identifies problems that are likely to arise;
- Determines budgets and estimates for all phases of the total project;

- Selects firms that offer the required services either from personal knowledge or from an appropriate directory such as those published by ACEC and its member organizations; and
- Gives the selected firms the project information set out above and invites them to offer their services.

The Consulting Firm:
- Responds with a letter of interest;
- Demonstrates an understanding of the project;
- Provides evidence of the firm's ability to perform the work;
- Submits profiles of the firm's principals and staff who will be assigned to the project;
- Gives references, including previous clients for whom similar projects have been carried out.

The Client Then:
- Evaluates the responses and selects a firm from the short list with which to begin negotiations;
- Makes reference to the fee schedules and list of alternative methods of remuneration published by the Association of Professional Engineers and by the Association of Consulting Engineers for each province or territory;
- Negotiates a fair fee and contract based on the most appropriate method of remuneration;
- Makes reference to the FIDIC Client/Consultant Model Services Agreement (1990).

Booklets describing the provincial fee guidelines (or recommended schedules) and performance standards may be obtained from provincial associations of professional engineers.

COMPENSATION FOR CONSULTING ENGINEERS

In the past, the engineer's fee was often calculated as a percentage of the construction costs. However, while this procedure is appropriate in some cases (mainly civil engineering), the main argument against it is that it penalizes the engineer for creating an economical design. A good compensation process should reward efficiency and innovation. The following paragraphs describe most of the common methods currently employed for establishing the consultant's fee, and are adapted from the FIDIC directory with permission.[10]

Per Diem Fixed daily rates are known as *per diem* payments. They are the most basic method of calculating remuneration for consulting engineering services and are generally used for those assignments where the scope of work cannot be accurately determined. Studies, investigations, field services, and report preparation all fall within this category. For example, some consulting engineering companies specialize in short-term overload assistance, sometimes called *contract engineering*. These consulting firms will, upon request, place their personnel within a client's firm to work side-by-side with the client's engineers. In this way, the client can absorb peak workloads without hiring and training new engineers who may become surplus when the peak has passed. The per diem payment is obviously the best compensation method in these cases. Direct out-of-pocket expenses, such as travel costs, are reimbursed in addition to the per diem payments.

Payroll Costs Times a Multiplier Payroll costs, multiplied by a factor to cover overhead and profit, are most often used for site investigations, preliminary design, process studies, plant layout, and detailed design. The multiplier is usually in the range of 2 to 3. Under this method, the client essentially pays the engineering costs as they occur, including a sufficient amount to cover overhead and profit. Direct out-of-pocket expenses are also reimbursed in addition to the payroll costs.

Lump Sum With this method, the consultant determines in advance a unit or lump sum fee that will cover costs, overhead, and profit. Many clients prefer this method of compensation because they know in advance the cost of engineering services that will be incurred. When the services to be performed are fairly well known, this is a simple process. The obvious disadvantage is that the consultant may incur a serious loss if costs have been underestimated.

Fee as a Percentage of Estimated or Actual Costs of Construction As mentioned above, the percentage of construction costs approach is used with less and less frequency. It has generally been applied to consulting engineering services where the primary task is preparing drawings, specifications, and construction contract documents. The method is increasingly unpopular with consultants because of the difficulty of relating design costs to rapidly changing construction technologies and to unpredictable market conditions in the construction industry.

The FIDIC directory also suggests some variations on these basic compensation procedures, and should be consulted if further details are required.

BECOMING A CONSULTING ENGINEER

An engineer starting up a private practice is much more likely to fail because of a lack of business ability than because of a lack of technical ability. Virtually all consultants warn the prospective newcomer about "that depressing first year." However, even if the engineer is well-prepared before taking the plunge into private practice, it may take two or three years to get fully established.

Some of the things the beginning consultant may neglect are such ordinary business matters as accounting, collections, overhead, taxes, and insurance. *Overhead*, for example, includes many items often overlooked. Employee benefits (vacations, sick leave, disability insurance, etc.) and business taxes may add from 10 to 15 percent to the direct costs. Rent, computer equipment, supplies, telephone and fax service, and secretarial help may come to as much as 30 or 40 percent of the direct costs. If there are more than six or eight employees, additional supervision may be required; this may add another 15 percent. Finally, there is the often belatedly recognized factor of *nonproductive time*, which may add another 10 percent. Thus, the direct engineering costs may have to be increased by as much as 65 to 80 percent of the original estimate, with no allowance made as yet for profit. Although these percentages were first estimated many years ago, they are still valid today.[11]

Even a determination to work extra hours cannot compensate for a lack of good business ability and adequate financial reserves, because extra hours are expected in any professional job, and consulting is no exception. In fact, it is essential to have an entrepreneurial spirit, since private practice makes you both boss and employee, and it is impossible to be the boss if you object to working long hours and if you intend to put all the business problems out of your mind when you leave the office. A special set of skills, education, and determination are required to make a success of private practice. The following list is one consultant's prescription for the recommended minimum requirements and personal characteristics for those who would like to enter private practice:[12]

1. Appropriate education, including humanities
2. Engineering registration
3. Confidence in one's professional ability
4. Broad prior experience in the responsible discharge of engineering work
5. Business acumen
6. Financial reserves to last at least one year (some say two to three years)
7. Ability to get along with people, especially clients and employees
8. A reputation for keeping your word
9. Good health and a willingness to work hard.

The basic problem of the new consultant is simple: to acquire that first project, it is necessary to demonstrate a minimum level of competence by pointing to projects that have been completed in the past. Given such circumstances, getting that first job can understandably be difficult. Yet every consultant in business today has had to get past this barrier.

Some consultants have made their start by quitting their jobs and taking one or more of their former employers' clients with them. However, such a practice is unethical and likely contrary to one's employment contract. Other aspirants have become junior partners with an established consultant. This is certainly ethical but offers a difficulty in that the senior party has to be convinced there is something to gain by taking on a new partner. Moreover, established consultants sometimes complain that the expectations of new graduates are often unrealistically high, and starting salaries in consulting companies are typically lower than those offered by high-tech firms. For those engineers who are willing to start at a lower salary and "pay their dues," there is some good news: within five or ten years, successful engineers in private practice will be earning more than their high-tech counterparts, who will be locked into more rigid salary structures and won't be appreciated for developing the managerial and sales skills so highly valued by consulting firms.

Consequently, although there are risks and problems in being a consulting engineer, there is also a possibility of proportionate rewards. Canada needs people with entrepreneurial ability to stimulate the generation of jobs, services, and material wealth. If you decide to enter into private practice, good luck!

TOPICS FOR STUDY AND DISCUSSION

1. Consider the list of nine personal characteristics on page 99. Rate your own ability under these nine headings on a scale from 0 to 10 (0 means you have serious doubts about your ability in the area, 5 means you have reasonable confidence in your ability in the area, and 10 means you have absolute confidence in your ability in the area). Sum your scores to get a rating out of 90. Give yourself an additional 5 points for each further qualification listed below.

   - You have adequate, current experience in computer software related to your field.
   - You own computer hardware that runs the software mentioned above.
   - You have published three or more technical papers.
   - You have already been involved in consulting.
   - You have a master's degree.
   - You have contacts in five or more local companies who might need your services.

- You have previous experience in making group presentations and writing technical proposals.
- You enjoy making important (and expensive) decisions under pressure.

Total your points; they should not exceed 130. If your total is 100 points or more and you have been scrupulously honest in your personal assessment, then you are probably ready to move into private practice. If your total is less than 80 points, then you probably need more experience, education, or determination.

2. Assume that you have carried out the evaluation in question 1 and have compared your qualifications and characteristics with those given for entering private practice. Recognizing that question 1 is not a perfect predictor of career success, do one of the following:

   - If your score was lower than 100, consider whether you see private practice as a career objective. If not, what other career paths (management, specialization, etc.) appeal to you? If private practice is your objective, do you need to change your personal characteristics or simply obtain more experience in your field? Make a list of steps that would make you more able to move into private practice.
   - If your score was higher than 100, prepare a business plan for establishing your private practice.

3. Imagine that you have decided to enter private practice and you are trying to become better known in your local area so that you can attract more clients and contracts. Advertising is a sensitive issue, since it must be consistent with the code of ethics. Read the sections in this text on advertising (consult the Index), and then devise at least five methods for becoming better known as a competent and ethical professional engineer that are clearly consistent with this text and with your provincial or territorial code of ethics.

4. Architects and engineers frequently work for the same firm, doing work that overlaps somewhat. Architects are supposed to design attractive structures and provide building arrangements that fulfil certain functions. Yet, in carrying out these responsibilities, they often get into such matters as spacing columns, which are structural in nature. Similarly, in providing for structural integrity, engineers may impose constraints on shape and function, which are the architects' realm. As a consequence, relations between these two professional groups have sometimes been strained, particularly over the issue of whether an architect or an engineer is to be in charge of the project. Consider both sides of this issue and prepare an argument for whether an architect or an engineer should be the ultimate boss. Does it make any difference whether the project is a

factory, an apartment building, a downtown office building, or a bridge? Consult your provincial or territorial Act (some Acts do define the architect/engineer boundary), and see if it agrees with your point of view.

5. There is a very fine distinction between the activities of an engineer who "provides" engineering services to the general public and an engineer who "offers" engineering services to the general public. For example, engineers in private practice do both, but engineers employed by the federal, provincial, and municipal governments may, in the course of their work, *provide* services to the general public, although, as government employees, they do not *offer* their services to the public. They are, however, covered by liability insurance (or equivalent coverage) as government employees. Should all engineers who provide services to the general public be required to hold a Certificate of Authorization in those provinces that require a Certificate, and demonstrate the experience and continuing competence requirements that may be associated with the Certificate? Explain and justify your response.

6. Some governments have suggested cost-reduction measures to reduce the number of government-employed engineers (at federal, provincial, and municipal levels) involved in inspection, approval, and quality-control activities. It is proposed that these duties be transferred to engineers in private practice. For example, instead of a government engineer inspecting a facility, as is frequently required by various laws, the owner of the facility would be required to hire an engineer to perform the inspection and prepare a compliance report. The government would merely perform a clerical task to ensure that a satisfactory inspection had taken place and that the facility complied with the law as shown in the report. Write a brief essay on the advantages and disadvantages of this proposal from the viewpoints of cost reduction and efficiency, maintenance and enforceability of standards, availability of service, and potential for conflict of interest.

NOTES

1. Association of Consulting Engineers of Canada (ACEC), *Directory of Member Firms* (Ottawa: ACEC, 1996).
2. J.M. MacEwing, "Professional Liability Perils," *Canadian Consulting Engineer* (September/October 1996): 8.
3. "Consulting Engineers Bemoan Liability Costs, Suits," *Engineering Times* (April 1988): 8.
4. R.T. Cosby, "Professional Liability — Are You Protected?" *Consulting/ Specifying Engineering* (April 1987): 104–8.

5. J.R. Clark, *Concerning Some Legal Responsibilities in the Practice of Architecture and Engineering* (Washington, DC: American Institute of Architects, 1961).

6. Professional Engineers Act, Regulation 941/90, RSO 1990, c. P.28, s. 74(3).

7. C.C. Hart, "Professional Liability Insurance: Some Interesting Statistics," *Gazette* 10, no. 3 (May/June 1990): 2.

8. C. Wallace, "Liability Insurance and Defence of Claims," *Canadian Consulting Engineer* (January/February 1997): 6.

9. Excerpts from ACEC, *Ingénieurs-Conseils CANADA Consulting Engineers* (Ottawa: ACEC, 1997), x, xi. Reprinted with permission of ACEC.

10. Based on International Federation of Consulting Engineers (FIDIC), *FIDIC International Directory of Consulting Engineers, 1997–1998* (Lausanne, Switzerland: FIDIC, 1997–1998). Used with permission of FIDIC.

11. A.J. Ryan, "Operating Your Practice," *American Engineer* (November 1955).

12. J.B. McGaughy, "So You Want to Open a Consulting Office — By Way of Qualifications," *American Engineer* (October 1955).

Part Three

Professional Ethics

CHAPTER SIX

Principles of Engineering Ethics

Ethical problems occur often in engineering. For example, an engineer may have to choose between risking the health of workers on a project or stopping the project to install safety equipment, thereby causing delays and increasing costs for the engineer's clients or employers. At what point does the severity of the risk and potential harm to workers outweigh the real loss to the client or employer that will result if the engineer stops the project to install safety equipment? In another instance, an engineer in a position of authority may have to decide whether a small gift is an innocent kindness or a serious attempt at bribery. As we shall see in subsequent chapters, many similar issues in engineering practice give rise to ethical problems, and it is useful to have a methodical approach for solving these problems.

To develop a method for solving *ethical* problems, let us first examine solution methods for *technical* problems. An engineer faced with a technical problem will generate one or more possible solutions and then analyze these solutions using axioms, theorems, and laws of mathematics, science, and engineering that are tested and true. It is therefore reassuring to know that, when faced with an ethical problem, there is a similar set of ethical theories that has been developed over the centuries. These ethical theories are the basis of the laws, regulations, and codes of ethics that guide the engineer in ethical decision making.

When an engineer is faced with a technical problem, there is usually more than one solution, and the goal is to select the best or optimum solution. Similarly, in dealing with ethical problems, there may be several possible solutions, and the goal is to determine the best solution from the ethical standpoint. This is not always easy, since ethical theories sometimes generate contradictions when applied. Alternate solutions frequently require diametrically opposite actions that are totally incompatible. This leads to a moral dilemma that may require breaking one ethical code to satisfy another. An example from daily life is easy to provide: we are all taught to be truthful and not to tell lies; however, if we hear gossip that would be hurtful to friends, relatives, or co-workers, we might alter the information to save their feelings. We would argue that a greater good is served by lying or denying the truth than by following precisely the ethical precept "Do not tell lies." When solving ethical problems in engineering, it is almost always necessary to evaluate several competing theories before making a decision; moral dilemmas will frequently result, and the best course of action must then be chosen.

In the next few paragraphs, we examine four important ethical theories that have evolved over the centuries, and use them to develop a methodical approach to solving ethical problems.

ETHICS AND PHILOSOPHY

Ethics — the study of right and wrong, good and evil, obligations and rights, justice, and social and political ideals — is one of four branches of philosophy, according to *The Canadian Encyclopedia Plus*.[1] The other three branches are *logic*, the study of the rules of reason; *epistemology*, the study of knowledge itself; and *metaphysics*, the study of very basic ideas such as existence, appearance, reality, and determinism. For our purposes, ethics can be defined more practically as defining, analyzing, evaluating, and resolving moral problems and developing moral criteria to guide human behaviour. Ethics has been a vital and important field of study since the dawn of civilization, and has had a written history for more than 2500 years. In fact, many ethical concepts that we commonly apply today are older than the basic concepts of engineering analysis (calculus, dynamics, stress analysis, etc.), which trace their origins to the seventeenth century.

Philosophy involves examining questions that are very fundamental, questions such as, "What is truth?" "How do we define justice?" and "What constitutes beauty?" Truth, justice, and beauty are concepts that we use every day. We cannot, in fact, recall when we first learned their meanings. It is therefore shocking to find that we cannot easily define these basic terms that we have used all our lives. Attempting to define these important but basic concepts is one of the fundamental motivations of philosophy.

In ethics, the goal is to differentiate between good and bad and between right and wrong. Therefore, the key question is "What is good?" and again we find it is difficult to define the basic term "good." Attempts to define goodness usually result in circular arguments, or definitions that describe the effects (or the lack) of goodness. For example, if "goodness" were defined as that property of an object, person, or act that creates pleasant, positive, or useful results, most people would probably agree with the definition. However, this is not a precise definition, since the terms "pleasant," "positive," and "useful" depend on subjective evaluations that vary from individual to individual. Therefore, this definition is not fully satisfactory as a basis for applying ethics. However, even if we cannot precisely define goodness, we must agree on some working definition if we are to apply criteria to differentiate it from its opposite, and the above definition is an initial attempt to do so.

FOUR ETHICAL THEORIES

Many prominent philosophers have devoted their lives to developing ethical theories, and a complete discussion of their thoughts would fill a thousand textbooks. It is presumptuous to think that this wealth of philosophical thought could be condensed into this single chapter. However, some review of ethical theories is essential for understanding the origin of codes of ethics and also for dealing with cases that "fall through the cracks" — cases that cause ethical problems but are not clearly addressed by the codes of ethics. This brief summary is merely an introduction, but it should be adequate to illustrate the basic concepts; perhaps it will inspire you to investigate the subject more deeply.

At least four moral theories or maxims have evolved over the centuries and are relevant to the application of ethics in engineering. Each theory, when it was proposed, was considered by its originators to be the basis for all ethical thought. All of the theories have stood the test of time and are useful aids to decision making. Although the theories may appear to differ significantly, and none of them is clearly universally superior to the others, it is startling to see how much they agree when applied to common ethical problems. The four theories are listed and described in detail on the next few pages. Each is identified by the name of its best-known proponent, although many earlier philosophers contributed to formulating the theories, and many contemporary philosophers have suggested modifications to improve them. The four moral theories are

- Mill's utilitarianism
- Kant's formalism, or duty ethics
- Locke's rights ethics
- Aristotle's virtue ethics

MILL'S UTILITARIANISM

This theory was stated most clearly by John Stuart Mill (1806–1873). Utilitarianism states that the best choice in a moral dilemma is that which produces the maximum benefit for the greatest number of people. This theory is probably the most common justification for ethical decisions in engineering, or indeed in modern society. Democratic government itself can be justified on utilitarian grounds, since it permits the maximum good (control over government) for the maximum number of people (the majority of voters). Although dictatorships or absolute monarchies may be more efficient, more convenient, or more stable, citizens who have lived in a truly democratic society would tolerate no other form of government.

The difficulty of applying the utilitarian principle lies in quantitatively calculating the "maximum benefit." Mill proposed that the *intensity* and *duration* of a benefit or pleasure (or, conversely, the intensity and duration of a pain to be avoided), and the *number* of people affected should be the three key factors. For example, in evaluating the benefit of automobile seat-belt legislation, the inconvenience is applied to all drivers and passengers, whereas the benefit (avoiding crippling or disfiguring injuries or death) accrues to only a few people, but the duration and intensity of the pain suffered in accidents more than outweigh the inconvenience of requiring everyone to wear seat belts.

In evaluating benefits, it is important that certain principles apply. The benefit to oneself must not be given any greater value or importance than the same benefit to any other individual. Benefits should, of course, be calculated without regard to discrimination on the basis of nationality, creed, colour, race, language, sex, and so on of the persons involved, and no preference should be given to any particular group. Moreover, the equality of distribution of the benefit is important when choosing a course of action on a utilitarian basis: an equal distribution of benefits is preferable to an unequal distribution. That is, the best course of action in an ethical dilemma is the choice that produces the maximum benefit for the greatest number of people, with the benefit most equally divided among those people.

The utilitarian theory is consistent with the concept of democracy, is easily understood, and, for simple cases, is easy to apply. For example, income tax is easily justified by utilitarian theory: a modest hardship (paying tax) is imposed equally on all residents (as a percentage of income), while yielding an immense benefit to those who need the hospitals, schools, and infrastructure of our society that are built mainly with tax funds. Although we may, from time to time, disagree with details of the taxation system (who should get tax exemptions, what priorities should be used in tax expenditures, etc.), the practice is rarely challenged as unethical.

Consequently, utilitarianism is a useful ethical theory for evaluating courses of action when an engineer is faced with an important moral choice.

KANT'S FORMALISM, OR DUTY ETHICS

The theory of duty ethics, or "formalism," was put forward by Immanuel Kant (1724–1804), who proposed that each person has a fundamental duty to act in a correct ethical manner. He evolved his theory from the belief or observation that each person's conscience imposed an absolute, "categorical" imperative (or unconditional command) on that person to follow those courses of action that would be acceptable as universal principles for everyone to follow. For example, everyone has a duty not to tell lies, since, if lying were to be done by everyone, then no promises could be trusted and our social fabric would be at risk of unravelling.

Kant believed that the most basic good was "good will," or actively seeking to follow the categorical imperative of one's conscience. This is in marked contrast to Mill, who believed that universal happiness was the ultimate good. In Kant's philosophy, happiness is the result of good will: the desire and intention to do one's duty. Kant emphasized that it was the *intention* to do one's duty that was significant, not the *actual* results or consequences. One should always do one's duty, even if the consequences in the short run are unpleasant, since this strengthens one's will. For example, even "white" lies should not be tolerated, since they weaken the resolve to follow one's conscience. Therefore, in solving a moral dilemma, the formalist theory states that one has a duty to follow rules that are generated from the conscience (the categorical imperative), and if a person strives to develop a good will, then happiness will be the result.

Examples of moral rules that result from applying this universal concept are easy to generate, and, not surprisingly, our happiness would certainly improve if everyone followed these rules: "Be honest," "Be fair," "Do not hurt others," "Keep your promises," "Obey the law," and so on. Kant also stated that a consequence of following the categorical imperative was an increased respect for humanity. Life should always be treated as an end or goal, but never as a means of achieving some other goal. Consequently, any engineering activity that endangers life by water or air pollution (regardless of the purpose or cause of the pollution) would be condemned as unethical. In Kant's philosophy, every engineer or engineering manager has an individual duty to prevent harm to human life and to consider the welfare of society to be paramount.

Kant's formalism therefore stresses the importance of following universal rules, the importance of humanity, and the significance of the intention of an act or rule, rather than the actual outcome in a specific case. The most significant problem with applying formalism is that duties based on the categorical imperative *never* have exceptions. We can all imagine cases where rules appear to conflict, resulting in moral dilemmas. Which rule we will follow requires deeper insight, and all four moral theories are useful aids to moral choice.

LOCKE'S RIGHTS ETHICS

The rights-based ethical theory comes mainly from the work of John Locke (1632–1704), and states that everyone has rights that arise from one's very existence as a human being. The right to life and the right to the maximum possible individual liberty and human dignity are fundamental, and other rights arise as a consequence. The rights of the individual must be recognized by others, who have a duty not to infringe on these rights. This is in contrast to the previously discussed duty-based ethical theory, which stated that duty was fundamental; in the rights-based theory, duties are a consequence of personal rights.

The writings of Locke had a significant effect on political thought in Britain in the 1690s and also influenced the French and the American revolutions. Basic human rights are embedded in Canadian law through the Canadian Charter of Rights and Freedoms. The Charter recognizes that everyone has

- fundamental freedom of conscience, religion, thought, belief, opinion, expression, peaceful assembly, and association
- legal rights to life, liberty, and security of the person and the right not to be deprived of these rights except in accordance with principles of fundamental justice
- equality rights before and under the law and the right to equal benefit and protection of the law

In evaluating ethical choices in engineering problems, it is important to recognize that individuals have these basic rights that, in general, should not be infringed upon. However, the above list does not contain every right that should exist, only the fundamental rights that have been hammered out in Parliament and in the courts of law over the last two centuries. There are, of course, more specific rights that stem from Locke's theory. For example, everyone has the right to a working environment that is free of sexual harassment or racial discrimination, and the employer has a duty to provide it. These rights would appear to be common courtesy, and few would challenge them. In specific cases, more rights may be derived, using respect for human dignity and individual liberty as their basis.

Rights-based ethical theory does have flaws and limits, however. For example, consider the earlier example of income tax: a few people, even today, challenge the concept of income tax, claiming that it infringes on the individual right to retain one's property. Clearly, the rights-based argument conflicts with the utilitarian argument, leading to a moral dilemma that has been resolved in favour of utilitarianism. Therefore, rights-based ethics has an important place in evaluating ethical problems and resolving moral dilemmas, but it is not sufficient in itself to deal with every case.

ARISTOTLE'S VIRTUE ETHICS

One of the earliest and most durable ethical theories was proposed by the ancient Greek philosopher Aristotle (384–322 B.C.), who observed that the goodness of an act, object, or person depended on the function or goal concerned. For example, a "good" chair is comfortable; a "good" knife cuts well. Similarly, happiness or goodness will result for humans if they can allow their specifically human qualities to function fully. The one quality that humans have, above all other animals, is the power of thought; therefore, Aristotle stated that true happiness would be achieved by developing qualities of character through thought, reason, deduction, and logic. He called these qualities of character "virtues" and visualized every virtue as a compromise between two extremes or vices. His guide to achieving virtue was to select the "golden mean" between the extremes of excess and deficiency. For example, modesty is the golden mean between the excess of vanity and the deficiency of humility; courage is the golden mean between excessive foolhardiness and the deficiency of cowardice; generosity is the golden mean between wastefulness and stinginess, and so forth.

Aristotle's concept of virtue as the golden mean can be applied to moral problems by examining the extremes of excess or deficiency and seeking the compromise, "happy medium," or golden mean between the extremes. This approach is frequently useful in resolving ethical dilemmas.

AGREEMENT AND CONTRADICTION IN MORAL THEORIES

The four moral theories just described have survived the test of centuries and can be considered true. However, while each is true for a wide range of applications, none of the theories is universally true or clearly superior to the other theories in every instance. Philosophers strive to find the single principle upon which all ethical thought is founded, but no single unifying concept has yet emerged. Each moral theory has a distinctive contribution; in some areas of ethical thought, all four theories are in complete agreement. Occasionally, however, they are in contradiction.

As an example of agreement between the theories, consider the Golden Rule: "Do unto others as you would have others do unto you." This is a clear statement of Kant's formalism: it imposes a duty on the individual to view human life as a goal, rather than a means to a goal. On the other hand, it could be considered a utilitarian principle, since it imposes an inconvenience on an individual while benefiting everyone with whom that person comes into contact. The proponents of rights-based ethics would agree with the Golden Rule, but would claim that the duty of the individual to act fairly comes from the rights of others to be treated fairly. Finally, the concept of "fairness" would be recognized as a "virtue" by Aristotle. The four moral theories are therefore in complete agreement that the Golden Rule is a good maxim for guiding human behaviour, as we would expect.

Similarly, if we were to examine almost any of the basic precepts of most religions, we would find that each one would be supported by all four moral theories. Consider, for example, the Ten Commandments from the Book of Exodus, which are part of the ethical basis of Judaeo-Christian religions. Each of the commandments clearly imposes a duty on the individual, while granting rights to others, requiring behaviour that would be virtuous and creating a stable environment that would yield the maximum benefit for all. An investigation of the basic precepts of all the great religions would show similar agreement.

However, although the moral theories show remarkable agreement on general principles, they occasionally give conflicting or contradictory guidance when applied to specific cases. Consider, for example, the case of engineers Smith and Jones, who are both employed in designing the control system for an electrical power generating plant. If Jones has definite knowledge that Smith has an addiction to alcohol or drugs that is seriously affecting Smith's mental stability and technical judgement, what is the proper course of action? The duty-based theory would state that Jones has a duty to the employer and should report Smith to management for reassignment or disciplinary action. The rights-based theory would state that Smith's health is a private matter; Smith has a right to privacy and Jones has no right to investigate it or discuss it with others. Clearly these two theories give contradictory guidance.

When there is conflict or contradiction between two moral theories, it is helpful to examine all four theories to see if a majority favour a particular course of action. In the case of Smith and Jones, the contradiction between the duty-based theory and the rights-based theory can be resolved by considering the utilitarian and virtue-based theories. The degree of danger to others and the degree of incapacitation resulting from Smith's drug or alcohol addiction are important factors that must be considered. The seriousness of the addiction and the possibility of recovery with and without intervention should also be considered. The utilitarian theory would then state that the risk of harm to others must be balanced against the harm that will be done to Smith if Jones invades Smith's privacy by exposing the addiction. The decision depends on the degree or intensity of several factors, so a decision cannot be made without knowledge of the specific case. If the virtue-based theory were applied to this case, similar knowledge would be required; the golden mean between the excess of harmful intervention and the deficiency of inaction would depend on a balance between factors specific to the individual case. However, since this information would be known by Jones, an informed ethical decision could be made.

When applying all four theories to the case does not resolve the dilemma, then the moral theory that is considered most appropriate must be selected and followed. This step requires a value judgement that may vary from person to person and is therefore not an absolute rule that always yields the same answer. Nevertheless, if the decision has been made in an orderly fashion, is consistent with at least one recognized moral theory, and has not been made lightly or wantonly, then the person who has made the decision will have a clear conscience. These ideas are expanded into a decision-making strategy later in this chapter.

A word of caution is in order here: we should always be alert to subconscious bias when selecting a course of action that benefits ourselves at the cost of someone else. A decision that renders a benefit to oneself is a conflict of interest, and professional engineers encounter this situation frequently. It is essential to avoid subconscious bias — in fact, sometimes it is essential to have an ethical decision verified by someone who does not have a conflict of interest. In important cases where life, safety, security, or personal reputation is at stake, a decision must not only be ethical, but also must be seen by others to be ethical.

The statements of the four moral theories and indications of where they may be in conflict are summarized in Table 6.1.

CODES OF ETHICS AS GUIDES TO CONDUCT

In order to put ethics into practice, most people need clearer day-to-day guidance than is provided by the general philosophical principles stated in the previous sections. Therefore, customs, conventions, laws, and ordinances that are consistent with the ethical theories, but give more specific guidance, have developed over the centuries. For example, criminal and civil law are probably the most important guides to ethical or moral conduct. Throughout the world there is remarkably close agreement in these laws, in spite of the different political systems, cultural influences, and moral attitudes that exist in different countries.

Table 6.1 A SUMMARY OF THE FOUR ETHICAL CODES

Mill's Utilitarianism	*Statement:* An action is morally correct if it produces the greatest benefit for the greatest number of people. The duration, intensity, and equality of distribution of the benefits should be considered.
	Conflict: A conflict of interest may arise when evaluating the benefits. It is important that a personal benefit be counted as equal to a similar benefit to someone else.
Kant's Duty-Based Ethics	*Statement:* Each person has a duty to follow those courses of action that would be acceptable as universal principles for everyone to follow.
	Conflict: Conflicts arise when following a universal principle may cause harm. For example, telling a "white" lie is not acceptable even if the truth causes harm.
Locke's Rights-Based Ethics	*Statement:* All persons are free and equal, and each has a right to life, health, liberty, possessions, and the product of his or her labour.
	Conflict: It is occasionally difficult to determine when one person's rights infringe on another person's rights.
Aristotle's Virtue-Based Ethics	*Statement:* Happiness is achieved by developing virtues, or qualities of character, through deduction and reason. An act is good if it is in accordance with reason. This usually means a course of action that is the golden mean between extremes of excess and deficiency.
	Conflict: The definition of "virtue" is occasionally vague and difficult to apply in specific cases. However, the concept of seeking a golden mean between two extremes is frequently useful in ethics.

The similarities are closest in criminal law: every country forbids theft, perjury, assault, and murder, although punishment may vary from country to country. These similarities are understandable, since laws are merely formal statements of the ethical theories discussed earlier and the four theories are generally in full agreement where basic standards of conduct are concerned.

Under the authority of the provincial and territorial Acts, Associations of Professional Engineers have been established and empowered to write and enforce regulations, by-laws, and codes of ethics that prescribe acceptable conduct for professional engineers. Infringements can lead to penalties enforced by the provincial justice system or by the Association of Professional Engineers, as described in Chapter 14.

The codes of ethics for each province and territory are contained in Appendix B. They give specific rules to guide the conduct of the engineer. The codes usually include statements of general principles, followed by instructions for specific conduct, and emphasize the duties that an engineer has to society, to employers, to clients, to colleagues, to subordinates, to the engineering profession, and to him- or herself. Although the various codes express these duties slightly differently, their intent and the results are very similar. The following paragraphs summarize the content that is common to each of the provincial and territorial engineering codes of ethics.

Duty to Society An engineer is required to consider his or her duty to the public, or society in general, as most important. This is consistent with the utilitarian concept that authority has been given to engineers under the provincial Act to use the title Professional Engineer, to define standards of admission, and to regulate professional behaviour. The purpose of awarding this authority is to create a greater benefit for society in general: protecting the average person from physical or financial harm by ensuring that professional engineers are competent, reliable, professional, and ethical. Engineers have a particular duty to protect the safety, health, and welfare of society if these are affected by the engineer's work.

Duty to Employers An engineer has a duty to his or her employer to act fairly and loyally and to keep the employer's business confidential. The engineer also has an obligation to disclose any conflict of interest that may arise in which the engineer may benefit by harming the employer's business.

Duty to Clients An engineer in private practice is employed by clients, and therefore has the same obligation to a client as the employee engineer has to the employer. Since the contract with a client is usually shorter than the typical employment contract, avoiding conflicts of interest and co-operating with other personnel involved with a project are special concerns.

Duty to Colleagues An engineer has a duty to act with courtesy and good will toward colleagues. This is a simple statement of the Golden Rule, which is supported by all four ethical theories. This duty benefits the people involved, their clients or employers, and the engineering profession in general. Clearly, a person who is awarded professional status should act professionally and should not permit personal or unrelated problems to intrude on the professional relationship. Most codes advise that it is unethical to review the work of a fellow engineer without that engineer's knowledge.

Duty to Employees and Subordinates An engineer has a duty to recognize the rights of others, particularly if they are employees or subordinates who are obligated to work with the engineer by contracts of employment.

Duty to the Engineering Profession An engineer has a duty to maintain the dignity and prestige of the engineering profession and not to bring the profession into disrepute by scandalous, dishonourable, or disgraceful conduct.

Duty to Oneself Finally, an engineer must ensure that the duties to others are balanced by the engineer's own rights. An engineer must insist on adequate payment, a satisfactory work environment, and the rights awarded to everyone through the Charter of Rights and Freedoms. An engineer also has a duty to strive for excellence and to maintain competence in the rapidly changing technical world.

You are urged to refer to Appendix B to review the code of ethics for your province or territory. The seven sets of duties just described will be evident, although described in more detail.

COMPARISON OF CODES OF ETHICS

In addition to the seven general duties just listed, individual provinces and territories may have additional requirements in their engineering codes of ethics. A brief overview of the different codes follows for each province and territory.

Alberta The Alberta code contains all of the duties described in the last section and also expects engineers, geologists, and geophysicists to have proper regard for the physical environment and to advise the provincial Registrar of any practices of other members of the Association that are contrary to the code of ethics. The Preamble to the code entreats engineers to serve in public affairs when their professional knowledge may be of benefit to the public and to demonstrate understanding for members-in-training under their supervision.

British Columbia The British Columbia code was recently revised. It was previously one of the most comprehensive codes, with over 40 clauses or subclauses giving very specific guidelines on ethical and professional topics. The present code is now clear and succinct, with ten clauses, similar to most other provincial codes, specifying the duties that engineers and geoscientists have to society, to clients or employers, to colleagues, to the engineering profession, and to themselves.

Manitoba The Manitoba code specifically states that breaches of the code may be considered unprofessional conduct and subject to disciplinary action under the Act. The Manitoba code is arranged as five basic, general "canons" to guide professional behaviour. Under each canon, a total of 35 very specific subclauses give useful advice on various aspects of professional practice. The Manitoba code also requires members to advise the Registrar of the Association about colleagues who are engaging in unethical, illegal, or unfair practices.

New Brunswick The New Brunswick code includes clauses prohibiting the use of free engineering designs from suppliers in return for specifying their products, requiring members to provide opportunities for professional development for engineers in their employ, and advising on the proper standards for professional advertising.

Newfoundland The Newfoundland code is arranged in three sections that specify the duties of the professional engineer or geoscientist to the public, to the client or employer, and to the profession. The code is brief (22 clauses) and clear.

Northwest Territories The code for the Northwest Territories is virtually identical to the Alberta code of ethics.

Nova Scotia The Nova Scotia code is fairly brief (28 clauses) and contains clauses typical of the other Acts, except for an admonition to refrain from conduct contrary to the public good, even if directed by the employer or client to act in such a manner, and a complementary instruction to employers not to direct employees to perform acts that are unprofessional or contrary to the public good.

Ontario The Ontario code of ethics is different from those in other provinces because it specifically is *not* enforceable under the Act. A separate regulation defining professional misconduct contains many of the clauses that appear in the codes of other provinces. One clause that appears to be unique to Ontario defines "permitting, counselling or assisting a person who is not a practitioner to engage in the practice of professional engineering" to be a form of professional misconduct. Two clauses in the code of ethics that are also unique to Ontario (but not enforceable under the Act) request the engineer to display his or her licence at the place of business and request moonlighting engineers to inform their clients that they are employed and to state any limitations on service that may result from this status.

Prince Edward Island The code of ethics for Prince Edward Island is relatively brief (24 clauses) and clear. All of its clauses are similar to those of other codes.

Quebec The Quebec code of ethics is one of the longest (52 clauses) and is arranged slightly differently from those of the other provinces. However, the basic clauses mentioned above are all represented in the code, and additional clauses are included. These clauses include the duty to show reasonable availability and diligence, to serve on Association committees unless the engineer has exceptional grounds for refusing, and to refrain from communicating with any person who has lodged a complaint against him or her. In addition, one section of the code describes criteria for setting fair and reasonable fees for service.

Saskatchewan The Saskatchewan code (30 clauses) has two clauses not common to other codes: engineers are required to constantly strive to broaden their knowledge and experience by keeping abreast of new techniques and developments. Also, engineers must not begin to offer services for a fee without first notifying the Council of the Association and receiving permission to do so.

Yukon Territory The code of ethics for the Yukon Territory is similar to Saskatchewan's, with the exception of three or four clauses. For example, the Yukon code does not contain the Saskatchewan clause requiring Council's permission to offer services for a fee.

The above overview shows that the codes for each province and territory are very similar, although not identical. They are useful guides to personal conduct. Adherence to the code of ethics is not voluntary, nor is it a lofty ideal that would be "nice but not essential" to achieve. With the exception of Ontario, the code of ethics is part of the Act that regulates engineering, and clear violations of the code can be bases for disciplinary action in the form of a reprimand, suspension, or expulsion from the profession, as described in Chapter 14. Fortunately, behaviour that is consistent with the basic ethical theories is rarely in conflict with the code of ethics.

In addition to the provincial and territorial codes, most engineering societies have developed codes of ethics. Three of these are included in Appendix E for comparison. The code promulgated by the National Society of Professional Engineers (NSPE) is very similar to the provincial codes and has been endorsed by many engineering societies. Since engineering societies do not have the regulatory power of provincial and territorial Associations, serious infractions of the society codes are punished by expulsion from the society.

A STRATEGY FOR SOLVING COMPLEX ETHICAL PROBLEMS

Most of the ethical problems that arise in everyday life are clear and simple and are solved by intuitive use of the ethical theories. Complex ethical problems can be much more challenging; intuitive methods usually do not work, and many people are in a quandary when trying to decide where to start and how to proceed to an acceptable solution. Engineers should have an advantage in resolving ethical dilemmas, since problem-solving and decision-making techniques are a routine part of engineering. This section explains the similarity between ethical problem solving and engineering design methods, and although this formal strategy would rarely be needed, it is reassuring to know that it exists and can be applied to complex cases.

THE ENGINEERING DESIGN PROCESS

The design for a new machine, structure, or electronic device does not spring fully developed into the mind of the designer. Rather, it is the result of a series of steps requiring inspiration and deduction in the right order at the right time. The design process usually begins with a vaguely perceived need or problem and ends with the manufacture of the structure or device that satisfies the need. The solution to an ethical problem can be developed by following a comparable series of steps. The typical steps or phases in the design process are as follows:

1. recognizing that a problem or need exists
2. gathering information and defining the problem to be solved or goal to be achieved
3. generating alternative solutions or methods to achieve the goal (synthesis)
4. evaluating benefits and costs of alternative solutions (analysis)
5. decision making and optimization
6. implementing the best solution

The design process begins by recognizing a need in the marketplace, the community, the factory, or the machine shop. For example, people living in the suburbs may want to travel to work in the city conveniently and believe that they need a new freeway or superhighway. It is important to gather data and define the problem precisely before committing resources to designing the highway. In our example, we must find out whether the travelling public really needs a new highway; would a new subway or bus line be more effective? When the need is investigated, it may be concluded that a new subway is the most desirable goal.

The detail design work would then begin with that goal in mind and alternative routes for the subway would be drawn up. Alternative sources for materials and alternative methods to make or buy trains and passenger cars would be considered. Each alternative would then be analyzed to obtain the cost estimates for acquiring property and constructing the line. Each route would be analyzed for safety, convenience to riders, and inconvenience to neighbouring residents. Finally, the optimum or best solution, which gives the maximum benefit at minimum cost, would be selected. The design would then be implemented, materials would be ordered, and construction would start.

APPLYING THE DESIGN PROCESS TO ETHICAL PROBLEMS

The design process described above is really a simple and straightforward problem-solving technique. We can develop a strategy for ethical problem solving based on this design process, as discussed below.

1. RECOGNIZING THE NEED OR PROBLEM
Ethical problems may be poorly defined and difficult to recognize, particularly in the early stages. For example, a manufacturing process once thought to be harmless may be suspected, over a period of years, to be the cause of cancer or toxicity in the workplace. Recognizing that a problem exists is an important first step.

2. GATHERING INFORMATION AND DEFINING THE PROBLEM
As a general rule, it is advisable to act on an ethical problem quickly and decisively, but it is equally important to have all the facts. Premature action will almost always offend someone and may create an even more serious problem. When the information has been collected and examined, the proper course of action may be quite different than what one initially perceived.

When the problem is clearly defined, the proper course of action is usually perfectly clear. The necessary action may be dictated by law, codes of ethics, or ethical theory, and one can skip directly to the implementation step. However, in some cases there may be conflicts that lead to a moral dilemma. For example, the code of ethics may give conflicting direction. In these cases, some inspiration is required, as explained in the next step.

3. GENERATING ALTERNATIVE SOLUTIONS (SYNTHESIS)
When a moral dilemma results — in which the engineer must choose between two courses of action, each of which is undesirable — then the engineer should strive to generate a new, positive, desirable course of action. This phase of the solution procedure requires creative thought, and therefore it is usually difficult. The new course of action may be a

compromise or a modification of one alternative to eliminate its negative aspects. For example, consider the case where an engineer receives a small cash gift from a contractor and is uncertain as to whether the gift is a favour to cover incidental expenses or a bribe. This creates a dilemma. If the gift is a conscientious favour, then to return it would offend the donor. On the other hand, if it is a bribe, then to keep it would incur an obligation. In this case, a creative compromise might be to give the cash to a charity (preferably in the name of the donor), thus avoiding offence without incurring an obligation. Other courses of action may exist, and some ingenuity is needed. Many of the creative methods used to generate alternatives in technical problems (such as brainstorming) can also be applied to generating alternatives in ethical problems.

4. EVALUATING ALTERNATIVES (ANALYSIS)

When two or more conflicting courses of action exist, they must be analyzed to see what consequences are likely to result before a decision can be made. The results of each possible course of action must be examined. What benefits accrue? For whom? What hardships are involved? Are the benefits and hardships equally distributed?

5. DECISION MAKING AND OPTIMIZATION

If the previous steps have been thoroughly followed, decision making simply involves comparing the consequences of each course of action with the code of ethics or moral theories discussed earlier and selecting the best or optimum course of action. If no solution appears to be acceptable, it may be necessary to go back to step 2 and verify that the problem has been properly defined.

In making decisions, a decision table is sometimes useful in summarizing the courses of action and their consequences. A decision table is merely a chart or matrix showing the various courses of action and the outcomes or consequences for the people concerned. The decision chart is particularly useful when numerical costs and benefits can be calculated or when numerical probabilities for good and bad events are known. The decision chart is then transformed into a type of balance sheet, with a numerical sum for each course of action.

For example, consider the simple case mentioned earlier, in which an engineer supervising a construction project has received a small cash gift from a contractor and is uncertain as to whether it is motivated by legitimate business or is an attempt to influence an engineering decision. The engineer does not want to keep the gift, since that would create an obligation to the contractor and could be seen as a bribe. On the other hand, the engineer does not want to return the gift and risk destroying a good working relationship by offending the contractor. After some thought, the engineer might generate an innovative third course of action: the cash gift could be used to install a temporary soft drink cooler on the work site that would benefit all of the workers and indirectly aid the project. The decision chart illustrating the various choices and outcomes is shown in Table 6.2.

The outcomes are the sums in the bottom row. The first two courses of action generate a negative sum; the third choice is the only positive course of action. In this simple example, the decision chart merely illustrates what we already knew: the best course of action is to use the gift for the benefit of the workers. The example illustrates the method, but is too simple to show its full value. The true usefulness of the decision chart lies in its ability to separate the courses of action and to identify the consequences of each possible course of action. The chart may stimulate the need to generate new courses of action, to review the information provided, or to restate the goal, so that an optimum course of action can be found.

In some cases, it may appear that a solution cannot be achieved; the arguments for conflicting alternatives may be so equally balanced that no course of action is clearly superior. In this case, the engineer should pose the following questions:

- Is the problem stated clearly?
- Has all the necessary information been obtained?
- Have I sought advice from the people concerned?
- Has an alternative or compromise solution been overlooked?
- Have all the consequences of each alternative choice been fully evaluated?
- Is a personal benefit or conflict of interest affecting my judgement?

If the above questions can be answered satisfactorily and there is still no optimum course of action, then it would be advisable to select the course of action that does not yield a benefit to the person making the

Table 6.2 SAMPLE DECISION CHART

	Possible Courses of Action		
Person Affected	Keep the Gift	Return the Gift	Use the Gift for Workers
Donor	No offence to donor (0)	Result may offend donor (−1)	No offence to donor (0)
Engineer	Result may incur an obligation (−1)	No obligation (0)	The project will benefit (+1)
Sum	−1	−1	+1

decision. If the choices are equally balanced and the possibility of personal benefit exists, then this choice will ensure that the decision is seen to be morally right.

If no personal benefit is involved, then the moral theory that is considered most appropriate must be selected and followed. Although this involves a personal value judgement, the person making the decision will have a clear conscience.

6. IMPLEMENTATION

Implementing the decision is the final step. Although the appropriate action will vary from case to case, it is usually advisable to act speedily and unequivocally when ethical decisions are needed, particularly if health, safety, or reputation is at stake.

SOME FINAL COMMENTS ON THIS STRATEGY

The strategy described above is a rather formal process that would be followed in detail only for very complex cases or cases in which a written report will be prepared to justify a specific decision. However, this strategy may be useful when an engineer is stuck for a solution to a simpler problem. In these circumstances, it is reassuring to know that many of the well-known methods of engineering — problem solving, generation of creative ideas, and decision making — can be applied to ethical problems as well. Engineers, in fact, should be much better prepared than the average person to resolve complex ethical problems as a result of their design education and problem-solving experience.

TOPICS FOR STUDY AND DISCUSSION

1. Examine the code of ethics for your province or territory in Appendix B. Does it include all seven of the basic duties described in this chapter?
2. Some codes of ethics include duties that do not appear in other codes. Examine the code of ethics for your province or territory to see if it contains clauses that require engineers to

   - Advertise in a dignified, professional manner.
   - Report infractions of the code of ethics to the Registrar of the Association.
   - Consider the difficulty of the task and the degree of responsibility in setting fees.
   - Refuse to pay commissions or reduce fees in order to obtain engineering work.

Should all of the above duties be included in your code of ethics, or should they appear elsewhere in the Act, or are they not appropriate for inclusion in either the code or the Act? Explain and justify your answer for each of the four clauses.

3. At least one of the provincial codes of ethics requires engineers to serve in public affairs when their professional knowledge may be of benefit to the public. This code also states that engineers, geologists, and geophysicists have a special obligation to demonstrate understanding, professionalism, and technical expertise to members-in-training under their supervision. Are these suitable duties for inclusion in the code of ethics? (Most provinces do not include them.) Should these duties be included in every code of ethics? Which of these duties is more fundamental? (That is, if only one of them could be included in the code, which would it be?) Discuss and explain your answers.

4. Compare the engineering code of ethics for your province with the code of ethics from a neighbouring province (see Appendix B), and with the code of ethics of one of the engineering societies (see Appendix E). What are the similarities and differences between the codes? Does your provincial code bear a similarity to the neighbouring provincial code or to the society code? Which of the three codes do you consider to be the best guide for professional conduct? Prepare a brief summary to explain and justify your answers.

5. The codes of ethics for the IEEE and ASME, two of the largest engineering societies in the world, are included in Appendix E. These codes are very brief. Compare them with the code of ethics for your province or territory. Do all three codes cover the seven basic duties described in this chapter? Since the code of ethics is a guide to personal conduct, is it preferable, in your opinion, to have a brief code that can be more easily remembered, or to have a more comprehensive code that gives more specific guidance? Explain your answer.

6. To illustrate the complexity of philosophical thought, try to write a brief answer to the question "What is the purpose of human life?" Since we are all human, we should be able to respond easily, but do not be surprised if you find this simple, basic question to be almost unanswerable. Is the answer to ensure the survival of our species? Is it to maximize economic gain? Is it to maximize individual pleasure? If so, what kind of pleasure? Is it to ensure a balanced natural ecosystem on Earth? Is it related to a higher power? Is it all of these things, or is it something different from any of these? Are the four moral theories discussed earlier in this text related in some way to the purpose of life? Prepare your answer in the form of a brief essay, discussing and explaining your viewpoint. (There is more than one right answer for this question.)

7. At the beginning of this chapter, it was stated that terms like "truth," "justice," "beauty," and "goodness" are very difficult to define. Is it possible to define any of these terms objectively, without using subjective opinions, so that your definition would be true for all cultures for all time? Consult textbooks on philosophy in order to prepare your answer, and don't be disappointed if you are unable to define completely objectively these well-known terms.

8. The six-step strategy for solving complex ethical problems discussed in this chapter is similar to the design process. This strategy may appear to be too formal and structured for enthusiastic engineers who want to dive into a problem and come up with a fast solution. For these engineers, there is another process, usually seen on bumper stickers, that outlines the following six phases of the design process:

   i. initial enthusiasm
   ii. unco-ordinated hard work
   iii. gradual disillusionment
   iv. evidence of chaos
   v. punishment of the participants
   vi. bestowing of honour on the uninvolved

   This process is easier to apply, but less likely to achieve a useful solution. Comment on these two six-step problem-solving methods. Is it really necessary to have a methodical process? Explain why or why not. Have any important steps been omitted? Are any of the steps unnecessary? Derive a method or mnemonic for remembering the six-step process described earlier in this chapter.

9. Some codes of ethics include clauses requiring engineers to treat one another with courtesy and good faith and forbidding them to make statements that maliciously injure the reputation or business of another engineer. Compare the benefits with the potential for abuse of such clauses. Although such clauses obviously have laudable aims, how does one determine when a violation of courtesy or good faith has occurred? Are such clauses enforceable? Is the prohibition against malicious statements a reasonable infringement on freedom of speech? How does one distinguish between malicious statements and statements that are critical of a fellow engineer but true, and perhaps necessary for technical reasons? For example, if a fellow engineer makes a serious calculation error, is it malicious to point it out? Is it better to encourage open debate than to make unenforceable rules about malicious lies?

10. Examine the code of ethics for your province to see if it includes a clause that would apply to sexual harassment. Which of the existing clauses forbids behaviour of this nature? If there is no appropriate clause in your code of ethics, write a clause that could be inserted.

FURTHER READING

The following sources were consulted in the preparation of this chapter and are recommended for additional reading.

P.L. Alger, N.A. Christensen, and S.P. Olmsted, *Ethical Problems in Engineering* (New York: John Wiley and Sons, 1965).

M.I. Mantell, *Ethics and Professionalism in Engineering* (Toronto: Collier Macmillan, 1964).

M.W. Martin and R. Schinzinger, *Ethics in Engineering*, 2nd ed. (New York: McGraw-Hill, 1989).

C. Morrison and P. Hughes, *Professional Engineering Practice: Ethical Aspects* (Toronto: McGraw-Hill Ryerson, 1988).

J.T. Stevenson, *Engineering Ethics: Practices and Principles* (Toronto: Canadian Scholars' Press, 1987).

NOTES

1. J.T. Stevenson, "Philosophy," *The Canadian Encyclopedia Plus* (Toronto: McClelland & Stewart, 1995).

CHAPTER SEVEN

Ethical Problems of Engineers in Industry

Engineers in industry are mainly employees and are occasionally subjected to pressures that constrain their ability to act professionally. A frequent source of pressure is the employer. However, other influences also may affect the engineer's ability to act professionally. Conflicts of interest, employer–employee negotiation procedures, trade secrets, and confidentiality are issues that the employee engineer will encounter sometime during his or her professional career.

In this chapter, we attempt to define the limits of the employer's authority and of the engineer's obligation, and to examine the miscellaneous factors that affect employee engineers in industry.

EMPLOYER AUTHORITY AND EMPLOYEE DUTIES

When an engineer accepts an offer of employment, a contract is created in which the engineer, as an employee, agrees to use his or her ability to achieve the employer's legitimate goals. The employer has a duty under the same contract to treat the engineer in a professional manner but also clearly acquires the authority to direct the engineer. The need for authority is obvious, particularly in large organizations where lack of direction could lead to chaos and bankruptcy. The employer has *management authority* to direct the resources of the company, whereas the engineer has *technical authority* to exercise the special knowledge and skill acquired through university education and practical engineering experience. In a well-run organization, the distinction between management and technical authority will be well defined; the individuals involved will show mutual respect and will co-operate to achieve the goals of the employer.

In many corporations, engineers are responsible for evaluating the technical feasibility of various courses of action, but management has the authority to decide which course will be followed. In general, the process works well and benefits the engineer, the managers, the shareholders, and society in general.

However, over the length of a professional career, an engineer may be directed to do something that, in his or her opinion, is morally wrong. For example, an engineer may be asked to alter calculations to show that a gear train has a slightly greater factor of safety against overload so that it meets the specifications requested by a client. An engineer may be asked to direct the improper disposal of industrial waste water that is suspected to contain toxic chemicals. In many well-publicized cases, the pressure on management to show a profit was converted into pressure on an engineer to act unethically. Although the problem does not occur frequently, it does exist and usually results in ethical dilemmas. At what point does the engineer's duty to follow his or her conscience exceed the obligation to the employer? To resolve an ethical dilemma, the procedure described in the previous chapter should be followed. A useful distinction can be drawn between the following categories or degrees of moral conflict:

Illegal Actions An engineer may be asked to perform an act that he or she considers unethical and that is also clearly contrary to the law. The law may be a criminal law, a civil or business law, a regulation made under the authority of an act (such as an environmental regulation), or an infringement of trademark, copyright, or industrial design legislation. In such a case, the engineer should advise the employer that the action is illegal and should resist any direction to break the law; the employer clearly does not have the authority to direct the engineer to break the law.

Actions Contrary to the Code of Ethics An engineer may be asked to perform an act that, while not clearly illegal, is a breach of the code of ethics of the provincial Association. In this case, the engineer should advise the employer of the appropriate section of the code and should decline to take any action on the employer's request. If the employer is not an engineer, he or she may not be sufficiently familiar with the code of ethics and its legal significance. If the employer *is* an engineer, then he or she is equally bound to follow the code. In either case, the employee engineer has a legal basis for insisting on ethical behaviour; an employer cannot direct the engineer to take action that clearly violates the code of ethics.

Actions Contrary to the Conscience of the Engineer An engineer may be asked to perform an act that, while not illegal nor clearly a violation of the code of ethics, nevertheless contravenes the engineer's conscience or moral code. These are, of course, the most difficult cases. For situations such as these, the decision-making procedure described in the previous chapter should be most useful. The engineer must gather all the relevant information and define the ethical problem as clearly as possible. In precisely what way does the required action offend the engineer's conscience? The engineer must attempt to see the problem from the employer's viewpoint. Alternative courses of action must be generated and examined in light of the basic ethical theories discussed earlier, and the optimum course of action should be selected. In these problems, the personal consequences to the engineer must, of course, be considered. Refusal to follow an employer's directive may result in disciplinary action or dismissal if the employer cannot be convinced otherwise. The possibility and consequences of dismissal and the remedies for wrongful dismissal should be considered by the engineer.

PROFESSIONAL EMPLOYEE GUIDELINES

In 1973, the U.S. National Society of Professional Engineers (NSPE) developed a set of *Guidelines to Professional Employment for Engineers and Scientists*. The factor that stimulated their development was the U.S. government's cutbacks on aerospace expenditures in the late 1960s, including the cancellation of the proposed supersonic transport aircraft (SST). Many American scientists and engineers wound up unemployed and suffered severe financial hardship in the years following the cutbacks.

The NSPE *Guidelines* are similar in many respects to the codes of ethics adopted by Canadian professional engineering Associations and U.S. technical societies. The *Guidelines* have, in fact, been adopted by at least 26 U.S. or international technical societies. However, unlike the codes of ethics established by provincial professional engineering Associations under the authority of provincial Acts, the *Guidelines* do not have any legally binding authority in either the United States or Canada. Nevertheless, they establish ethical practices for employment of engineers (and scientists) that are consistent with, but much more specific than, the provincial codes of ethics. A copy of the *Guidelines* is included as Appendix F.

As described in Chapter 3, the emergence of these *Guidelines* was a significant contribution to the engineering profession and recognizes that, first, most engineers are employees of someone else, and second, the usual codes of ethics apply primarily to engineers in private practice and sometimes are not very helpful to engineers in industry, government, and education.

The NSPE *Guidelines* contain far more injunctions for employers than for employees. In brief, the employee is expected to be loyal to the employer's objectives, safeguard the public welfare, avoid conflicts of interest, and pursue professional development programs. The employer is expected to keep professional employees informed of the organization's objectives and policies, establish equitable compensation plans, minimize new hiring during layoffs, provide for early vesting of pension rights, assist in professional development programs, provide timely notice in the event of termination, and assist in relocation efforts following termination.

The NSPE *Guidelines* are not legally binding, but they have had a constructive impact upon employee–employer relationships. Certain items are clues to the problems that in the 1970s beset aerospace engineers who suddenly found themselves out of work. For example, the emphasis on early vesting of pension rights reflects the fact that many people found, after several years of service with employers, that they had acquired little or no right to the contributions made by the employers to the employees' pension funds. Many — perhaps most — pension plans provide that the employees' rights to employers' contributions become vested only after many years of service. Thus, an employee who changes jobs every few years could easily reach retirement age with no pension except what the government might provide.

A related concern is reflected in the *Guidelines'* reference to limiting new hiring during periods of layoffs. This guideline was stimulated by the realization on the part of some engineers that the companies that had just laid them off were continuing to hire brand new engineering graduates. The companies explained that they had new needs that could not be met by the people who had been let go, but the unemployed engineers were more likely to believe that their former employers were trying to save money by hiring less expensive people. Some companies replied by saying that the desire to save money could not be a correct explanation, because new graduates could hardly be considered cheap, were certainly expensive to recruit, and generally could not be very productive for a year or two anyway.

The NSPE *Guidelines* are useful for employee engineers and give more specific advice and guidance than the provincial codes of ethics. Although the Guidelines do not have the legal authority of the codes of ethics, they are consistent with the codes and form an excellent basis for ethical decision making.

PROFESSIONAL ENGINEERS AND LABOUR UNIONS

Engineers have a right to fairness in negotiating pay scales and other conditions of employment. Occasionally, employers will fail to provide these basic conditions, and the engineer is usually faced with the dilemma of resigning or taking part in collective action against the employer. This creates a moral dilemma, since the engineer has an obligation to the employer but also has an obligation to himself or herself, and to the engineering profession as a whole, not to accept unprofessional working conditions or inadequate pay.

It has been well established, in both Canada and the United States, that engineers are entitled to take collective action and even to form or join unions, if desired. Each province has a labour board and a labour relations act that can provide advice and can assist engineers who are contemplating collective action. Engineers who are part of company management are not permitted to organize collectively, since it would be illogical to do so. However, employee engineers are under no such prohibition.

As a general policy, engineers should try to resolve problems with employers through negotiated contracts that do not involve formal unionization, for the simple reason that it usually will be less work. Requesting the provincial labour board to create a formal union will require the employer to negotiate in good faith but will also create a lot of formal procedures and bureaucracy. On the other hand, in every province, labour legislation guarantees the rights of employees to form unions and to negotiate in good faith with employers, and this is an effective procedure when every other route is closed.

Although employee engineers are free to join existing labour unions within their company or industry, it is usually advisable for them to form a collective group composed entirely of engineers, if possible. This guarantees that the goals of the group will always be consistent with the wishes of the engineers. Otherwise, there is a risk that the engineers will be a minority in the union and may be obliged to support labour action that is not in their best interests.

When the question of unionization is discussed by engineers, an emotional and intemperate debate frequently results. Engineers are professional people and deserve professional employment. However, in the modern world, the majority of engineers are also employees, and when policies for negotiating pay or other terms and conditions of employment do not exist, they are entitled to take the same collective action as other employees. The need to resort to collective action does not generally indicate a failure of the engineer to act ethically; it usually shows a failure of the employer to act ethically by establishing fair policies and negotiating procedures within the company.

The Canadian Society of Professional Engineers (CSPE) has been formed with the goal of establishing a professional group, modelled on medical associations and bar associations, that works collectively for professionals while not incurring the problems usually associated with unions. CSPE is active mainly in Ontario.

UNETHICAL MANAGERS

In very rare instances, the management of a company may appear to be unethical. An engineer who finds evidence of dishonesty, fraud, misrepresentation, pollution, or similar unethical acts should take immediate action to inform management of the problem and to suggest remedial action. A dilemma arises for the engineer if management is unresponsive to arguments based on ethics. This puts the engineer's duty to the employer in direct conflict with the duty to the public welfare, which should be considered paramount, as all codes of ethics clearly state.

The engineer must therefore make a serious assessment of the situation and balance the seriousness of the risk to public welfare against the likelihood of overcoming management resistance to remedial action. The engineer should act quickly on the problem, since delay may be interpreted by management as agreeing to or condoning the unethical action. The engineer is generally faced with three possible courses of action:

- First, the engineer could continue to work for the company while trying to correct company policy. This is probably the friendliest action and would be possible if the dishonest actions were minor and if management is open to improvement and change. In most cases, this is the effective and preferred choice.
- Second, the engineer could continue to work for the company while alerting external regulatory agencies that the company is acting dishonestly. This is commonly called whistleblowing, and it is an unpleasant and usually unfriendly act. However, in rare cases where the engineer has full knowledge (preferably documented) of a clear and serious hazard to the public, where supervisors and management have refused to take action to correct the problem, and where attempts to correct the situation have failed, then whistleblowing may be the last resort. However, it should be noted that this course of action is not recommended until all other possible courses of action have been exhausted. Whistleblowing and the problems associated with it are discussed in Chapter 10.
- Third, the engineer could resign in protest. This course of action may be necessary in serious cases where complicity may be suspected if the engineer remains with the company. In this case, the engineer should consult a lawyer before resigning. There may be grounds for considering such a forced resignation as equivalent to wrongful dismissal.

INTRODUCTION TO CASE STUDIES

Chapters 7, 8, and 9 present eighteen case studies in which the reader is presented with a moral dilemma stemming from an engineering problem and asked to make a decision that can be substantiated using a code of ethics or basic ethical concepts. By their nature, case studies are slightly artificial, the information is restricted to a brief summary, the opportunity to gather more information is not available, and the opportunity to propose creative alternatives is therefore limited. Nevertheless, case studies are useful exercises in ethical decision making regardless of their shortcomings.

All of these case studies are based on real cases or reports of real cases, but names and some details have been changed to provide complete anonymity. Any similarity to real people in comparable situations is entirely coincidental.

CASE STUDY 7.1
ACCEPTING A JOB OFFER

STATEMENT OF THE PROBLEM

During a period of economic recession, an electrical engineering student, Joan Furlong, is nearing graduation and seeks a permanent position with an electronics company in digital circuit design and analysis. She is interviewed by several electronics and power companies. Her résumé clearly states her qualifications, job objective, and interests, which are mainly in digital circuit design. As her graduation day approaches, she receives an offer from the Algonquin Power Company to work on scheduling maintenance activities at their substations. The salary is good, so she writes immediately and accepts. About two weeks later, she receives a letter from Ace Microelectronics offering her a position on a new project in digital circuit design; the salary is approximately equal to the Algonquin offer, although the employment may end when the project ends. Furlong is uncertain what to do, as she realizes that she sincerely wants to work in digital circuit design and not in scheduling of maintenance activities. She identifies three possible courses of action.

First, she could write to Algonquin Power, tell them that her plans have changed, and apologize for the inconvenience. She is aware of the code of ethics of her provincial Association, but she is not yet a member of the Association and does not feel bound to follow the code. Although she is a student member of the Institute of Electrical and Electronics Engineers (IEEE), the IEEE code does not appear to have any clause that pertains to this particular case.

Second, she could write to Algonquin Power, as above, but offer to reimburse them for the recruitment expenses that they have paid on her behalf.

Third, she could write to Ace Microelectronics and advise them that she has unfortunately already accepted an offer from Algonquin Power but might be in a position to join them on a later project, in a few years' time, when her obligation to Algonquin Power is satisfied.

QUESTION

Which of the above three alternatives is best, from the ethical viewpoint?

AUTHORS' RECOMMENDED SOLUTION

This problem is not clearly defined in codes of ethics, although almost every code states that an engineer has an obligation to act with "good faith" or "good will" toward clients, employees, and employers.

The first action is clearly unethical. Furlong *does* have an obligation to Algonquin Power that cannot be erased with a simple apology. The power company has probably sent rejection letters to the other applicants for the position and may stand to lose more than just the recruitment costs if its maintenance program is delayed. (The seriousness of a failure to recruit personnel is rarely appreciated by people outside the company.) The argument that Furlong is not bound by the Association's code of ethics is spurious, legalistic, and unacceptable as a justification for her actions.

On the surface, the third course of action might seem to be the most ethical: Furlong made a promise to Algonquin Power, and she has a duty to fulfil it. However, it is clear that the job is not in the area that she wanted, and she chose it mainly for security. Consequently, although she may grow to enjoy her job, the probability is that she will not enjoy it, will regret the missed opportunity, will not be as productive an employee as the company would want, and may leave in a few years. This course of action is ethically correct but not ideal.

The second course of action is the mean, or compromise, between the evils of the other two courses of action. It is possible that Algonquin Power may request the return of expenses paid during the recruitment, but that would be a small price for Furlong to pay for realigning her career path to suit her goals. This choice acknowledges her ethical duty, maximizes the benefits, and tries to ensure that the person who benefits most from this course of action (Furlong) alleviates at least part of the losses incurred by Algonquin Power.

CASE STUDY 7.2
SLOW PROMOTION AND JOB DISSATISFACTION

STATEMENT OF THE PROBLEM

John Smith is a licensed professional engineer in the jigs and fixtures group of the Dominion Press and Stamping Company, which is a fairly large, privately owned manufacturing company. He was first hired by Dominion during his final summer at university. They offered him a part-time job during his final year of university and a permanent position after graduation. Smith learned a great deal during his first year of permanent employment, which was almost entirely devoted to training. He was sent away on a two-month computer course and then given on-the-job familiarization with each of the three divisions of the company (design, manufacturing, and sales) during the rest of his first year. He elected to stay in the design division and has worked there productively for three years.

Unfortunately, he now feels that he has reached a dead end. He has mastered all the skills necessary to carry out his present job, but he sees that further promotions are blocked by colleagues with more seniority. Moreover, the design job has become routine, and his request to transfer to the manufacturing division was denied because there was no opening and because he was considered too valuable in his present job. Smith feels very guilty about resigning to apply for a job in another company, since he is sincerely grateful for the experience he has obtained over the past four years with Dominion, has an excellent working relationship with his co-workers, and knows the company will have trouble replacing him. He also feels it may be unethical to change jobs now that he has been fully trained and is of maximum benefit to the company. On the other hand, he is aware of several vacant positions with rival manufacturing companies.

QUESTION
Is it ethical for Smith to seek employment elsewhere?

AUTHORS' RECOMMENDED SOLUTION
This problem occurs fairly often among engineers in their first full-time job in industry. Training periods for junior engineers may take one or two years, and then, when the engineer is finally a productive member of the company, he or she wants to move on to greener pastures. A dilemma arises, since the employee clearly has an obligation yet also has the right to seek advancement in his or her career, commensurate with talent, effort, and education. This problem cannot easily be resolved by examining the provincial code of ethics; the code stipulates that engineers must act with "fairness and loyalty" to employers, but precisely what is "fair" in this instance requires further examination.

When an employee accepts a training benefit, an implied contract is created. A company invests in a new engineer during the training period, and that investment is returned during the period of useful employment that follows. The cost and extent of the training period is therefore a factor in determining how much useful employment would be a fair return. Using this concept of return on investment, it seems that we can reduce an ethical decision to a simple computation. However, it is not easy to determine how much expense was actually incurred during the training period. Did the engineer receive a full salary? Did he or she carry out useful duties while being trained? Was the company reimbursed for training or for work done during training? What obligation does a company have to carry out technical training? The calculation may be rather approximate and subjective, but the concept of fairness hinges on showing that the obligation incurred by training has been satisfied by adequate useful service.

In the case of Smith, common sense would dictate that if most of his first year consisted of training, then he would be obligated to remain for at least an equal one-year period. However, a company would usually expect to amortize the initial training costs over more than one year. As a comparison, most military postings are four years long, most town councillors are elected for three- or four-year terms, and most university departments appoint department chairs to four-year terms. Each of these jobs would include some initial familiarization or training.

When the initial training is particularly lengthy or expensive, then the company should stipulate, in the contract of employment, what time period would be expected to amortize the cost. For example, university students who want to become engineering officers in the Canadian Forces may receive full room, board, and tuition expenses during their university education; however, in return they must usually sign employment contracts for a minimum of five years' service after graduation.

Therefore, since Smith has completed three years of satisfactory service after the initial year of training and no minimum period was specified in his contract, he has probably satisfied any indebtedness he might have incurred. If he is unable to advance or transfer within the company, then he should feel no guilt about moving on to a job with another company that offers more challenge and interest.

CASE STUDY 7.3
PART-TIME EMPLOYMENT (MOONLIGHTING)

STATEMENT OF THE PROBLEM

Philip Fortescue is a licensed professional engineer who has worked for Federal Structural Design for ten years. Unfortunately, for reasons that are not clear to either Fortescue or his employer, the company has not had many large contracts, and Fortescue's salary is very low. His pay raises have rarely exceeded cost-of-living increases over the ten years of his employment. As a result, he has been forced to take on extra employment in his spare time, and he secretly brings the work to his office in the evening, where he uses the CAD (computer-aided design) system on the computer. He is careful to ensure that the paper, pens, and photocopying are paid out of his own pocket, and he argues that the computer would be sitting idle in the evening anyway, so his employer is suffering no loss. In fact, Fortescue argues that his evening work benefits his employer, since it permits him to continue to work for Federal Structural Design in spite of his low salary.

QUESTION
Is it ethical for Fortescue to carry on his part-time employment in this manner?

AUTHORS' RECOMMENDED SOLUTION
The question of an engineer accepting part-time employment (or moonlighting, as it is commonly called) occurs frequently, and guidelines have evolved over the years. In general, it is not unethical for an employee to work for more than one employer, although it requires determination and stamina.

However, as every code of ethics indicates, the employee engineer must show fairness and loyalty to the employer. This means that the part-time employer should not compete for the full-time employer's contracts, that the time and effort spent on moonlighting should not reduce the employee's efficiency during the usual workday, and that the employer must be fully informed in order to verify the situation.

It is clear that Fortescue is not acting ethically in this case, since he has not informed his employer of his part-time employment. Therefore, the question of whether his part-time work competes or conflicts with his full-time job cannot be verified. The fact that an engineer would remain with an employer for ten years with no promotion or significant increase in salary is curious and implies that there are other, unstated factors that influence this case. It is not clear whether Fortescue is exploiting the employer's facilities and contacts to generate a large part-time income, or whether the employer is exploiting Fortescue by forcing him to carry two jobs to survive financially. However, the engineer's secrecy about his part-time employment is clearly unethical. Moreover, if Fortescue is offering services to the general public when he moonlights, then in some provinces (such as Ontario) he must obtain a Certificate of Authorization, and in almost every province the question of liability insurance must be addressed. Fortescue may be putting his career at more risk than he realizes.

CASE STUDY 7.4
ENGINEERS AS MEMBERS OF LABOUR UNIONS

STATEMENT OF THE PROBLEM
Jeanne Giroux is a licensed professional engineer working for Acme Automotive Manufacturing, a medium-size company that makes parts

for cars and trucks. She works in the design engineering office (four engineers, six designers) and supervises modifications to parts, including stress analysis, detail drawing, and prototype testing — a wide range of duties. In her job, she has frequent contact with the machinists and other tradespeople in the manufacturing plant and observes that the shop union is very effective in negotiating terms and conditions of employment. The union steward informs her that the employer is required by law to bargain in good faith with the union and that there are procedures for mediation and arbitration in case of stalemates in the negotiations.

In contrast, the design and engineering staff have had very low pay raises in recent years, are limited to short vacations, and are required to work overtime (and occasional Saturdays) without additional pay when there is a crisis. Although all four of the company's engineers are licensed, the provincial Association of Professional Engineers cannot assist them in negotiating with the employer. The Association has, however, provided them with a survey of engineering salaries. Giroux notes that the mean salary for the four staff engineers is 20 percent below the median salary for the appropriate group in the survey. The company administration manual does not mention procedures for staff pay raises. As a comparison, the sales staff in the company (none of whom are engineers) are paid partly on a commission basis and always receive higher pay than the engineers.

Giroux has been assured by the union steward that the design and engineering staff could be included in the bargaining group if at least six of the ten employees sign application forms. Giroux believes that all ten would probably join, if asked.

QUESTION
Is it ethical for engineers to join the shop labour union?

AUTHORS' RECOMMENDED SOLUTION
Yes, it is ethical and legal for engineers to join labour unions, provided that the engineers do not exert managerial control in the company. From the details of this case, it would appear that the design engineering staff are not considered part of management. However, although it may be ethical and legal to join the union, it may not be appropriate; labour relations acts in some provinces specifically permit professional engineers to form a bargaining unit composed entirely of professional engineers unless the majority of engineers wish to be included in a bargaining unit with other employees. Moreover, it is not essential to have the support of an established union to form a bargaining unit. This can be done directly through the labour board of the provincial government in most provinces. A majority of members of the bargaining group may request certification of the group as a bargaining unit; the elected repre-

sentatives then negotiate contracts for the entire group. However, establishing a union (certification) requires a great deal of organization and paperwork; for small groups like the one described in this case, it might be more appropriate to make management aware of their dissatisfaction and of the routes open to resolve the problem, and to suggest the negotiation of individual contracts, a collective contract, or a salary negotiation procedure for the design engineering staff. If the employer is uncooperative, it would be advisable to obtain legal advice and to examine unionization as a last resort.

CASE STUDY 7.5
FALSE OR MISLEADING ENGINEERING DATA IN ADVERTISING

STATEMENT OF THE PROBLEM
Audrey Adams is a licensed mechanical engineer with marine experience working for a manufacturer of fibreglass pleasure boats. She has conducted buoyancy tests on all the boats manufactured by the company and has rated the hull capacity of each according to the procedure specified by Transport Canada. She observes in the company's sales literature that a boat hull rated for a maximum of five people is consistently shown in photographs with six people on board. The sales literature appears to be otherwise correct. Adams is aware that the boat would be safe in still water with six people, but could be flooded and sink in rough water. Adams believes the sales literature is misleading and possibly hazardous.

QUESTION
What action should Adams take?

AUTHORS' RECOMMENDED SOLUTION
The code of ethics for every provincial Association and technical society states that the welfare of the general public must be considered most important. In this case, there is a potential hazard to the public, and the engineer has an ethical duty to take action to reduce or eliminate this hazard.

Adams's first step should be to inform the engineering manager about the problem. This would typically be done in an internal memorandum describing the errors in the sales literature. If these are simple errors or oversights by the sales personnel, then they will be easy to rectify. Most companies are honest and would take immediate action to correct the sales literature by issuing new data to customers.

In rare instances, company management may be dishonest. For example, if Adams, while investigating the problem, should discover that test results had been altered or that the manufactured boats were being sold with incorrect capacities stamped on the serial nameplates, then the problem would be much more serious. In this case, company management would be guilty of misrepresentation, which could be a criminal act. If a serious failure should occur (such as an overloaded boat sinking, resulting in a loss of life), the erroneous literature and incorrect capacities would become public knowledge, and the engineer could be the subject of investigation for possible unethical acts, incompetence, or collusion with management in the misrepresentation.

Therefore, if an engineer recognizes that she or he is working for a dishonest company, a decision must be made quickly to dissociate her- or himself from any unethical activity, to work for change within the company, and, in extremely rare cases, to act as a whistleblower or to resign in protest. Fortunately, extreme action is rarely necessary.

CASE STUDY 7.6
DISCLOSING PROPRIETARY INFORMATION

STATEMENT OF THE PROBLEM

This incident allegedly occurred in the aircraft industry. An aeronautical research engineer from Company A conducted tests of a new aircraft tail assembly configuration in his company's wind tunnel and knew that devastating vibrations could occur in the configuration under certain circumstances, leading to destruction of the aircraft. Later, at a professional meeting, Company A's engineer hears an engineer from Company B, a competitor, describe a tail assembly configuration for one of Company B's new aircraft that runs the risk of producing the same destructive vibrations that Company A's engineer discovered in his tests. Presumably there is an obligation, as a matter of both ethics and law, to maintain company confidentiality regarding Company A's proprietary knowledge. On the other hand, engineers have a duty to safeguard public safety and welfare. If the engineer from Company A remains silent, Company B might not discover the destructive vibrations until a dreadful crash occurs, killing many people. As an added complication, assume that the simple statement by Company A's engineer that the tail design "will not work" will not be believed by Company B's engineers.

QUESTION

What would you do if you were the engineer from Company A?

AUTHORS' RECOMMENDED SOLUTION

On the one hand, under the code of ethics you have an obligation to your employer to maintain the confidentiality of proprietary information. Your company paid a lot of money to test the tail assembly, and it would not be fair to your employer to turn this information over to Company B. Moreover, if Company B is as diligent in its testing and analysis as Company A, then it will discover the vibration problem in due course, and you need take no action.

On the other hand, you also have an obligation under the code of ethics to consider the welfare of society as paramount and to report a condition that endangers public safety. If Company B should fail to discover the design flaw and a prototype aircraft later crashes as a result of this flaw, your lack of adherence to the code of ethics will be painfully clear.

Therefore, you have an ethical dilemma with two undesirable courses of action. In this case, the duty to society must prevail over the loss of advantage, and Company B should be informed of the potential problem. You would, of course, notify your employer before contacting Company B. You should also keep in mind the objective of this exchange of information. It would be unethical to convey the information in a way that harmed the reputation of Company B. However, it is *not* your duty to release detailed data or to save money for your competitor. Your goal is merely to ensure that public safety is not endangered by Company B's defective design. Therefore, in consultation with your employer, you would determine what minimum information would achieve this purpose, and convey it in a direct and unambiguous way.

TOPICS FOR STUDY AND DISCUSSION

1. Consider the circumstances of Case Study 7.6, in which it has been recommended that Company A inform Company B that its aircraft tail assembly design has a serious vibration problem. If the problem was *not* dangerous to passenger safety but merely increased fuel consumption and created noise in the cabin, would Company A still be ethically bound to inform Company B? Explain your answer.

2. The engineer who accepts an offer of employment creates a contract in which the engineer agrees to use his or her ability to achieve the employer's legitimate goals. Consider the case where you have been hired to design electrical or mechanical components for manufacturing machinery. During a recession, the employer decides to diversify into new areas to attract more business. What would your position be, both ethically and personally, if the employer asked you to participate in the design of

   - bottling equipment for the beer and liquor industry?
   - manufacturing equipment for the tobacco and cigarette industry?
   - medical equipment to make abortions safer and more convenient?

- pill-making machines for the birth-control or pharmaceutical industries?
- security locks for the prison system?
- equipment for nuclear power plants?

3. As an employee of a large Canadian manufacturing corporation, you have been assigned assistant to the chief engineer on a six-person team that is to establish a branch plant in an underdeveloped country. Your task is to supervise the installation and commissioning of the manufacturing equipment. The local people who will be running the equipment are rural people with little or no education. As soon as you arrive on the site and familiarize yourself with the plan, which is well under way, you realize that the manufacturing line to be installed was removed from service in Canada because it created toxic waste. The waste must be disposed of by special incineration equipment that does not exist in this foreign country. Although the manufacturing line would not be permitted in Canada, the underdeveloped country does not have environmental laws that would prevent its installation and operation. You have some concerns about this project and discuss them with the chief engineer. He is sympathetic but points out that the manufacturing line ran in Canada for over ten years before pollution laws stopped it, and no deaths were attributed to it. Moreover, the local people will be much better off when the line is running and there is useful employment for all concerned.

   What guidance does your provincial code of ethics give on this problem (see Appendix B)? Does the code apply to activities conducted in a foreign country? What alternative courses of action are open to you? Which course is best from the ethical standpoint?

4. Assume that you are working as a professional engineer in a small consulting company that gives its employee engineers considerable latitude in scheduling tasks, meeting deadlines, and reporting expenses. You are approached by the company president, who states that your professional attitude and attention to high standards have been recognized by senior management. The president also expresses concern about the lax attitudes of your colleagues, who appear to be abusing the freedom awarded them. Would it be ethical for the president to offer, and for you to accept

   - a secret assignment to monitor the behaviour of your colleagues and your immediate superior and report back to him?
   - a promotion to head engineer to replace your immediate superior, on the basis that the head engineer is not competent as a manager and should be replaced?

   Discuss, explain, and justify your answers.

CHAPTER EIGHT

Ethical Problems of Engineers in Management

Engineers in management positions may experience some or all of the ethical dilemmas discussed in the previous chapter. Managers, however, control the resources of the corporation to a much greater degree than their subordinates. In particular, managers are in a position to hire, fire, delegate, and direct other employees. Conflicts of interest may occur. Managers may also be required to negotiate and make agreements with other businesses, and the potential for ethical problems arises in these dealings as well.

Specific problems vary from company to company. Also, certain types of problems are more common in certain industries. That is, the engineer in management may encounter slightly different pressures and conflicts in working for an industrial manufacturer, a consulting engineering firm, or a government agency. The manager must be alert to the potential for conflict of interest and must avoid becoming involved, or even appearing to be involved, in unethical practices.

ADHERENCE TO THE PROVINCIAL ACT

One of the most obvious responsibilities of the professional engineer in a management position is to ensure that the Professional Engineering Act is being obeyed within the manager's area of responsibility. The two most common infringements of the Act in most provinces concern the use of unlicensed personnel to carry out the work of professional engineers and the misuse of engineering titles. It is not only unethical and unprofessional to continue such practices, it is illegal: these practices are contrary to the Professional Engineering Act in every province and territory.

UNLICENSED PERSONNEL

Using unlicensed personnel is a serious infringement, since it can potentially harm the client, employer, or general public. If personnel are carrying out professional engineering work as defined in the Professional Engineering Act for that province or territory, the work must be done by, or under the direct supervision of, a professional engineer. Where this is not the case, the situation must be rectified. This problem is not uncommon in many smaller industries in Canada, usually due to ignorance of the law.

Correcting these practices may require determination and tact. If a manager tries to restrict a practice that has endured for years, hard feelings and antagonism are likely to result. The manager must risk offending the employee who is involved in order to obey the Act. A certain amount of diplomacy may be required, depending on the circumstances of the case and the flagrancy of the violation. In cases where the employee would be eligible for a licence, then the appropriate action is to insist on proper registration. When this is not possible, the employee must be put under the supervision of a professional engineer, perhaps the manager, in order to continue to perform useful, but now properly regulated, work.

MISUSE OF ENGINEERING TITLES

Misusing engineering titles is usually a less serious infringement. Many companies have positions with the word "engineer" or "engineering" in the title, such as a "design engineer" who is really a designer, or an "engineering manager" who is really a manager, giving the erroneous impression that the person holding the position is a licensed professional engineer. If the tasks performed do not require a licence, the title must be changed to eliminate the ambiguity. Again, this is a situation where tact and diplomacy may be needed to alter practices that may go back many years. It may take some creative thinking to develop new job titles that are elegant and accurate but do not contravene the Act.

REVIEWING WORK AND EVALUATING COMPETENCE

Engineers, whether they are employees or managers, are required by law to practise only within their limits of competence. Engineers should not undertake — and managers should not assign — work that is not within the competence of the engineer. In some cases, preparation or review may be needed to gain or regain the needed competence. In these cases, the question of fairness to the employer may be raised. An extensive period of education at the employer's expense to gain competence in a specific field is not uncommon, but the expense to the employer should be approved before a commitment is made.

WORK REVIEW FOR ACCURACY

Most engineers have their work routinely reviewed by a second engineer for accuracy. In some industries — particularly the aircraft, aerospace, and nuclear power industries, where errors could be extremely costly and could have serious liability implications — the review of calculations is a common and expected procedure. In these industries, an important decision should never be made on the basis of a single engineer's unchecked calculations. These reviews are always carried out with the knowledge of the person who did the original work, and the purpose of the review is always to guarantee safety, improve quality, or reduce liability.

WORK REVIEW TO ASSESS COMPETENCE

It is common practice to evaluate the performance of all employees on a regular basis, and the engineering manager is generally responsible for this evaluation. It may be necessary to review an engineer's work to evaluate competence at other times as well. However, a manager should never ask a professional engineer to review the work of another engineer without the knowledge of the engineer who prepared the work. This precept is included in most codes of ethics, is simple common courtesy, and should apply to any professional employee. Since the work of a professional, by definition, requires specialized knowledge, additional information, data, or explanations might be required. A professional reputation is a valuable asset that requires years of study and experience to build. The review of any professional's work must not be done in a careless or cavalier way that could inadvertently damage someone's professional reputation. A secret review is like a trial in absentia, and this is generally contrary to our system of natural justice. Consequently, although reviews of an engineer's work for accuracy or competence are common and do not require the *permission* of the engineer, such reviews must not be done without *informing* the engineer. Engineering managers should be particularly sensitive to the need for this common courtesy.

DISCRIMINATION IN ENGINEERING EMPLOYMENT

Since the engineering manager plays a key role in hiring, evaluating the performance of, and dismissing engineers, he or she is in the front line of the battle against discrimination in engineering. Discrimination should not be a problem in Canada, since the Charter of Rights and Freedoms prohibits discrimination on the basis of race, national or ethnic origin, colour, religion, sex, age, or mental or physical disability. Although progress in overcoming discrimination is evident, there is still a serious underrepresentation of certain groups in engineering, such as women, Aboriginal peoples, people with disabilities, and others. The problem is particularly obvious where women are concerned, since they are a majority in the general population but are definitely a minority in the engineering profession.

Women and minority groups have a legal right to be treated fairly; although they would not expect preference over their colleagues, artificial obstacles must not be created for them. This topic is discussed in more detail in Chapter 13, "Fairness and Equity in Engineering."

SPECIAL PROBLEMS RELATED TO COMPUTERS

In the last 25 years, computers have been put to work in every engineering office. Computer-aided design (CAD) workstations are now the focal point of design decisions, since they permit configurations to be visualized, simulations to be run, and detailed analyses to be performed even in the very earliest stages of the design process. It is hard to remember — or even to imagine — what the engineering design process was like before the computer revolution. Almost every engineering office has hardware and software for CAD, automated drafting, computer-aided analysis, dynamic simulation, project management, spreadsheets, and word processing. The rapid dissemination of information and data through the Internet is already happening, and expert systems and artificial intelligence are under development and will be seen in the very near future. The use of computers in engineering thus will continue to grow.

Computers permit unbelievable visual power in design and incredible speed in analysis, and have therefore removed barriers to inspiration and reduced the tedium of routine calculations and repetitive drawing. However, computers create new responsibilities and problems for the engineering manager, such as security, backup of data, infringement of copyright, the possibility of engineering errors caused by flaws or "bugs" in the computer programs, and incorrect transmission or unreliable data from the Internet. Managers must be alert to these problems.

COMPUTER SECURITY AND BACKUP

Computer hardware and software, including stored data, represent a large financial investment; in a small engineering office, they may be one of the major assets of the practice. It is the responsibility of the manager to be alert to any risks, problems, or damage that threaten this investment. Maintaining the equipment and providing for alternative facilities in the event of failures are the most obvious responsibilities. The manager should have a disaster plan for the possibility, however remote, of complete malfunction or destruction of the computers and for possible corruption or loss of the programs and data. The simplest protection is to have critical data and programs duplicated on backup disks and stored in a safe, secure location. Many books are available to provide further advice on this subject.

PREVENTING SOFTWARE PIRACY

One of the most frequently encountered conflicts of interest that arise today in engineering concerns protecting copyright for engineering programs and data. It is so easy to violate copyright that the practice, whether through ignorance or by intent, is widespread.

The purchase of a computer program permits use of the program but does not include the right to duplicate the program except for backup. Some programs are leased (particularly large ones) and may include a monitor program that checks the date given by the computer and disables the program if the lease has expired and no extension is indicated (usually in the form of a secret password provided by the leasing supplier). It is contrary to the Canadian Copyright Act to duplicate programs for personal use or to disable the password protection on leased programs. This may create an ethical dilemma for an engineering manager, since unauthorized use of programs duplicated elsewhere or unauthorized use of leased programs (and many similar practices) may be illegal or unethical, but they certainly improve the efficiency and productivity of the office. This dilemma has only one solution: the unethical practices must not be allowed, and it is the responsibility of the engineering manager to protect the copyright of software.

This issue has been clarified in the Ontario *Guideline to Professional Practice*. Two circumstances arise in which it is not an infringement of copyright to make duplicate copies of programs:

> The first exception provides that it shall not be infringement for a person in a lawful possession of a copy of a computer program to modify, adapt, or convert a reproduction of the copy into another program to suit needs, provided that:
>
> i) the modified program is *essential* for the compatibility of the computer program with a particular computer;
> ii) the modified program is used only for the person's own needs;
> iii) not more than one modified copy is used by the person at any given time; and

iv) the modified copy is destroyed when the person ceases to be entitled to possession of the copy (i.e. upon expiry of a software licence).

The second exception provides that a person who is in *lawful possession* of
a copy of a computer program or of a modified reproduction of a program
may make a single backup copy of the program, provided that the backup
copy is destroyed when the person ceases to be the owner of the copy of
the computer program.

The intention of these exceptions is to give the authorized user of software a limited right to change the software to ensure compatibility of the
software with the authorized user's computer system, and to allow for the
protection and security of the original program.[1]

LIABILITY FOR ERRORS IN COMPUTER PROGRAMS

Computer programs, even those that are widely distributed and highly
regarded, can occasionally have hidden flaws, or "bugs," that cause incorrect calculations. The question of liability arises if the errors should
result in an incorrect engineering decision. Disclaimers in the documentation for computer programs usually state quite clearly that, in the
event of malfunction, the liability of the seller of the computer program
is limited to the cost of the program. The PEO guideline *The Use of
Computer Software Tools by Professional Engineers and the Development of
Computer Software Affecting Public Safety and Welfare* is much more specific. Under the heading "Use of Computer Software Tools by Professional Engineers," the guideline states:

> The practice of professional engineering has become increasingly reliant
> on computers, and engineers use many computer programs that incorpo-
> rate engineering principles and matters. Many of these programs are
> based upon or include assumptions, limitations, interpretations and judg-
> ments on engineering matters that were made by or on behalf of an engi-
> neer when the program was first developed. Therefore, it is often difficult
> to determine, just by using a program or by being given a description of
> its function, the engineering principles and matters it incorporates.
>
> The engineer must have a suitable knowledge of the engineering prin-
> ciples involved in the work being conducted, and is responsible for the
> appropriate application of these principles. When using computer pro-
> grams to assist in this work, engineers should be aware of the engineering
> principles and matters they include, and are responsible for the interpre-
> tation and correct application of the results provided by the programs.
>
> Engineers are responsible for verifying that results obtained by using
> software are accurate and acceptable. Given the increasing flexibility of
> computer software, the engineer should ensure that professional engineer-

ing verification of the software's performance exists. In the absence of such verification, the engineer should establish and conduct suitable tests to determine whether the software performs what it is required to do.[2]

Clearly, the engineer must be alert to the possibility that errors may exist in design software and, as for any design data or design aid, must perform independent checks to ensure the validity of such design assistance. It is not acceptable professional practice to assume that computer output is accurate unless you have verified both the program and the output by independent validation. Briefly, the engineer must ensure that the program is appropriate for the application, that it is accurate when used properly (as established by validation tests), and that it is correctly used by properly trained personnel. Most importantly, remember the old adage in engineering that an engineer should never base a really important decision on a single calculation (or unvalidated computer output). Alternative and independent computations must be made to validate the original results.

Therefore, proper practice requires the engineer to make adequate independent checks to evaluate the program in general, and to validate the specific output for the project in question, before using the results as the basis for key design decisions. It is the manager's responsibility to ensure that this is done consistently by the engineers under his or her direction.

HIRING AND DISMISSAL

The engineering manager usually hires and dismisses engineering staff when required. The manager should therefore be aware of the following aspects of hiring and dismissal.

EMPLOYMENT CONTRACTS AND POLICIES

The best method for employing professional engineers is through clear-cut employment contracts. These contracts eliminate uncertainty by specifying the duration (either fixed-length or indefinite), the remuneration and how it will increase with time and duties, vacation entitlement and statutory holidays, what would constitute just cause for termination, and terms and amounts of severance pay and other payments. When a company has too many employees for each engineer to have a personal contract, then the company must have clear policies that deal with these issues. The NSPE *Guidelines to Professional Employment for Engineers and Scientists* (in Appendix F) provide a good explanation of the topics that should be included in the company policies for the benefit of the professional employees.

TERMINATING EMPLOYMENT FOR JUST CAUSE

A manager must take responsibility for terminating or discharging employees when their services are no longer required. Such terminations must be in accordance with the employment contract or published company policies. In addition, employees may be discharged for *just cause*, which is defined below.

Those matters which would allow an employer to terminate an employee, without notice or severance pay, are as follows:

1. serious misconduct;
2. habitual neglect of duty;
3. serious incompetence, not just management dissatisfaction with performance;
4. conduct incompatible with his or her duties or prejudicial to the company's business;
5. wilful disobedience to a lawful and reasonable order of a superior in a matter of substance;
6. theft, fraud or dishonesty;
7. continual insolence and insubordination;
8. excessive absenteeism despite corrective counselling;
9. permanent illness; and
10. inadequate job performance over an extended period as a result of drug or alcohol abuse and failure to accept or respond to the company's attempt to rehabilitate.

If one of these elements of misconduct exists, and is ascertained even after the employee has been discharged, the company can rely on that misconduct and not pay the employee any severance allowance.[3]

WRONGFUL DISMISSAL

When an employee without an employment contract is dismissed and the reason does not constitute just cause, as described above, then there is a risk of *wrongful dismissal*. These cases may have to be resolved in a court of law, and legal advice is extremely useful for both the employee and the manager.

In a comprehensive article on wrongful dismissal, lawyer Howard Levitt described six situations that could also be considered wrongful dismissal even though the employee is not technically dismissed. These are forced resignation, demotion, a downward change in reporting function, a unilateral change in responsibilities, a forced transfer, and serious misconduct of the employer toward the employee.[4]

In summary, it is important for a manager to be alert to the myriad difficulties and complications that are associated with supervising the work of other people. The manager needs leadership ability, sensitivity, and a professional attitude. A knowledge of the law or access to legal advice is also beneficial and, when needed, should be obtained *before* taking hard decisions, not after the fact.

CASE STUDY 8.1
THE UNLICENSED ENGINEER

STATEMENT OF THE PROBLEM

Assume that you are the manager of the engineering design department for a fairly large consulting engineering firm. As part of your job, you hire and dismiss department staff members, including engineers, designers, CAD operators, and clerical workers. Six months ago, you hired Jorges Xavier, who had recently moved to your area from another province. During the employment interview, you emphasized that it was essential that he be licensed, and the letter of appointment sent to him stipulated that he was being hired as a Professional Engineer. After Xavier started work, you had a sign placed on his door and had business cards printed, both of which had the P.Eng. designation after his name.

You are startled to receive a complaint from a client who claims that Xavier is not a licensed professional engineer. The client is furious that you and your company would send unqualified people to work on her project. You contact the provincial Association of Professional Engineers, and they confirm that Xavier has *not* been awarded a licence. Now *you* are furious.

QUESTION

Who is responsible for this problem? Can you fire Xavier for just cause? Would it make any difference if

- *Xavier is licensed in another province but has neglected to apply for a transfer of licence?*
- *Xavier has applied to transfer his licence, but it is still being processed by the provincial Association?*
- *Xavier has never been licensed in any province?*

AUTHORS' RECOMMENDED SOLUTION

It is not presently possible in all cases to transfer a licence from one province to another. However, a person who has been licensed in one province will generally be accepted as qualified for licensing in another province, and although additional requirements (such as writing the Professional Practice Exam) may be required, they can generally be satisfied fairly easily in a matter of months. This case involves the code of ethics and a more fundamental problem: a breach of the Professional Engineering Act.

There can be little doubt that Xavier is guilty of practising professional engineering without a licence. He has used the business cards that clearly say P.Eng. without protest or correction, and he is not licensed in the province where he is working. Consequently, he is responsible to the provincial Association for any infraction of the Act, although the fact that you, as manager, had the business cards prepared could be considered a mitigating factor. You will be guilty of a breach of the code of ethics if you permit Xavier to continue practising engineering.

It is essential to determine what work Xavier has done for the client. If he has been in a junior or training position during his first six months with the firm and his work has been supervised by another engineer, as would usually be the case, there is no problem. There has been no risk to the client or the public, and no damage has occurred.

However, it would probably be advisable to discuss the case with a lawyer if Xavier has been involved in making independent decisions on engineering projects, since the engineering firm would undoubtedly be liable for any problems that arise from those decisions. In any discussion of liability, it would be made clear that it was your responsibility, as manager, to verify the qualifications of your subordinates.

The appropriateness of dismissal and the basis for it are slightly different, depending on which of three possible situations applies. If Xavier has failed to apply for a licence in six months of employment but has a valid licence from another province, then it could be argued that this constitutes either serious misconduct or habitual neglect of duty, which are both recognized as breaches of the code of ethics and just cause for dismissal.

If Xavier has applied for a licence but it has merely been delayed, and he has a valid licence from another province, then he has probably complied with your requirements and dismissal would probably be unjust. If Xavier has never been licensed in another province, then he has been dishonest in his employment interview with you, and such fundamental dishonesty would be just cause for dismissal.

Xavier clearly contravened the Act when he used the P.Eng. designation while not licensed. However, you, as manager, must bear much of the responsibility for any embarrassment or liability that the firm suffers: although you stated the requirement for a licence clearly, you did not follow up to verify that Xavier had his licence. A company involved in offering services to the public has a duty to verify that its engineers keep their licences up to date.

Xavier is of course at fault for using the P.Eng. designation, and he may also be subject to a charge under the Act, as discussed above, particularly if he does not have a valid licence from another province. Such a charge would be prosecuted in the provincial court, but under the authority of the Act, the Association would initiate the charge.

CASE STUDY 8.2
DISMISSAL OF OFFENSIVE ENGINEER

STATEMENT OF THE PROBLEM

Assume that you are a licensed professional engineer employed as manager of an engineering department with ten employee engineers and eighteen designers and CAD operators. You are summoned into the office of the vice-president of development, your direct superior, who instructs you to fire one of your engineers who disgraced himself by talking in a loud and offensive way to the vice-president at a company picnic, held the previous week. You attended the picnic and recall that the discussion began about rival sports teams and was unrelated to company business. However, after a few beers, the vice-president and the engineer exchanged bitter insults.

You point out to the vice-president that the engineer's work has always been satisfactory. The vice-president states that the engineer's behaviour was offensive and insubordinate, and insubordination is grounds for dismissal. He hints that he would also interpret your failure to co-operate as insubordination.

QUESTION

You know that the incident has gravely offended the vice-president and has created some awkwardness in the department, which you believe would be eliminated if the engineer was fired. And you think the vice-president will jeopardize your salary increase and possibly your job if you don't co-operate. What should you do, and how should you do it?

AUTHORS' RECOMMENDED SOLUTION

This problem is typical of the pressure applied to middle-level managers. The first step is to get the facts and define the problem: was the behaviour of the engineer just cause for dismissal? Since your action could result in a wrongful dismissal, it is a very serious matter and also involves issues of law. However, you must make a decision quickly, and you probably will not have the luxury of discussing the case with a lawyer or labour consultant. Since you witnessed the incident and did not initiate the dismissal action, you evidently did not consider it a serious case of insubordination. Moreover, for the engineer's behaviour to be considered sufficiently scandalous to justify dismissal, there must be a record of continual insolence and insubordination or wilful disobedience on a matter of substance, not merely an altercation at a social event. There does not appear to be sufficient basis to justify dismissal of the engineer, and under the code of ethics, you have a duty to the employee as well as to the employer.

Therefore, the ethical issue is quite clear: the engineer's behaviour does *not* constitute just cause. Although it might solve some of your immediate problems, it would be unethical and unwise to dismiss him. A reprimand to the engineer might, however, be appropriate, but the best action would probably be to do nothing more. If the vice-president was speaking in a moment of anger, he may later realize the error and be relieved that you did not follow his instructions. However, should he later approach you and insist that you follow his instructions, you should ask to have the order in writing. In your reply to the written order, you would decline to carry out his instructions, justify your decision, and mention the reprimand that you have given the engineer. This would probably end the matter. However, should it be pursued further, the written record would be essential to show senior management that you were protecting the company against a legal case for wrongful dismissal. The documents might also be useful as the basis for a complaint of unethical conduct against the vice-president or as evidence in your own suit for wrongful dismissal, however unlikely that may be.

CASE STUDY 8.3
CONFLICT OF INTEREST

STATEMENT OF THE PROBLEM

You are an engineering manager in a fairly large company, and you have been asked to sit on a ten-member standards committee that sets performance and safety specifications for the automotive equipment that your company manufactures. The committee comprises three industry representatives such as yourself, three government representatives, and three engineering professors, and it is chaired by a representative from an engineering society associated with the automotive industry.

One of the other industry representatives has proposed a revision to the specification for a component that you manufacture. The change will make a fairly modest improvement in quality but will require specialized manufacturing expertise and equipment. During the meeting on the specification, you realize that the revision, if approved, will be very beneficial to your company since the company has the necessary expertise, but it will create hardships for some of your competitors.

You believe that the person proposing the revision would also benefit in a similar way. You are uncertain whether you should bring these points to the attention of the committee. You did not propose this revision, but it does improve the quality of the product, and any benefit that your company receives would come strictly by chance.

QUESTION

Do you have an ethical obligation to inform the committee that your company may benefit from this revision? Do you have an obligation to point out that the person proposing the revision may also stand to benefit?

AUTHORS' RECOMMENDED SOLUTION

This is a clear conflict of interest, and you must disclose it to the committee. The code of ethics states that the engineer must put the welfare of society above narrow personal interest. In fact, the main function of a standards committee is to serve the public welfare, not the parochial interests of its members. The committee may nevertheless decide, after discussion, in favour of the revision. Although it would probably be acceptable for you to express an opinion on the revision, once the conflict has been disclosed, it would be inappropriate for you to participate in a formal vote, particularly if there is controversy over the revision.

You do not have an obligation to speak about the member who is proposing the revision unless you believe that there is deliberate fraud, which does not appear to be the case. In fact, your disclosure of a conflict would place an expectation on the other industry representatives to declare their positions.

Conflicts of interest are common in such committees for the simple reason that the best-informed people are those involved in the design and manufacture of the components concerned. However, this makes it particularly important to be alert to unfair and unethical advantages that may result from such positions of trust.

CASE STUDY 8.4
ERRORS IN PLANS AND SPECIFICATIONS

STATEMENT OF THE PROBLEM

You are the engineering manager for the Acme Assembly company, which designs, fabricates, and assembles machinery. You have received a contract to construct twenty gearboxes that have been designed by Delta Designs, a company that is occasionally a competitor. However, Delta is extremely busy and does not have the capacity for this work at the present time.

One of your engineers notices that the sizes of shafts and gears on the drawings appear to be rather small for the torque and power ratings of the gearboxes, and rough calculations seem to confirm that assessment. You call the chief engineer at Delta Designs, and he states that he is too busy to double-check the drawings. He has full confidence in his designers and says you should get on with the job. He also points out that you are employed in this contract as fabricator, not as designer, and should not be reviewing his work.

QUESTION

Do you have an ethical obligation to pursue this apparent discrepancy further? Would it make any difference if failure of the gearboxes could result in injury or death rather than just financial loss?

AUTHORS' RECOMMENDED SOLUTION

Under the code of ethics, an engineer has an obligation to a client to ensure that the client is fully aware of the consequences of failing to follow the engineer's advice. In this case, a single telephone call probably would not be deemed to satisfy this requirement, either ethically or legally. You should follow up the telephone call with a letter describing your concerns and request written instructions to proceed.

If the chief engineer at Delta Designs should instruct you, in writing, to proceed with the fabrication, you would do so unless you consider the flaws in the design to be obvious and serious. This might indicate a problem of incompetence, negligence, or fraud on the part of Delta's chief engineer. If these suspicions are supported by other evidence, it would be appropriate to ask your provincial Association for advice on possible disciplinary action.

The potential for injury or death in the case of failure is important, because failure to safeguard the safety of the public could be considered professional misconduct on your part. In the case where serious injury or death is possible, the chief engineer's complaint about your reviewing his work is irrelevant. A review of the design is appropriate in these circumstances, and through your diligence, you have sought to protect the public and to safeguard his reputation as well.

CASE STUDY 8.5
MANIPULATION OF DATA

STATEMENT OF THE PROBLEM

You are a professional geologist responsible for all exploration and ore assays in a mine. You report directly to the chief executive officer of the mine, who is an accountant by training. You have just finished evaluating initial ore assays for a newly opened part of the mine, and they show much lower ore content than hoped or expected. The CEO is very disappointed at the news, even though you reassure him that the results are preliminary and that more thorough results will be available in a week or so. The CEO had hoped to present good news about the exploration to shareholders at a meeting to be held in the next few days.

The CEO asks you to keep the poor results confidential and not to report or discuss them until after the shareholders' meeting, even with people within the company.

QUESTION

Is it ethical to keep this information confidential from the shareholders, who are the owners of the company?

AUTHORS' RECOMMENDED SOLUTION

In the mining, oil, and gas industries, geological data are extremely sensitive information and can be the sole basis for major decisions. The financial welfare of the company may depend on keeping such data confidential. In this case, you have initial results and they have been classed as preliminary. Releasing preliminary data showing low ore content could provoke a loss of confidence in the company. Therefore, providing that there is no concern about fraudulent intent, it would be ethical to keep these reports confidential until the results are known with certainty.

CASE STUDY 8.6
PROFESSIONAL ACCOUNTABILITY[5]

STATEMENT OF THE PROBLEM

Ethel Eager, P.Eng., is a mechanical engineer working in the production department of a well-known specialty chemicals company. The company makes consumer products in Canada for the North American market. It also has plants in the United States, which compete with Canadian plants for North American production mandates.

Eager started out five years ago in a junior production position, reporting to Cam Complacent, P.Eng., the production supervisor. When Eager started at the Canadian plant, it was highly successful. However, over the five years of her employment, the plant has become steadily less competitive versus other firms and its sister plants in the United States. When Complacent retired recently, Eager was promoted to fill his job.

Having passed her Professional Practice Exam during this time, Eager is aware of the importance of professional ethics in engineering. Over the past five years, she has noticed several unusual practices and events in the plant and in the office. For example, supplies often run out before forecast, inventory is invariably balanced by assuming losses, and there are frequent shortages for customer shipments. In the human resources area, she has noticed a tendency to "horseplay" on the graveyard shift, and what she would consider to be instances of ethnic and gender harassment. There also appears to be lax attention to the hours employees worked, coupled with high overtime.

These situations disturbed her, and Eager had approached her boss, Cam Complacent, about them several times. Each time, he played down her concerns and said being "easy" on these subjects helped to keep morale and productivity up. Although Eager was personally convinced that some employees were cheating their employer by taking products home and misrepresenting their hours of work, as a junior employee until recently, she had decided to take her manager's advice and keep quiet, since there didn't appear to be a significant problem.

However, shortly after replacing Complacent, Eager was informed, early one Monday morning, that there had been a major theft at the plant on the weekend. A truck had pulled up to the warehouse, not been challenged, loaded up, and disappeared. Fortunately, the police soon caught the two thieves, who turned out to be employees, one of them a relative of a senior employee. Indeed, the police soon found that a network of employees was involved, and wanted to interview Eager about further investigations.

Meanwhile, Eager received a fax from the company's vice-president for North American manufacturing wanting to review why the Toronto plant's costs were high and productivity was low relative to the company's other plants. The fax concluded with the following: "Understand major theft has occurred. Will be in Toronto tomorrow to review your situation." The future of Eager and her plant looks grim.

QUESTION

Should Eager be held accountable for the employees' actions? What lessons, if any, could be learned from this case?

AUTHORS' RECOMMENDED SOLUTION

As a middle manager and a professional engineer, Eager is accountable to her management, possibly to the police, and to her profession, because she knowingly allowed a dishonest environment to flourish. All the stakeholders involved — Eager, her managers, her peers, her employees, and even her suppliers — have suffered or will suffer. Because she is a professional engineer, Eager has a duty under the provincial code of ethics to all of these stakeholders to act at all times with devotion to high ideals of personal honour and professional integrity. She also has a duty to expose, before the proper tribunals, unprofessional or unethical conduct by another engineer.

Although there are mitigating circumstances in this case (Eager's relative inexperience and her employer's lack of an ethics program, for example), Eager has learned two valuable lessons:

- the meaning of accountability
- that there are no small ethics problems

In hindsight, Eager now knows that turning a blind eye to the problems at her plant was wrong. She should have explained to Complacent that her professional duty included dealing with her concerns, the benefits of dealing with them, and the consequences of ignoring them. If this approach proved ineffective, she could then have suggested to Complacent that together they discuss the subject with more senior management, or that they obtain advice from their provincial Association on meeting their ethical duties in these circumstances. If Complacent was unwilling to consider this approach, as a last resort, Eager could have considered going alone to senior management or obtaining advice from the provincial Association.

TOPICS FOR STUDY AND DISCUSSION

1. Azim Khan, P.Eng., is assistant manager of the engineering department for a large Canadian city. He has been assigned to supervise the construction of a new sewage treatment plant since he participated in designing the plant. The construction contract has been awarded, after a competitive bidding process, to the Zenith Construction Company. About ten days before construction is to begin, Khan finds a gift-wrapped case of rye whisky on his doorstep (approximate value: $300). A card attached to the box says, "To an esteemed colleague. We look forward to a long and professional relationship — Zenith Construction." Is it ethical for Khan to accept this gift? Justify your answer using your provincial code of ethics.

2. You are the manager of a new project, and one of your first responsibilities is to make a realistic estimate of the time the project will take and the cost that will ensue. Your calculations result in very high estimates, so high that you fear the project may be discontinued. Some older engineers on the project say that many earlier projects would have been cancelled if the true extent of their final costs had been known early in the game. Moreover, they argue that no one can ever be really sure of what something is going to cost; after all, these are only *estimates*. In the earlier projects, a very optimistic face was put on the cost estimates, and, even though the final costs exceeded the estimates, the projects were successful.

   The older engineers urge you to reduce your estimates so the project will not be cancelled. But you have put a lot of careful work into your estimates and believe your figures are as correct as any estimate of the future can ever be. Therefore, if you reduce the estimates, you know you will be lying. Furthermore, you know your own reputation in the company will be flawed if it becomes apparent that you shaved your estimates. However, you fear that some of the people in your project team may be laid off if your project is cancelled. You are caught in a dilemma, and, as a manager, you must decide one way or the other. Explain how you would try to solve this ethical dilemma. Summarize the process and your decision. For assistance, review Chapter 6.

3. Renée Langlois is a professional engineer who has recently been appointed president of a large dredging company. She is approached by senior executives of three competing dredging companies and asked to co-operate in bidding on federal government dredging contracts. If she submits high bids on the next three contracts, then the other companies will submit high bids on the fourth contract and she will be assured of getting it. This proposal sounds good to Langlois, since she will be able to plan more effectively if she is assured of receiving the fourth contract. Is it ethical for Langlois to agree to this suggestion? If not, what action should be taken? If she agrees to this suggestion, does she run any greater risk than the other executives, assuming that only Langlois is a professional engineer?

NOTES

1. Professional Engineers Ontario (PEO), *Guideline to Professional Practice* (Toronto: PEO, 1988, revised 1996), 21. Reprinted with permission of PEO.
2. Professional Engineers Ontario (PEO), *The Use of Computer Software Tools by Professional Engineers and the Development of Computer Software Affecting Public Safety and Welfare* (Toronto: PEO, 1993), 4. Reprinted with permission of PEO.
3. Howard A. Levitt, *The Law of Dismissal in Canada*, as quoted in *CSPEAKER*, Canadian Society of Professional Engineers (CSPE) (September 1981): 1–4. Reprinted with permission of Howard A. Levitt.
4. Ibid.
5. Case Study 8.6 is adapted from James G. Ridler, P.Eng., "Accountability: At the Core of Professional Engineering," *Engineering Dimensions* 18, no. 1 (January/February 1997): 40–41. Used with permission of James G. Ridler and PEO.

CHAPTER NINE

Ethical Problems of Engineers in Private Practice

The engineer in private practice may face ethical problems in addition to those described in the previous two chapters. The frequent interaction with clients, suppliers, contractors, subcontractors, and government agencies leads, occasionally, to conflict of interest or ethical dilemmas. For example, some clients or competing firms may not feel that they are bound by ethical rules in awarding engineering contracts. Engineers must thus answer the question of whether they should adopt unethical practices to obtain contracts, even though the contracts themselves are competently and conscientiously carried out, or refuse to become ensnared in unethical practices, and thereby risk losing work. It is not hard to find spectacular examples of people yielding to the use of unethical methods; our newspapers document these cases daily. However, it is not our purpose to dwell on past inequities or to cause embarrassment to those who are guilty or alleged to be guilty of acting unethically; our purpose is to alert the engineer to the possibility of unethical practices and to urge the consistent adoption of fair, legal, and ethical methods. In this chapter, we examine some of the ethical problems that may arise when obtaining and carrying out engineering contracts.

THE CLIENT–CONSULTANT RELATIONSHIP

It is important for an engineer in private practice, or consultant, to develop a good working relationship with the client. In many cases, the consultant is engaged by a client to monitor an engineering project, such as the design or construction of a structure or factory, for example. This usually creates a three-way relationship between the client (owner), the contractor (designer or builder), and the consultant (engineer). The client typically needs the advice of the consultant to ensure that the work of the contractor is adequately performed and of good quality.

The relationship between the client and the consultant will, of course, depend on the personalities of the individuals, the type of project, and the problems encountered, but it is useful for the engineer in private practice to recognize that there is a spectrum of client–consultant relationships. D.G. Johnson describes three points along this relationship spectrum:[1]

> **The "independent" model:** The client explains the problem and then turns over decision-making power to the consultant, who takes charge of the problem and makes decisions for the client. The consultant does not provide technical knowledge to the client, but acts in place of the client, keeping the clients' interests in mind, in a paternalistic but independent way. This is one end of the client–consultant spectrum, and it is generally unacceptable, since it robs the client of the ability to make any choices.

> **The "balanced" model:** The consultant interacts with the client, by providing engineering advice and evaluating the risks and benefits of various alternatives, but the client makes the choice of the action to follow. This relationship is similar to the ideal patient–physician relationship, where the professional may have the knowledge and expertise to solve the client's problem, but the client must be informed of the possible choices, and their benefits and risks, before making a decision to proceed with treatment. In a balanced relationship, the client and consultant must treat each other as equals. The consultant has a responsibility to provide engineering expertise to the client, but the client retains power to make the key decisions. The balanced relationship is the approximate mid-point of the spectrum, and is generally the optimum client–consultant relationship.

> **The "agent" model:** The consultant is simply an agent or "order-taker" for the client, and contacts the client for instructions before acting. This is the other end of the client–consultant spectrum, and it is also generally unacceptable, since the client does not make full use of the engineer's knowledge. This relationship may also be seen as demeaning by the consultant.

In any project, the client–consultant relationship will be situated somewhere along the spectrum described above. The precise point will depend on the personalities and relative knowledge of the participants and the types of problems encountered. The key goal is to keep information flowing between client and consultant to ensure that both are fully aware of the crucial areas of the work.

ADVERTISING FOR ENGINEERING WORK

An engineer in private practice may not be known to the many corporations seeking engineering services and may feel the need to advertise. This brings up a thorny issue that has plagued all the professions in North America. Every province has some restrictions, usually in the code of ethics or in practice guidelines, on advertising engineering services. The purpose of the restrictions is to ensure fairness and honesty in competitive evaluation of professional qualifications and experience.

Advertising fills our newspapers, magazines, radio programs, and television screens, and it seems obviously demeaning and unprofessional to have engineering services promoted in the same way as soap powder or chewing gum. However, advertising that communicates facts and data about the availability, experience, and areas of expertise of an engineer in private practice is fair and unobjectionable. In the past, the "calling card" or "business card" form of advertising was the only acceptable advertising method, and the back pages of many engineering publications continue to be filled with these advertisements. In recent years, the restrictions have eased slightly, but many rules still apply.

Most codes of ethics state that the engineer "shall not advertise his or her professional services in self-laudatory language or in any other manner derogatory to the dignity of the profession." Alberta and the Northwest Territories require engineers to advertise through "factual representation without exaggeration"; Saskatchewan and Yukon state that an engineer shall "not advertise in a misleading manner or in a manner injurious to the dignity of the profession." Quebec and Ontario do not mention advertising in their codes of ethics, but each has a regulation under the Act that gives specific instructions. In Quebec, Regulation 10 under the Act gives very precise rules concerning the information that may be conveyed on business cards, stationery, newspapers, magazines, directories, and signs on work premises, offices, and vehicles.

In Ontario, regulations made under the authority of the provincial Act permit advertising provided that it is done in a professional and dignified manner; factually, without exaggeration; and does not directly or indirectly criticize another licensed engineer or the employer of another licensed engineer. In addition, the regulation expressly forbids the use of the engineer's seal or the Association's seal in any form of advertising, including the use of these seals on business cards or letterhead.[2] The seal has a legal significance (explained later in this chapter) that is totally incompatible with advertising. On the other hand, the Association's decorative logo, which includes the Association's working name, may be used on business cards and letterhead, but only to signify membership in the Association. In several provinces, a Certificate of Authorization (or equivalent credential) is required for an engineer to provide services to the general public, so presumably anyone advertising for engineering work in those provinces would be expected to possess the Certificate.

In addition, Professional Engineers Ontario provides general guidelines to test proposed advertising and ensure that it does not contravene the regulations. The following advertising guidelines are reproduced with permission.

Advertising may be considered inappropriate if it:

i) claims a greater degree or extent of responsibility for a specified project or projects than is the fact;

ii) fails to give appropriate indications of cooperation by associated firms or individuals involved in specified projects;

iii) implies, by word or picture, engineering responsibility for proprietary product or equipment design;

iv) denigrates or belittles another professional's projects, firms or individuals;

v) exaggerates claims as to the performance of the project or;

vi) illustrates portions of the project for which the advertiser has no responsibility, without appropriate disclaimer, thus implying greater responsibility than is factual.[3]

In general, advertising that is factual and truthful and that communicates information about an engineer in private practice concerning qualifications, experience, location, or availability in a dignified manner is acceptable.

ENGINEERING COMPETENCE

Engineering competence gained through education or experience is an important and valuable asset. The client is, in fact, paying for that competence when the engineer is hired to work on the client's project. An engineer who accepts an assignment that is beyond his or her level of competence is guilty of unethical conduct and could be guilty of unprofessional conduct or incompetence, which could be the basis for disciplinary action.

This does not mean that an engineer must be a world-class expert in every phase of a proposed project before accepting it. However, the engineer must be sufficiently familiar with the subject matter to know that he or she can become competent through study or research in a reasonable period of time or that a colleague or consultant can be hired without delaying the project or incurring unnecessary expense to the client. The essential criterion is that the client's project must not be put at risk because of the engineer's lack of competence.

Each engineer must be the monitor of his or her level of competence. This means making use of opportunities to expand one's knowledge and experience and maintaining one's engineering competence, as discussed in more detail in Chapter 15. It also means being realistic about evaluating one's own abilities, a difficult task at the best of times. However, no one knows the limits of one's knowledge better than oneself.

USE OF THE ENGINEER'S SEAL

Each provincial Act provides for engineers to obtain and use a seal of a specific design for that province. The seal is usually an inked rubber stamp indicating that the person named on the stamp is licensed in that province (or territory). The terms *seal* and *stamp* are interchangeable. The Act (or regulation) typically requires that all final drawings, specifications, plans, reports, and other documents involving the practice of professional engineering, when issued, must be dated and bear the signature and seal of a professional engineer. The use of the seal is not optional; it is a standard requirement of the provincial Act. Most documents typically state that they were *prepared* or *checked and approved* by the person who sealed them.

The seal has legal significance, since it typically indicates that the person accepts responsibility for the accuracy of the documents and, indirectly, implies that the person has a thorough knowledge of the project to which the documents relate. Conversely, an engineer who knowingly signs or seals documents that are not based on such thorough knowledge may be guilty of professional misconduct and may also be liable for fraud or negligence if the misrepresentation results in someone suffering damages.

In a case cited recently, the British Columbia Supreme Court decided that an engineer was liable in a dispute over an improperly designed residence foundation. The court stated, "By affixing his seal to the drawings and by his letter to the defendant municipality ... the defendant [engineer] ... certified that the foundation drawings conformed to all the structural requirements of the 1980 National Building Code."[4] This statement clearly indicates that the court considered the seal on the drawings a guarantee of their accuracy and conformance with codes. However, in a somewhat different type of case, the Supreme Court of Canada ruled that "The seal attests that a qualified engineer prepared the drawing. It is not a guarantee of accuracy. The affixation of the seal, without more, is insufficient to found liability for negligent misrepresentation."[5]

Although the contradiction in these court rulings may indicate that there is still some debate over fine legal points, the best strategy for the professional engineer is to avoid the courts entirely. Do not affix your seal to a document until you are willing to accept full responsibility for it, based on detailed knowledge of the document and the project to which it applies, and you are completely satisfied with the accuracy of the document.

"CHECKING" ENGINEERING DOCUMENTS

It is important to define what "checking" means. The term *checked and approved* usually means that the documents were prepared by someone under the direct supervision of the person who signed and sealed them. However, as mentioned in Chapter 2, Alberta recently reviewed and revised its standards of practice to permit engineers, geologists, and geophysicists to seal the work of others (not under direct supervision), provided that the seal is preceded by a thorough review of the work and that the engineer who seals the documents accepts full responsibility. The Practice Standards Committee of the Association of Professional Engineers, Geologists and Geophysicists of Alberta (APEGGA) examined the legislation of other associations, and, while the wording of the legislation was similar to the Alberta Act, seven associations permitted or encouraged the interpretation that stamping the work of others is acceptable when accompanied by a thorough review and acceptance of responsibility.[6] For example, Ontario regulations require engineers to "sign, date and affix the ... seal to every final drawing, specification, plan, report or other drawing prepared or checked" by the engineer.[7]

Therefore, if you have performed sufficient review and analysis of the work that you are willing to accept full responsibility for it, then it is acceptable to seal the work in most provinces. However, for some projects, a proper "review" would require complete duplication of the analysis; obviously, if the work is completely redone by you, then it would be appropriate for you to assume responsibility for it. It is important to avoid the grey areas where an engineer could be led into a trap. For example, it is fairly common for an unlicensed person to ask an engineer to seal documents with the primary purpose of aiding the person to practise without a licence or to avoid the scrutiny or cost of a full engineering analysis. Regardless of the urgency of such requests, do not assume responsibility for work that is beyond your area of expertise, or for work that you have not thoroughly reviewed for accuracy.

PREPARATION AND APPROVAL

If one engineer has prepared a document or drawing and another engineer must approve it, then *both* seals should appear on it whenever possible. If for any reason this is not possible or expected, then only the approving engineer should seal it, indicating that he or she takes the responsibility for the document or drawing. Where final drawings cover more than one engineering discipline, it is typically recommended that the drawings be sealed by the approving engineer (typically a chief engineer or project leader) and by the design engineer for each discipline. The seals of the design engineers should be "qualified" by an explanatory note to indicate clearly each engineer's area of responsibility.

PRELIMINARY DOCUMENTS

Preliminary documents, drawings, or specifications are usually not sealed but are clearly marked "preliminary" or "not for construction." Only the final drawings are sealed. Similarly, an engineer should not seal a document that has no engineering content.

Occasionally, to satisfy the requirements of a regulatory agency, a preliminary document may need to be sealed. In this case, the comment "preliminary" or "not for construction" should be included prominently.

REPORTS

Individual pages of a report or drawings included in a report need not be sealed providing the report as a whole has been signed, sealed, and dated.

SEALING OF DETAIL DRAWINGS

The engineer generally has responsibility for a project as a whole, and the engineer's seal must appear on the major reports, specifications, or drawings that describe the project. Usually the engineer is not expected to seal every detail drawing, although the drawings must be prepared under the engineer's control and supervision, and he or she assumes responsibility for them, whether sealed or not. The process for the special case of structural steel is described in the Ontario *Guideline to Professional Practice:*

> In the case of structural steel, the steel supplier provides shop drawings for review by the structural engineer. The steel supplier has selected standard connections from the handbook published by the Canadian Institute

of Steel Construction according to the moments and forces given by the engineer for each connection. The selection of "standard connections" is not considered to be part of professional engineering practice and such shop drawings need not be sealed. Shop drawings depicting special connections do require sealing.

Some design engineers require a seal on all shop drawings and erection diagrams. This is their prerogative. Alternatively, a letter signed by a professional engineer stating that the shop drawings have been prepared under his or her supervision may be acceptable.[8]

SEALING OF MASTERS AND PRINTS

The master drawings must of course be complete and unambiguous, since they are usually the major reference for describing the concepts and the details of the structure, machine, process, or whatever is being designed. It is important that an engineer in private practice have an effective procedure for controlling the issuing of preliminary and final drawings so that security and confidentiality of the client are protected and no confusion arises. The appropriate time to seal a drawing is when it is approved and released for fabrication or construction. Modifications to final drawings must be rigidly controlled and documented. This control is aided by sealing only the prints and not the master. In this way, the prints can be checked for modifications when sealed.

SEALING OF "SOFT" DRAWINGS

Computer-aided design has simplified the easy, rapid production of drawings, but it has resulted in problems of control over modifications and general security. Erroneous data or unauthorized modifications are usually difficult to detect. The development of electronic seals using passwords is evolving but is not fully implemented in design offices. It is therefore important for an engineer in private practice to implement some form of control within the firm to prevent unauthorized copying or modification of computer files containing design drawings, reports, specifications, and so on. Until standards and procedures develop, it is recommended that "final" drawings in computer file form be protected by password or by storing them on magnetic media in secure locations and that seals be applied, by hand, only to prints made from these files. The prints or hard copies of the files would then be the master document. Copies of the engineer's seal generally should *not* be reproduced in any form, whether by computer or by stick-on labels, since this is equivalent to losing secure control of the seal.

CONFIDENTIALITY

The engineer has a clear obligation, under the code of ethics, to keep the affairs of the client confidential. However, in some cases, a client may request the engineer to sign a confidentiality agreement to prevent the engineer from disclosing the client's affairs to any third party unless authorized to do so by the client. Engineers usually do not hesitate to sign such agreements, since they have no intention of disclosing information improperly and the agreement for confidentiality is redundant. Two types of ethical dilemma may arise from the need for confidentiality.

An engineer in private practice may encounter a potential conflict of interest when accepting work from a new client who is a competitor of a former client. The engineer should not accept a contract that requires disclosure of a previous client's affairs, whether technical, business, or personal. This is particularly applicable to proprietary information or "trade secrets" that could cause financial loss to the former client if disclosed. If a confidentiality agreement has been signed with the former client, it should be reviewed. However, the code of ethics requires the engineer to keep the former client's proprietary information confidential, and in some cases, it may not be advisable to accept work from the new client even if a confidentiality agreement was not signed.

Another less-common problem with confidentiality agreements arises in environmental projects. In some cases, the engineer may be compelled to reveal information under the code of ethics (where public danger is involved) or by law, as may occur in court proceedings or under regulations of the Environmental Protection Act. For example, consider the case where an engineer advises the client to remedy an environmental hazard, but the client refuses to do so. If the engineer has signed a confidentiality agreement, then any whistleblowing done by the engineer could be interpreted as a breach of contract. This creates a serious ethical dilemma for the engineer, who must choose between breaching a contract or following the code of ethics. A compromise, suggested by the Ontario *Guideline to Professional Practice*, is to include a clause in the confidentiality agreement stating that, if the client should fail to act on certain hazards within a specified period of time, the engineer is entitled to fulfil any reporting requirements that are specified in law after first notifying the client.[9] However, an engineer practising in the environmental area would be well advised to get personal legal advice on the proper wording of such agreements.

CONFLICT OF INTEREST

An engineer in private practice may occasionally encounter conflicts of interest. As mentioned earlier in this text, a "conflict of interest" occurs in any professional relationship where the professional has an interest that interferes with the service owed to the client. For example, if an engineer recommends that a client purchase goods or services from a company in which the engineer has partial ownership (of which the client is not fully aware), the engineer has created a serious conflict of interest that is contrary to the code of ethics. Conflicts may be much simpler: an engineer may be tempted to suggest that the client adopt a course of action where the main benefit is to reduce the engineer's workload. Unless there is a similar reduction in fee, the engineer has a conflict of interest that must be disclosed fully to the client. In every instance of conflict or potential conflict of interest, the engineer must make a full disclosure to the client of the engineer's personal interest, whatever that may be. Although the client may agree that the conflict is insignificant, the client is then making a fully informed choice.

A client who learns after the fact that an engineer benefited personally and secretly from a decision that was ostensibly based on technical factors would be justified in contacting the provincial Association to lodge a complaint of professional misconduct.

REVIEWING THE WORK OF ANOTHER ENGINEER

The sensitive problem of reviewing the work of another engineer was discussed in an earlier chapter with respect to an employee. The question of review is equally delicate when engineers are in private practice. In all instances, the welfare of the client or the general public must come before the personal wishes of the engineer. Moreover, it must be emphasized that an engineer should be *informed* when his or her work is to be reviewed, but it is *not* necessary to seek or obtain the engineer's permission for the review. The Ontario *Guideline to Professional Practice* summarizes the situation as follows:

> The Code of Ethics permits engineers to be engaged to review the work of another professional engineer when the connection of that engineer with the project has been terminated. Before undertaking the review, reviewers should know how the information will be used. Even when satisfied that the connection between the parties has been terminated, reviewers should, with the agreement of the client, inform the other engineer that a review is contemplated. They should recognize that the client has the right to withhold approval to inform the engineer, but [should] satisfy themselves that the reasons for the owner's decision are valid before proceeding with the review.
>
> If a client asks an engineer to review the work of another engineer who is still engaged on a project, either through an employment contract or an agreement to provide professional services, the reviewer

should undertake the assignment only with the knowledge of the other engineer. Failure to notify the engineer under this circumstance constitutes a breach of the Code of Ethics. On the other hand, should a second engineer be engaged by another person (say, a building department) to provide professional engineering services on the same project, he or she would have no obligation to advise the original client of the commission.

Senior engineers are often asked to review a design prepared by another engineer. (Most engineers are expected to have their work routinely reviewed as part of an ongoing quality control and professional development process.) If reviewers find that design changes are necessary, they should inform the design engineer of these findings and the reasons for the recommended changes. During the design stage, reviewers (who are acting as the client's agent in this case) and engineers may agree on changes to the engineers' proposal. However, design engineers must not agree to any change or alternative suggested by reviewers that could result in an unworkable installation, be in conflict with the relevant codes, or create a risk of damage or injury.

Reviewers must administer the design contract and evaluate engineers' work at arm's-length, so that the engineer of record maintains full responsibility for the design.

. . .

It is emphasized that the acceptance of mutually agreed-upon changes does not relieve the original design engineer of responsibility for the design or work under review.

Once the review has been completed, there is no obligation or right for the reviewers to disclose their findings to the other engineer. In fact, in most cases, disclosure of the findings would not be permitted by the client. Reviewers' contractual obligations are to the client. However, reviewers should seek the client's approval to inform the engineer of the general nature of the findings, and if appropriate, should try to resolve any technical differences.[10]

DESIGN CALCULATIONS

A client may request that an engineer submit calculations that were done to support a recommendation. This amounts to a review of the engineer's work, but obviously it is done with the full knowledge and co-operation of the engineer. The client has an ethical right to review these calculations and to make a copy for a permanent record. However, the time necessary to prepare the calculations in a format understandable to the client should be included as part of the contracted service.

Occasionally the computation techniques or the data on which the computation is made may be proprietary and the engineer may not wish to divulge them. In this case, the conditions for reviewing the calculations should be negotiated beforehand and the extent of disclosure should be understood in advance. The usual procedure is to provide the proprietary data to the client with the clear understanding that they will be kept confidential.

COMPETITIVE BIDDING FOR SERVICES

A detailed procedure for selecting an engineer in private practice was given in Chapter 5. The procedure involves three stages and separates the process of selecting the best-qualified engineer (or firm) from the process of negotiating the fees. This prevents many of the problems that arise when engineers are selected on a competitive basis by lowest bid.

However, it should be stressed that seeking professional services by lowest bid is not illegal or unethical, and no one should be dissuaded from the procedure by the misguided belief that competition is harmful. Quite the opposite: ingenuity thrives on healthy competition. However, there is a danger in competitive bidding, as explained in the Ontario *Guideline to Professional Practice:* "With professional services there are ultimately only two elements which a client is retaining, i.e. the engineer's knowledge and time. Shortchanging on a professional engineering fee will result in the substitution of less skilled engineers or less time put into the assignment, thus potentially shortchanging the project."[11]

However, some competitive activities in obtaining contracts are considered to be unfair and therefore unethical. For example, any agreement to pay a kickback, gift, commission, or consideration, either openly or secretly, would be considered an unfair and unethical method of obtaining contracts. Many codes also describe supplanting a colleague as unethical, where *supplanting* is defined as intervening in the client–engineer relationship of a colleague and, through inducements or persuasion, convincing the client to fire the engineer and hire the intruding engineer.

NEGLIGENCE AND CIVIL LIABILITY

The engineer in private practice generally has two sources of concern that can give rise to civil liability: breach of contract and negligence. Both of these are usually inadvertent events and are to be distinguished from professional misconduct and incompetence, which are discussed in Chapter 14. A *breach of contract* is a failure to complete the obligations specified in a contract, whereas *negligence* is a failure to exercise due care in the performance of engineering. Although it is possible to obtain protection for breach of contract by incorporating a practice and protection against negligence by purchasing liability insurance, it is not possible to avoid disciplinary action for negligence, incompetence, or professional misconduct. The Ontario *Guideline to Professional Practice* summarizes the situation as follows:

> An individual engineer can protect personal assets against an action for damages for *breach of contract* by incorporating the practice. After incorporation, it is the company that is the contracting party and not the individual. As far as protection from liability for *negligence* there is nothing available to an engineer other than careful, thorough engineering and insurance.[12]

ACEC CODE OF CONSULTING ENGINEERING PRACTICE

The Association of Consulting Engineers of Canada (ACEC) has a code of practice that applies to member firms of ACEC and requires them to fulfil their duties with honesty, justice, and courtesy toward the public as a whole, clients, other consulting engineers, and employees.

ACEC CODE OF CONSULTING ENGINEERING PRACTICE

Members of the Association of Consulting Engineers of Canada shall fulfil their duties with honesty, justice and courtesy towards Society, Clients, other Consulting Engineers and Employees.

Society

Members shall practice their profession with concern for the social and economic well-being of Society.

Members shall conform with all applicable laws, by-laws and regulations.

Members shall satisfy themselves that their designs and recommendations are safe and sound and, if their engineering judgement is overruled, shall report the possible consequences to clients, owners and, if necessary, the appropriate public authorities.

Members expressing engineering opinions to the public shall do so in complete, objective, truthful and accurate manner.

Members should participate in civic affairs and work for the benefit of their community and should encourage their employees to do likewise.

Clients

Members shall discharge their professional responsibilities with integrity and complete loyalty to the terms of their assignments.

Members shall accept only those assignments for which they are competent or for which they associate with other competent experts.

Members shall disclose any conflicts of interest to their clients.

Members shall respect the confidentiality of all information obtained from their clients.

Members shall obtain remuneration for their professional services solely through fees commensurate with the services rendered.

Other Consulting Engineers

Members shall relate to other consulting engineers with integrity, and in a manner that will enhance the professional stature of consulting engineering.

Members shall respect the clientele of other consulting engineers and shall not attempt to supplant them when definite steps have been taken towards their employment.

Members shall compete fairly with their fellow consulting engineers, offering professional services on the basis of qualifications and experience.

Members engaged by a client to review the work of another consulting engineer, shall inform that engineer of their commission, and shall avoid statements which may maliciously impugn the reputation or business of the engineer.

Employees

Members shall treat their employees with integrity, provide for their proper compensation and require that they conform to high ethical standards in their work.

Members shall encourage their employees to enhance their professional qualifications and development.

Members shall not request their employees to take responsibility for work for which they are not qualified.[13]

FIDIC CODE OF ETHICS

The International Federation of Consulting Engineers (FIDIC) has adopted a code of ethics that guides the conduct of individual consulting engineers and member firms of consulting engineers. In addition, several FIDIC policy documents on consulting engineering practice are available on its Web site, including copyright of engineering documents, the role of the expert witness, professional risks, design/build ("turnkey") contracts, professional liability, and so on. The code of ethics is the most relevant FIDIC document for our purposes; it appears to be in complete agreement with the ACEC code of ethics and other codes discussed in this textbook, and is reproduced below with permission.

FIDIC CODE OF ETHICS

The International Federation of Consulting Engineers recognises that the work of the consulting engineering industry is critical to the achievement of sustainable development of society and the environment.

To be fully effective not only must engineers constantly improve their knowledge and skills, but also society must respect the integrity and trust the judgement of members of the profession and remunerate them fairly.

All member associations of FIDIC subscribe to and believe that the following principles are fundamental to the behaviour of their members if society is to have that necessary confidence in its advisors.

Responsibility to Society and the Consulting Industry

The consulting engineer shall:

- Accept the responsibility of the consulting industry to society.
- Seek solutions that are compatible with the principles of sustainable development.
- At all times uphold the dignity, standing and reputation of the consulting industry.

Competence

The consulting engineer shall:

- Maintain knowledge and skills at levels consistent with development in technology, legislation and management, and apply due skill, care and diligence in the services rendered to the client.
- Perform services only when competent to perform them.

Integrity

The consulting engineer shall:

- Act at all times in the legitimate interest of the client and provide all services with integrity and faithfulness.

Impartiality

The consulting engineer shall:

- Be impartial in the provision of professional advice, judgement or decision.
- Inform the client of any potential conflict of interest that might arise in the performance of services to the client.
- Not accept remuneration which prejudices independent judgement.

Fairness to Others

The consulting engineer shall:

- Promote the concept of "Quality-Based Selection" (QBS).

- Neither carelessly nor intentionally do anything to injure the reputation or business of others.
- Neither directly nor indirectly attempt to take the place of another consulting engineer, already appointed for a specific work.
- Not take over the work of another consulting engineer before notifying the consulting engineer in question, and without being advised in writing by the client of the termination of the prior appointment for that work.
- In the event of being asked to review the work of another, behave in accordance with appropriate conduct and courtesy.

Corruption
The consulting engineer shall:

- Neither offer nor accept remuneration of any kind which in perception or in effect either a) seeks to influence the process of selection or compensation of consulting engineers and/or their clients or b) seeks to affect the consulting engineer's impartial judgement.
- Co-operate fully with any legitimately constituted investigative body which makes inquiry into the administration of any contract for services or construction.[14]

CASE STUDY 9.1
APPROVAL OF ENGINEERING PLANS BY TOWN COUNCIL

STATEMENT OF THE PROBLEM
Edward Beck is a consulting engineer in a small town. He has been elected to sit on the town council as a councillor, a part-time job that he does mainly as a form of public service. Beck has also been hired by a developer to draw up plans for the street layout and water and sewage facilities for a new residential subdivision in the town. The developer's submission to town council includes Beck's drawings and specifications. Later, in a town council meeting, Beck votes to approve the subdivision. During the discussion, Beck does not publicly state his relationship with the developer, nor does he conceal it. His signature and seal are on some of the plans submitted to council. Everyone knows that he is the only engineer in town who does this type of work, and he is certain that they would prefer to see local people hired for this project.

QUESTION
In voting to approve this project, has Beck acted unethically?

AUTHORS' RECOMMENDED SOLUTION
This situation sometimes occurs in small towns with few engineers, where a conflict of interest cannot be avoided. Engineers certainly should

not be disqualified from projects because they are performing a public service as members of town councils. However, in this case, it is not enough that "everyone knows" that Beck has a business relationship with the developer. Beck created a serious conflict of interest when he voted to approve plans that he himself prepared. He should have made a clear, unequivocal statement of his involvement in the project and his relationship with the developer, then withdrawn from the debate and abstained from the vote. By participating in a formal vote without declaring the conflict of interest, Beck has exposed himself to the possibility of a complaint to the provincial Association and possible disciplinary action.

CASE STUDY 9.2
ADVERTISING PRODUCTS AND ENGINEERING SERVICES

STATEMENT OF THE PROBLEM
Alonzo Firenze is a consulting engineer to the Acme Amphibious Transporter Company, which manufactures small amphibious recreational vehicles with a moulded plastic hull. Each vehicle is driven by eight low-pressure balloon tires and can manoeuvre quickly and safely on land and water. In preparing a television campaign to increase sales of the vehicle, the television producer suggests that Firenze appear on camera and endorse the safety aspects of the vehicle as a professional engineer and safety expert. The television producer points out that Firenze has conducted extensive tests, studies, and surveys on the vehicle and can speak with authority. In addition, Firenze is very photogenic and would like the exposure to the general public.

QUESTION
Would it be unethical for Firenze to appear in the television commercial and make a statement endorsing the recreational vehicle?

AUTHORS' RECOMMENDED SOLUTION
Although participation in a television news or documentary program on vehicle safety would be considered a suitable professional activity (and perhaps more engineers should be seen in these roles), endorsing a product in a television commercial would be seen as unprofessional, would lower the public esteem for the engineering profession, and would therefore be unethical. The key distinction is that the purpose of a commercial is to increase sales, and it is a sad commentary on the television industry that commercial advertisements have a rather sordid history of half-truths and appeals to emotion rather than logic. Moreover, an engineer who publicly praises a product manufactured by a corporation that employs him or her clearly has a conflict of interest, and the endorsement would lack the expected professional detachment.

A second point at issue is Firenze's interest in appearing because of the personal exposure he will receive. This is not the proper format for advertising professional services. Moreover, although Firenze is willing to participate and is convinced of the safety of the product, how would he respond in future in different circumstances? For example, what response would Firenze give if the employer expected an endorsement as part of his employment contract, but he was not confident of the product's safety? This illustrates the risk: the profession and Firenze's reputation could both suffer by subordinating professional standards to the pressures of the marketplace.

CASE STUDY 9.3
CONTINGENCY FEE ARRANGEMENTS

STATEMENT OF THE PROBLEM

As an engineer in private practice, you are considering whether to offer your services on a contingency basis, an arrangement in which you would be paid a percentage of some outcome.

Two clients wish to retain you. Client A wants to retain you to act as an expert witness in a lawsuit against a third party. The lawsuit, if successful, should result in the award of a very large sum as a settlement. Client B has shown a hesitant interest in retaining you to recommend changes to the energy usage of a manufacturing process. After an initial study of the problem, you believe that the energy savings could be immense. You believe Client B would be more responsive if fees were contingent on the savings.

QUESTION

Would it be ethical to offer your services on a contingency basis to either of these clients, with the understanding that you would be paid a percentage of the legal settlement (Client A) or a percentage of the value of the energy savings (Client B)?

AUTHORS' RECOMMENDED SOLUTION

These two cases seem similar but are distinctly different.

An expert witness is permitted to express opinions, whereas a nonexpert witness must confine his or her testimony to known facts. Therefore, an engineer testifying as an expert witness must have an impartial attitude toward the outcome of a case. However, as a recipient of a percentage of the potential settlement from Client A, you would have a conflict of interest and your testimony would be suspect. Therefore, it would be unethical to accept this case on a contingency basis. You should bill Client A for time and expenses or on a flat-rate basis so that the reimbursement is independent of the outcome of the case.

The case of Client B is somewhat different, since there is no need for impartiality. In fact, your bias toward reducing energy consumption could be very beneficial to the client. Also, you have a duty to yourself and to your colleagues to charge an adequate fee. From your study, you evidently believe that this fee will be adequate. Therefore, the proposal to base the fee on a contingency is not unethical. However, a word of warning: there might be a perception of unethical behaviour unless the results can be measured accurately and impartially and can be achieved without degrading the client's product or facilities. Therefore, although this method of setting a fee is not unethical, it has some risks associated with it. You would be well advised to use one of the more common billing methods (as described in Chapter 5) unless the client expresses a preference for the contingency method and the savings can be clearly and unequivocally measured. A further word of warning to engineers in British Columbia: the code of ethics in that province specifically forbids contingent fee arrangements that depend on a "finding of economic feasibility" such as this case describes.

CASE STUDY 9.4
ADHERENCE TO PLANS AND CONTROL OF SEALED DRAWINGS

STATEMENT OF THE PROBLEM
A professional engineer in private practice is engaged by a building contractor to prepare drawings for the forms and scaffolding needed to construct a reinforced concrete bridge. The forms and scaffolding must sustain the weight of about 1400 tonnes of concrete until the concrete is cured. The engineer prepares the drawings and signs and seals the original, which he gives to the contractor. The contractor later engages the engineer to inspect the completed structure; the engineer finds that the contractor has made several major deviations from the plans. He is not sure whether the structure is safe or unsafe. The contractor has stated that time is of the essence, and concrete is to be poured in the next 48 hours. The engineer feels an obligation to the contractor because of their previous professional relationship and hopes that it will continue.

QUESTION
What should the engineer do?

AUTHORS' RECOMMENDED SOLUTION
Two issues are at stake here. When the engineer passed the sealed original drawings to the contractor, control was lost. Changes could have been made to the original that, if unsafe, could have created problems for the engineer. As a general rule, only prints should be signed and sealed so

modifications will be evident. In this case, apparently no changes were made. However, the contractor did not construct the forms and scaffolds according to the plans, and the engineer is now faced with the unpleasant task of informing the contractor that the deviations from the plans must be evaluated to ensure that they are safe. This will undoubtedly require some calculations and perhaps a second inspection. The engineer should notify the contractor in writing that concrete must not be poured until the review and reinspection is complete and that the structure could constitute a hazard to workers and the general public. The strength analysis should be carried out as quickly as possible, but if the 48-hour deadline cannot be met, then the project must not proceed until all safety concerns have been satisfied. It is perhaps useful to point out that the contractor could have consulted the engineer about the changes earlier in the construction so that the delay could have been avoided.

CASE STUDY 9.5
FEE REDUCTION FOR SIMILAR WORK

STATEMENT OF THE PROBLEM
Susan Johnson is a professional engineer in private practice. She is hired by Client A to design a small explosion-proof building for storing flammable paints, chemicals, and explosives. The work is carried out in her design office, and copies of the plans are provided to her client. After construction is complete, she is approached by Client B, who has seen the building and has a similar requirement. Client B suggests that the fee should be substantially reduced since the design is already finished and only minor changes would be required.

QUESTION
Would it be ethical for Johnson to reduce her fees as suggested? Would it be good business practice?

AUTHORS' RECOMMENDED SOLUTION
Establishing fair and reasonable fees depends on five factors:

- level of knowledge and qualifications required
- difficulty and scope of the assignment
- responsibility that the engineer must assume
- urgency with which the work must be accomplished (will overtime or extra personnel be required?)
- time required (number of person-hours)

Of these five factors, only the last one (time required) is likely to be reduced because a similar project was successfully completed. The client benefits from receiving a design that has been tested and is likely more dependable and easier to construct. The level of knowledge, the qualifications, and, most important, the responsibility that the engineer must assume are unchanged. In assessing the responsibility, one might ask: "If the engineer reduces the fee for the design by 10 percent, say, and the building collapsed, would the client expect reimbursement for only 90 percent of the building cost?" Another question that might be asked is, "Are savings, as a result of prior experience in the client's business, passed on to the client's customers?" The answer to both of these questions is undoubtedly "No," so it would be unfair to Johnson and poor business practice to accept a substantial reduction in fee for providing the drawings for this structure. Moreover, even if Johnson should make a significant savings on the reduced time, it would still be unethical to pass these savings on to Client B, since the original design was commissioned and funded by Client A and a reduced fee for Client B amounts to a subsidy paid by an innovative client to a follower. If Client A was requesting a second building of the same design, then it would be appropriate in this case to pass on some of the savings in time. (Indeed, Client A may claim to own the copyright for the drawings unless retaining copyright ownership was specified in the original agreement.)

CASE STUDY 9.6
ALLEGED COLLUSION IN FEE SETTING

STATEMENT OF THE PROBLEM
A large corporation wants to expand its manufacturing facilities and interviews three consulting engineer firms to design and supervise the construction of the new plant. Each consulting engineering firm states in its proposal that fees would follow the schedule proposed by the provincial Association. Later, the corporation decides that it could reduce the cost of the project by conducting some initial studies itself and by providing its own engineers to assist in supervising the construction of the plant. The corporation asks each consulting firm to quote on how much its consulting fees would be reduced if the corporation provides this assistance.

The three consulting firms meet, discuss the corporation's request, and then submit the same amount as a fee reduction. The corporation complains to the provincial Association that the engineers are colluding in their bids and that this is unethical, if not illegal, conduct.

QUESTION
Is it unethical for the three consulting firms to agree on the fee reduction to be allowed for the assistance?

AUTHORS' RECOMMENDED SOLUTION

The contracting procedure recommended by the Association of Consulting Engineers of Canada (ACEC) for engaging consulting engineers is fairly well established, as described in Chapter 5. The procedure provides for competition on the basis of qualifications, experience, scheduling, and service but discourages competition based on price alone. Although competitive bidding is not illegal or unethical, corporations must make the basis for selection clear when the request for proposal is issued.

In this case, the corporation appears to have been following the ACEC procedure in the early stages, and the engineers responded appropriately. It appears that the corporation misunderstood the procedure. Price competition is not part of the ACEC contracting procedure, although fee negotiation is appropriate once a consulting firm has been selected. In this case, the co-operative action by the firms is not unethical. At no time was it made clear that the corporation wanted competition on a fee basis, and the complaint after the fact appears to reveal a misunderstanding on the part of the corporation.

TOPICS FOR STUDY AND DISCUSSION

1. Assume that you are a consulting engineer in a partnership arrangement. Your partner suggests that your business cards and stationery should contain some discreet advertising and suggests that the following be printed on them:

   - a stylish logo
   - your engineering seal, reduced in size to fit
   - the slogan "The best in the business!"

   Which of these advertising components would conform to the constraints on advertising defined in your provincial or territorial code of ethics (or provincial advertising regulation)? Explain your answer with reference to your Act or code of ethics.

2. You have been hired as a machine design consultant to a soap manufacturer to suggest methods of speeding up a liquid detergent production line. During your work, you have access to confidential company documents and discover that the company is adding very small quantities of a known cancer-causing substance to the detergent but is not listing it as an ingredient. You know that the substance has been banned. This confidential information is totally irrelevant to the job you were hired to perform, and you discovered it entirely by chance. Do you have an obligation to act on this information? If so, what action would you take? Explain the reasons for your decision.

3. A rural town in a resort area has been instructed by the provincial government to replace an old wooden bridge for safety reasons. The town council hires a consulting engineer, Ali, to design a concrete bridge to replace the unsafe wooden structure. Because of poor soil conditions, pilings are required, and the resulting design will clearly be very expensive to construct. One of the town councillors discusses the matter with a neighbour, Baker, who is also a consulting engineer and has a summer cottage in the town. Baker suggests that a culvert might serve the same purpose and would be much cheaper than the bridge in view of the soil problems. Since Ali is a concrete specialist and is not capable of carrying out the redesign for the steel culvert, he is paid for his work on the concrete bridge design and replaced by consulting engineer Gambon, who designs a large culvert structure. The culvert is subsequently constructed at a fraction of the predicted cost of the concrete bridge. Ali is disappointed that construction of the concrete bridge did not proceed (and fees for supervising the construction therefore were not paid). Ali alleges that there was unethical conduct by Baker or Gambon or both. In this example, were the actions of any of the engineers unethical? Would the replacement of Ali by Gambon be considered "supplanting" as defined in this chapter?

NOTES

1. D.G. Johnson, "Engineering Ethics," in I.J. Gabelman, ed., *The New Engineer's Guide to Career Growth and Professional Awareness* (Piscataway, NJ: IEEE, 1996), 173. © 1996 IEEE. Reprinted with permission of IEEE.
2. Professional Engineers Act, Regulation 941, Section 75, RSO 1990, c. P.28.
3. Professional Engineers Ontario (PEO), "Advertising," *Guideline to Professional Practice* (Toronto: PEO, 1988, revised 1996), 21. Reprinted with permission of PEO.
4. Quoted in J.M. MacEwing, "Legal Significance of the Engineer's Seal," *Canadian Consulting Engineer* (July/August 1996): 8.
5. Ibid.
6. Association of Professional Engineers, Geologists and Geophysicists of Alberta (APEGGA), "Stamping Policy Undergoes Change," *The Pegg* (November 1995).
7. Professional Engineers Act, Regulation 941, Section 53, RSO 1990, c. P.28.
8. Professional Engineers Ontario (PEO), "Use of the Seal," *Guideline to Professional Practice*, 19. Reprinted with permission of PEO.
9. Professional Engineers Ontario (PEO), "Recommended Confidentiality Agreement," *Guideline to Professional Practice*, 15.
10. Professional Engineers Ontario (PEO), "Rules of Practice," *Guideline to Professional Practice*, 7. Reprinted with permission of PEO.

11. Professional Engineers Ontario (PEO), "Selection of an Engineer," *Guideline to Professional Practice,* 9. Reprinted with permission of PEO.

12. Professional Engineers Ontario (PEO), "Contractual Liability," *Guideline to Professional Practice,* 20. Reprinted with permission of PEO.

13. Association of Consulting Engineers of Canada (ACEC), *Directory of Member Firms* (Ottawa: ACEC, 1996), xxvii. Reprinted with permission of ACEC.

14. International Federation of Consulting Engineers (FIDIC), *FIDIC International Directory of Consulting Engineering, 1997–1998,* Section 5 (Lausanne, Switzerland: FIDIC, 1997–1998), 8. Reprinted with permission of FIDIC.

CHAPTER TEN

The Engineer's Duty to Society and the Environment

This chapter discusses the engineer's duty to society, with special reference to the importance of controlling environmental hazards, and also describes the appropriate action that should be taken in those rare instances where the engineer's duty to society takes precedence over obligations to clients, employers, and colleagues.

THE DUTY TO SOCIETY

Each provincial code of ethics requires engineers and geoscientists to consider their duty to society paramount. That is, they must put the greater good of society ahead of narrow personal gain. The need to invoke this duty probably arises most frequently in the area of protecting the environment. Engineers and geoscientists are, to a great degree, the caretakers of the environment — whether or not they want this role — since many of their activities have an environmental impact. They design the infrastructure of our society: the buildings, bridges, tunnels, and roads that connect us; the water, gas, and electrical networks that support us; and the sewage and waste disposal systems that are critically important to the modern lifestyle. Engineers and geoscientists also discover the resources and design and operate the processing and manufacturing plants upon which our standard of living depends. Decisions made by engineers and geoscientists frequently have a direct effect on the environment and society in general, and these decisions must be made ethically and with consideration of the duty to society.

CANADA'S ENVIRONMENTAL HEALTH

There is no question that engineering, science, and technology have been of immense benefit to Canada and to humanity in general. It is said that medicine has given people health, that the humanities have given people pleasure, and that engineering and technology have given people the time to enjoy both. In fact, most people enjoy a range of activities for personal fulfilment that would have been absolutely incredible only a generation or two ago. And since we are now in the midst of a computer revolution, this trend is likely to continue. However, while engineering, science, and technology have generally been good for society, two obvious environmental problems have accompanied the industrialization of our society:

- **Proliferation of Machine-Made Hazards** Because of the wider use of manufactured goods and the heavier use of public facilities, some risks to the public have actually increased. Even "safe" structures, highways, cars, aircraft, and appliances pose a risk of death or injury simply because we use them more often and more widely than previous generations did. In addition, manufacturing toxic chemicals (such as herbicides and pesticides), generating power by nuclear energy, and similar hazardous processes were unknown 50 years ago. Engineers must be alert to these hazards and must reduce or eliminate them wherever possible.
- **Degradation of the Environment** The lifestyle of industrialized nations requires a high energy usage to maintain it. The consumption of fossil fuels and the careless disposal of waste products have caused a gradual deterioration of the environment. This deterioration is evident in increased water pollution, acid rain, the "greenhouse effect" (global warming), and the growing problem of waste disposal, to mention only a few of the most obvious aspects.

The extent of the problem is evident from the following brief summary of some well-known Canadian environmental tragedies.

The 1970s was a period of awakening to the reality of the costly effects of pollution resulting from industrial by-products. This recognition was based on scientific investigations and frightening, sometimes painful, experience. For example, in northern Ontario hundreds of residents along the English-Wabigoon river system have lost their jobs as commercial fishermen, their community stability and their expectation of a long and healthy life because of invisible, poisonous mercury in the fish. The mercury was dumped as waste material by a local paper mill before 1970, but will remain a danger in the water and fish for another century. In New

Brunswick, although several children have died, residents continue to be exposed to aerial spraying with a pesticide designed to kill forest insects and protect the forest industry. At Port Hope, Ontario, thousands of tonnes of slightly radioactive wastes, dumped years earlier, were removed from beneath houses, schools, and stores to lessen the danger of cancer developing among citizens. In Toronto, public clinics are held to test the blood of residents of some neighbourhoods for low levels of lead dust from nearby factories, which could lead to nervous disorders and learning disabilities.

. . .

In 1981, Canadian scientists discovered, for the first time, very small amounts of one of the most deadly man-made chemicals, Dioxin 2, 3, 7, 8-TCDD, in the Great Lakes. Dioxin is a by-product of agricultural chemical production and the degradation of industrial wastes. As little as a droplet of the pure chemical is deadly in thousands of litres of water. Researchers are only beginning to consider the problems which would be involved in the protection and purification of Great Lakes water for human consumption if Dioxin concentrations increase.[1]

Although many Canadians believed that some of these long-standing problems had improved in recent years, the shocking results of a survey released in July 1997 showed that some provinces were among the worst polluters in North America. The report, *Taking Stock: North American Pollutant Releases and Transfers*,[2] was prepared by the Commission for Environmental Cooperation (CEC), an agency set up by the governments of Canada, Mexico, and the United States as part of the North American Free Trade Agreement. One of the major tasks of CEC is to monitor the release of pollutants into the environment and their transfer to other jurisdictions in North America.

The comparison of Canadian and U.S. pollution, based on 1994 data, showed that Canadian factories, refineries, smelters, pulp operations, and other industries released 2.5 times as much pollution per plant as U.S. facilities. In the ranking by province and state, the top polluters were Texas and Tennessee, but Ontario was third, Quebec was twelfth, and Alberta was seventeenth. Texas has almost twice the number of facilities and reported twice the amount of pollution released as Tennessee and Ontario. Figure 10.1 shows the top twenty polluting provinces and states, and Table 10.1 summarizes the data for the ten provinces.

The CEC pollution data and rankings have been criticized as inaccurate by some groups, since the 1997 report was based on 1994 data, and many industries (particularly pulp and paper industries) have improved their controls and efficiency since that time. Moreover, Mexico was not included in the report because it is in the process of establish-

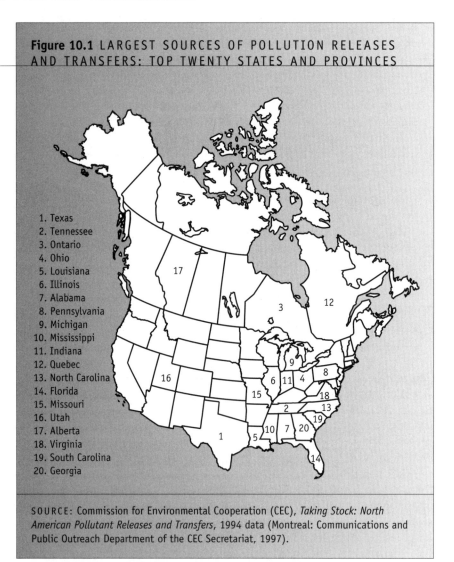

Figure 10.1 LARGEST SOURCES OF POLLUTION RELEASES AND TRANSFERS: TOP TWENTY STATES AND PROVINCES

1. Texas
2. Tennessee
3. Ontario
4. Ohio
5. Louisiana
6. Illinois
7. Alabama
8. Pennsylvania
9. Michigan
10. Mississippi
11. Indiana
12. Quebec
13. North Carolina
14. Florida
15. Missouri
16. Utah
17. Alberta
18. Virginia
19. South Carolina
20. Georgia

SOURCE: Commission for Environmental Cooperation (CEC), *Taking Stock: North American Pollutant Releases and Transfers,* 1994 data (Montreal: Communications and Public Outreach Department of the CEC Secretariat, 1997).

ing reporting procedures, and two large Canadian plants inadvertently submitted data in kilograms that were later assumed to be in tonnes, thus inflating the Quebec and Ontario data slightly. (A later report from CEC said that the errors did not change the report's rankings.) Conversely, U.S. plants are required to report more substances than Canadian plants, and the United States has stricter and more recently developed laws to control pollution.

Table 10.1 RANKING OF PROVINCES IN ORDER OF POLLUTION RELEASES AND TRANSFERS

Province	Population (1994)	Land Area (km²)	No. of Facilities Releasing	Total Releases (kg)	Rank (out of 64 in N.A.)	Rank (per capita)	Rank (per km²)
Ontario	10 928 000	1 068 586	767	78 803 309	3	15	35
Quebec	7 281 000	1 540 689	315	52 809 233	12	13	48
Alberta	2 716 000	661 194	87	30 314 399	17	8	45
British Columbia	3 668 000	947 806	85	7 369 917	43	48	55
New Brunswick	759 000	73 440	20	5 499 023	44	14	34
Nova Scotia	937 000	55 491	18	5 396 854	45	25	32
Manitoba	1 131 000	649 953	37	4 006 097	47	39	57
Saskatchewan	1 016 000	652 334	17	1 258 184	55	54	61
Prince Edward Island	134 000	5 659	2	38 789	61	60	56
Newfoundland	582 000	405 721	3	15 122	62	63	63

SOURCE: Commission for Environmental Cooperation (CEC), *Taking Stock: North American Pollutant Releases and Transfers,* 1994 data (Montreal: Communications and Public Outreach Department of the CEC Secretariat, 1997).

GUIDELINES FOR ENVIRONMENTAL PROTECTION

While the care of our environment is everyone's responsibility, frequently professional engineers, geologists, geoscientists, and geophysicists must assume the key caretaker role because their decisions often have a great impact on the environment. To help engineers in carrying out this responsibility, a guideline on environmental practice has been developed by the Association of Professional Engineers, Geologists and Geophysicists of Alberta (APEGGA),[3] and the basic guideline has also been adopted (with some changes to the explanatory instructions) by Professional Engineers Ontario (PEO).[4] The guideline should be considered a complement to the code of ethics, and it commits professional engineers, geologists, and geophysicists to protecting the environment and safeguarding the well-being of the public. The nine key parts of the guidelines are listed below with permission.[5]

Professional engineers, geologists, and geophysicists:

1. *Education:* Shall develop and maintain a reasonable level of understanding of environmental issues related to their field of expertise.
2. *Interdisciplinary Approach:* Shall use appropriate expertise of specialists in areas where the member's knowledge alone is not adequate to address environmental issues.

3. *Professional Judgement:* Shall apply professional and responsible judgement in their environmental considerations.

4. *Integration:* Shall ensure that environmental planning and management are integrated into all their activities which are likely to have adverse environmental impact.

5. *Environmental Cost Assessment:* Shall include the costs of environmental protection and/or remediation among the essential factors used for evaluating the life-cycle economic viability of projects for which they are responsible.

6. *Pollution Prevention and Waste Management:* Shall recognize the value of waste minimization, and endeavour to implement the elimination and/or reduction of waste at the production source.

7. *Cooperation with Public Authorities:* Shall cooperate with public authorities in an open manner, and strive to respond to environmental concerns in a timely fashion.

8. *Compliance with Legislation:* Shall comply with legislation, and when the benefits to society justify the costs, encourage additional environmental protection.

9. *Cooperation with Others:* Are encouraged to work actively with others to improve environmental understanding and practices.

RECOGNIZING AND REDUCING ENVIRONMENTAL HAZARDS

The above guideline calls on engineers to be alert to environmental hazards and work to stop and reverse the trend toward environmental degradation. As an aid to achieving the first step, education, the following paragraphs contain a brief overview of some environmental challenges. A much more detailed (and mathematical) description and analysis of these problems are given in the comprehensive 1996 text *Environmental Science: The Natural Environment and Human Impact.*[6]

WASTE DISPOSAL

The most common activity that causes degradation of the environment is the indiscriminate disposal of wastes — whether solid, liquid, or gaseous — as a by-product of manufacturing, processing, or construction activities. Controlling waste disposal is usually within the authority and responsibility of an engineer, or should be. Thus, the engineer may be responsible for finding a solution that is least harmful to the environment.

In the early part of the twentieth century, little was known about the insidious effects of heavy metals, dioxin, asbestos, pesticides, and other toxic substances, and people believed that the environment was a vast "sink" that could accept any amount of waste without becoming contaminated. Waste was tipped into dumps as cheaply and quickly as possible, with little regard for its effects on the environment.

However, as we enter the twenty-first century, waste disposal from both industrial and domestic sources is becoming a crisis in many parts of Canada. Small towns are beginning to realize that the "town dump" is a source of disease, a fire hazard, and a danger to groundwater, and dumps are gradually being replaced by closely monitored landfills, where the low-hazard solid waste is covered daily with a layer of soil, thus reducing odour and pest problems. Some landfills are lined with plastic to prevent, or at least reduce, waste fluids from entering the groundwater. Such "sanitary" landfills are a great improvement over dumps.

However, some major cities are unable to find locations for landfills and are incinerating low-hazard waste in specially built facilities. Incineration has many advantages; obviously, the volume of waste is significantly reduced, with the accompanying reduction in land requirements. However, gas and particulate emissions from the incinerators may still be objectionable. Moreover, if chlorine-based organic compounds are burned, trace amounts of toxic chemicals may be dispersed. The best way to solve the waste disposal problem is to reduce waste volume through more efficient use of resources, coupled with the reuse or recycling of waste materials wherever possible. For example, waste automobile tires have recently been used as fuel to heat cement kilns, and even the ash is consumed since it becomes part of the product.[7] Composting, pyrolysis, and density-based separation of organic and nonorganic materials have also been shown to be successful in various applications. More research is needed in this area of reusing and recycling waste materials.

Environmental protection legislation has been passed in most provinces, and recycling programs have been started. These have reduced some of the solid waste, but there is still a risk that some liquid waste — particularly toxic, flammable, explosive, radioactive, or hazardously reactive chemicals — are still being dumped illegally because of the shortage of proper incineration and high-hazard disposal areas. Hazardous liquid waste can pose a serious threat to health if it leaks out of a dump site into the underground water table. In the past, some companies and individuals have disposed of hazardous waste dangerously. The names Love Canal and Minamata are now synonymous with unethical disposal of industrial waste and subsequent human tragedy. In the United States alone, 32000 hazardous waste disposal sites had been identified as of 1991, but very few have been cleaned up.[8] (See Chapter 11 for a case history of Love Canal, Minamata, and other environmental tragedies.) The solution to this problem requires both technical ability and political awareness.

AIR POLLUTION

There are many components to air pollution, but the most publicized are sulphur oxides and nitrogen oxides. Sulphur oxides arise principally from the burning of fossil fuels like coal and petroleum, although many other kinds of industrial activity also produce them. SO_2 is a foul-smelling gas that reacts with oxygen in the atmosphere to form SO_3, which then combines immediately with water to yield sulphuric acid in the form of droplets. The highest SO_2 values have been reported in the northeastern region of North America and in Europe, where large quantities of high-sulphur fossil fuels have been burned. For most large cities, the pollution from SO_2 has improved in recent years, mostly because of the shift from high-sulphur coal to low-sulphur natural gas.

Sulphur oxides are detrimental to plant life, and they corrode metals, discolour fabrics, and cause the deterioration of building materials. Severe plant damage can be observed many miles downwind from certain smelting operations. A combination of sulphur oxides and air particles seems to be especially damaging to human health, partly because of the action of small particles in conveying sulphuric acid into the lungs, and partly because of the chemical role of air particles in converting SO_2 to sulphuric acid. SO_2 is therefore a serious irritant to the lungs, and dramatic episodes such as the "London smogs" have been attributed to the combination of SO_2 with particulates. The London smog of 1952 proved the link between atmospheric pollution in smog and increased mortality.[9] The worst smog in Canada occurred in southern Ontario in 1962 and lasted five days. It is believed to have been a London smog, but was called the "Grey Cup smog" because it caused the 1962 Grey Cup football game to be postponed because of poor visibility.[10]

Even in the absence of sulphur, combustion of fossil fuels causes serious air pollution in urban atmospheres. Exhaust gases usually contain unburned hydrocarbons (HC), carbon monoxide (CO), nitrogen oxides (NO_x), and the "normal" combustion products such as CO_2 and H_2O. In the atmosphere, many of these products react chemically to produce new contaminants. These processes are stimulated by sunlight, and the products are thus given the general name *photochemical oxidants*. Two of the principal photochemical oxidants are ozone (a lung irritant) and peroxyacetyl nitrate (PAN, a lung and eye irritant). Ozone is also continually being created in the atmosphere by natural processes, but not to a degree great enough to constitute a pollution hazard.

Nitrogen oxides are also a problem in air pollution. There are several known oxides of nitrogen, but the important ones from the standpoint of air pollution are nitric oxide (NO) and nitrogen dioxide (NO_2). The term NO_x is commonly used to refer to nitrogen oxides collectively. NO_x is a product of almost any combustion process that uses air, since nitrogen is the chief component of air. To a great degree, the formation of

NO_x is the result of high combustion temperatures, principally from motor vehicles, which account for as much as 50 to 60 percent of NO_x in the atmosphere in industrialized urban areas. The Los Angeles type of photochemical smog is caused mainly by NO emissions from cars; occurs on warm, sunny days when traffic is heavy; and reaches a peak in early afternoon. It therefore differs from the London smog, which forms on cold winter nights as a result of SO_2 produced from coal combustion.[11]

The nitrogen oxides participate actively in photochemical reactions with hydrocarbons, thus helping to produce photochemical smog. NO_2 plays a double role in air pollution as a component in the formation of photochemical smog and as a toxicant in its own right. NO is much less toxic than NO_2, but NO is readily converted into NO_2 in the atmosphere by reacting with oxygen and, in the presence of water, becomes nitric acid, which is extremely toxic to any growing organism. The adverse effects of air pollution on humans and animals include serious lung disorders, reduced oxygen in the blood, eye and skin irritation, and damage to internal organs.[12] Damage to painted surfaces, cars, and buildings is mainly the result of acid rain, which is discussed in detail later in this chapter.

Most provinces and the federal government have Clean Air Acts that specify emission standards and ambient air-quality standards. Air pollution control is mainly a provincial responsibility, although the federal government regulates trains, ships, and gasoline.[13] These government regulations must, of course, be followed in any engineering practice, and where possible, reducing emissions of SO_x and NO_x should always be a prime objective.

ACID RAIN

The problem of acid rain captured Canadians' attention in the 1980s. Both sulphur and nitrogen oxides, as mentioned previously, are implicated, because they form sulphuric and nitric acids in the atmosphere and cause rainfall to become more acidic than otherwise would be the case. Neutral water ideally has a pH of 7.0, but "normal" rainfall in remote areas that are unpolluted has a pH of about 5.6 because of the presence of small amounts of acid of natural origin.[14] Rain is typically called "acid rain" when the pH falls below 5.0, and in many areas of Canada and the northeastern United States, the rain has a pH value as low as 4.0.

When acid rain falls, it harms mainly fish, trees, farms, buildings, and cars, but it occasionally harms human health as well. Aquatic life begins to be affected when the pH falls below 5.0, and most fish are killed when a pH of 4.5 is reached. The result is hundreds of lakes in northeastern North America (and in Scandinavia) that are devoid of fish, and thousands of other lakes that are threatened.[15] Lesions on plants caused by simulated acid rain have been observed when the pH drops lower than 3.4, although subtle effects may be occurring at higher pH levels. Vegetation and humans may both be harmed, because the acidity leaches magnesium, aluminum, and heavy metals out of the soil and concentrates them in drinking water. Fish are particularly susceptible to dissolved aluminum, and this may be a risk to humans, as well. Acid rain does very serious harm to limestone buildings and monuments, since it dissolves the key chemicals in limestone.

The areas with the most acid rain lie downwind from the areas that produce most of the sulphur and nitrogen oxides, so the downwind people obviously feel that the upwind people should take corrective action, which is costly. In 1981, estimates of sulphur oxide emissions were 20 million tonnes in the United States and 4.8 million tonnes in Canada.[16] Sulphur oxides produced from burning coal (especially high-sulphur coal) are the main source of acid rain. These pollutants also result from smelting nickel, lead, and copper ores, which is done extensively in Ontario and several northern U.S. states. A secondary cause of acid rain is NO_x from power stations and automotive emissions, mainly on the U.S. west coast.

Acid rain is an international problem, since pollutants know no borders. In Canada's case, although the heavier flow of pollutants is believed to be from the United States into Canada, because of the greater industrial activity of the States, the smelters in the Sudbury area of Ontario are major sources of the sulphur dioxides that cause acid rain. The Canadian and U.S. federal governments have made agreements to control acid rain, but engineers (particularly those in the power generation and smelting industries) must monitor this problem and work to alleviate it where possible. The most effective way to reduce acid rain is to reduce acid emissions, although low pH levels in lakes can also be reversed by adding lime to neutralize the acid. The remedies are costly, but an early study shows that, in addition to the benefits of a cleaner environment, the economic benefits are about equal to the cost of the controls.[17]

WATER POLLUTION

In a historical sense, some rivers are less polluted now than they were in the nineteenth century, when there was a serious water pollution crisis. For example, in the mid-nineteenth century, 20 000 people in London were killed by cholera. As Donald Carr says, "in the Western world this was the greatest pollution disaster of history."[18] Typhoid and cholera epidemics stemming from water contaminated by sewage were widespread.

Water pollution may occur from at least six sources:

- disease-causing bacteria
- organic waste decaying in the water, reducing the dissolved oxygen content
- fertilizers that stimulate plant growth and also depress oxygen levels
- toxic materials, such as heavy metals and chlorinated hydrocarbons (DDT, PCB)
- acidification, as mentioned earlier
- waste heat, which can also reduce dissolved oxygen levels[19]

Threats of cholera and typhoid are almost unknown in Canada today as a result of sewage treatment and the use of chlorine to kill bacteria in drinking water. Nevertheless, a comprehensive survey of U.S. rivers in the early 1980s showed that we must be vigilant in monitoring water quality. Although improvements had occurred in fecal bacteria and lead concentrations, water quality had deteriorated with respect to nitrate, sodium, chloride, arsenic, and cadmium concentrations. It is believed that the nitrate came mostly from the enormous increase in fertilizer use that occurred during the previous decade and that the sodium and chloride came from the even greater increase in the use of salt on the highways in winter. The arsenic and cadmium apparently came from fossil-fuel combustion products and primary metals manufacturing.[20] The increased presence of nitrate and sodium in the surface waters was especially troubling, because these come mostly from so-called "non-point" sources (i.e., from all over the place) instead of from "point" sources like sewage treatment plants, power plants, and factories. Many strategies have been developed to monitor and control point sources, but reducing pollution from non-point sources requires new approaches.

It is not commonly known that agriculture is one of the biggest polluters and one of the biggest users of water. As water passes across farmland, it picks up pesticide residues and dissolved salts, fertilizers, and nitrogen from animal manure. The pollutants wind up in our surface water. Nutrients such as nitrogen and phosphorous promote the growth of algae, which consumes the dissolved oxygen in the water, making it unlivable for most other aquatic life.[21]

Every summer, many beaches in Canada are closed to swimmers because of high bacteria and fecal content in the water. This is a scandal for a country that boasts the largest endowment of fresh water in the world. Lake Erie has been one success story, however. In the 1970s, it became popular for the news media to say that "Lake Erie is dead" because of heavy algae growth and a major decline in the fish population. The lake was not dead, of course, and in fact has been slowly recovering as a result of water pollution control programs. Full recovery will be in the distant future, however.

Laws to control pollution have been passed at all levels of government, and as we move into the twenty-first century, society will more and more adopt the principle that the polluter pays for cleaning up these problems. Where industries discharge waste into public waterways, the professional engineer has a duty to society to ensure that environmental laws are followed. Techniques for treating pollution are occasionally effective, but these "end of the pipe" remedies are not adequate in most cases. The need for waste disposal must be eliminated by improving the efficiency and cleanliness of production processes. The professional engineer must play a key role in initiating these improvements and getting the problem of water pollution under control.

GLOBAL WARMING AND OZONE DEPLETION

There is evidence that the global mean surface air temperature rose between 0.3°C and 0.6°C in the 100-year period between 1890 and 1990. Although this may seem to be a small change, it is estimated that the end of the last Ice Age was accompanied by a temperature increase of only about 2°C. So, if this trend continues, even a slight warming of the entire earth could cause severe climatic changes. Although the cause of global warming is not known with certainty, evidence points toward the *greenhouse effect*, which is caused by the emission of the "greenhouse" gases: carbon dioxide (CO_2), methane (CH_4), nitrogen dioxide (NO_2), ozone (O_3), and the chlorofluorocarbons ($CFCl_3$ and CF_2Cl_2, called CFCs). Chlorofluorocarbons were first manufactured in great quantities in the 1940s, when wartime factories converted to producing household appliances. Production of CFCs increased because they are superb refrigerant gases (commonly called "freons"). Only recently have we learned that they are much more potent greenhouse gases than CO_2. One molecule of CFCs has the same greenhouse effect as 10 000 molecules of CO_2.

Since the end of the nineteenth century, the amount of CO_2 in the atmosphere has increased by about 23 percent, to 340 parts per million (ppm). The increase is attributed to two factors: the burning of fossil fuels and deforestation in the tropics. Carbon dioxide is vital to the earth's heat balance, because it causes the trapping of a certain amount of heat in the atmosphere, creating a greenhouse effect. CO_2 has a great influence on global temperature: if there were substantially less CO_2 in the atmosphere than there is, very little heat would be retained and the surface of the earth would be coated with ice. Conversely, if the amount of CO_2 were to double, the earth's average temperature might increase by 3 or 4°C, with serious consequences. The doubling time is dependent upon the rate at which we burn fossil fuel, but estimates have ranged from 88 to 220 years.[22] Based on observed trends and computer models of the climatic system, many researchers have declared with confidence that global warming will continue into the twenty-first century. In particular, some of the predictions are as follows:

- Global temperatures in the year 2050 will be higher than at any time in the last 150 000 years.
- The rate of change of global temperature will be larger than any rate seen in the last 10 000 years.
- There will be significant rises in sea levels, leading to flooding of low-lying areas.
- Increased evaporation may cause significant decreases in the levels of lakes and reservoirs.
- Global warming will not be uniform in either space or time. Warming will be more intense over land than over sea, and will be greatest in the northern hemisphere in winter.[23]

The most evident aspect of the greenhouse effect is the rise in sea level. The rise of 15 cm in this century has already caused some beaches on the Atlantic coast to erode at the rate of .9 to 1.5 m (3 to 5 feet) per year.[24] It has been predicted that the sea level in the next 100 years could rise another 8 to 25 cm if the average global temperature increases by 1.5 to 5.4 °C, because of the melting of glaciers. The huge glaciers in Antarctica, which make up 85 percent of the total ice on earth, are not involved in this calculation, because it is believed that they are currently subtracting water from the earth's oceans and thus might remain in balance.[25] The Arctic ice cap is also not involved at all, since it is largely floating; its melting would not substantially change the sea level.

A somewhat more worrisome scenario emerges from the computerized climate models that show global warming would drastically shift existing climatic patterns. The hydrological cycle might be intensified, with higher temperatures and greater rainfall at northern latitudes; simultaneously, rainfall at mid-latitudes might decrease. Although Canada might prosper, the wheat fields of the United States might turn into dust bowls. Those who work with the climate models hasten to remind everyone that these models are not yet very accurate: global warming appears to be predicted with reasonable confidence, but its accompanying regional effects are uncertain. However, the floods in Quebec (1996) and in Manitoba (1997), the ice storm in Ontario and Quebec (1998), and anecdotal evidence of wild and unexpected climatic swings in other parts of the world tend to support the predictions of shifting climate patterns as a result of global warming.

Chlorofluorocarbons play another sinister role in environmental degradation. It has been well established that CFCs are primarily responsible for depleting the ozone layer. Ozone in the stratosphere helps to screen out damaging ultraviolet rays (ozone at ground level is a pollutant and irritant). CFCs combine with ozone to create gaps in the stratospheric ozone layer, thus permitting ultraviolet rays to reach the earth's surface, where they harm plants and cause increased incidences of skin cancer. This was first observed in 1985 by scientists from the British Antarctic Survey, and measurements since then show that ozone depletion is becoming more severe with time.[26] International agreements and, in Canada, provincial laws have been passed to phase out the use of CFCs, which will not only help protect stratospheric ozone but will also lessen the greenhouse effect. Regrettably, a criminal "black market" trading in illicit CFCs has also sprung up, mainly for servicing out-of-date refrigeration plants. This illegal trade is hindering the move toward safer refrigerants.[27]

What should engineers do about all this? As a start, any operation that produces needless carbon dioxide or other greenhouse gases should be discouraged. More importantly, refrigeration lines containing CFCs — which have been used in refrigeration systems for decades — should never be voided into the atmosphere and, of course, any criminal trade in CFCs should be reported to the police.

ENERGY CONSERVATION AND NUCLEAR POWER

During the 1970s, two oil shortages caused long line-ups at gasoline stations. By the 1980s, there was an oil glut. Gasoline became cheaper, and we lost interest in energy conservation. However, as we enter the twenty-first century, analysts warn that a new and different form of energy crisis looms. Experts predict that coal reserves will last another three or four centuries, but oil and natural gas reserves will be severely depleted in the near future. One source predicts that the world's oil reserves will be 90 percent depleted by the year 2020,[28] whereas a more recent prediction quotes the year 2050.[29] None of the alternative energy sources (solar, wind, wave, and geothermal) has made more than a minor addition to the global energy supply. Nuclear fission and nuclear fusion (should it ever be developed) appear to be essential if we are to maintain our present standard of living and extend it to the citizens of the underdeveloped nations.[30] In the period from 1980 to 1989, the proportion of the world's energy needs satisfied by nuclear energy doubled from 2.5 to 5.0 percent.[31]

Atomic Energy of Canada Ltd. (AECL), the designer of the CANDU nuclear reactor, argues that the CANDU is safer and more reliable than the U.S. light-water reactors, since it is fuelled by natural uranium and moderated by heavy water. In spite of recently publicized management problems[32] that have plagued Ontario Hydro, Canada's largest producer of energy using CANDU reactors, the CANDU consistently wins international honours for technical excellence.[33] As a result, Canada delivers more energy from nuclear sources per capita than any other country in the world. Nuclear energy has the advantage of using fuel that is compact and plentiful, although public concerns over operating safety and disposal of radioactive wastes have created public resistance to building new nuclear generating plants. The risk of a meltdown causing massive damage to surrounding residents is small, but this small potential for devastation generates real fear. The design and management of nuclear plants must therefore be done accurately, professionally, and ethically and seen to be so by the general public. AECL and Ontario Hydro must pay specific attention to public fear and work to demonstrate (and improve even further) the safety of the CANDU reactor, since the predicted future depletion of oil and gas reserves may cause a greater dependence on nuclear energy.

Another serious worry connected with nuclear power concerns the long-term disposal of highly radioactive waste. It will be necessary to keep the waste out of circulation for thousands of years because of the extremely long half-lives of some of the elements, such as plutonium. At present, the plan is to store such waste in stable geological underground layers from which water has been absent for millions of years. As with most matters related to nuclear energy, waste disposal is the subject of bitter debate. One issue has to do with the level of certainty regarding the future. Supporters of nuclear energy admit that there can be no absolute guarantee that humans will not be exposed to the waste for thousands of years into the future. However, they say that the risk of future exposure is very small. Opponents of nuclear energy have insisted on a guaranteed method for keeping nuclear waste permanently out of contact. Since they readily agree that no such guarantee can be made, they say that this is reason enough to phase nuclear energy out of existence.

One alternative to nuclear energy is coal-fired power, with its associated problems of air pollution. In the aftermath of the Three Mile Island and Chernobyl disasters, an argument ensued concerning whether coal power or nuclear power is more dangerous. One writer asserted that their dangers are similar, even if one accepts the high estimate of 39 000 future cancer deaths from Chernobyl. He estimated that the death toll from the use of coal in the former USSR is between 5000 and 50 000 per year. Many of these deaths result from mining and transporting coal, because using coal requires handling 100 times as much material as using uranium for an equivalent energy output. In the United States, 100 or more coal miners die each year, and nearly 600 of the 1900 deaths in railroad accidents each year are the result of transporting coal. But the big killer is air pollution, although it is impossible to say with certainty that any given death is specifically caused by coal burning. Nevertheless, it has been estimated that 50 000 people in the United States die each year due to air pollution, mostly resulting from the burning of coal. The estimate of 5000 to 50 000 deaths from burning coal in the former USSR is derived by extrapolating these data, assuming similar populations and similar pollution conditions.[34]

The debate over the safety of our energy sources will not be resolved in this textbook; however, the issue illustrates the importance of professional ethics, conserving energy, increasing efficiency, and avoiding waste. The energy problems of the twenty-first century will be upon us much too soon, and some hard decisions will have to be made. These decisions must not be made in a moral vacuum: ethical decision making must prevail.

EXPONENTIAL POPULATION GROWTH

Population growth is generally viewed as an achievement, not a problem. However, the consumption of nonrenewable resources and the resulting environmental degradation are proportional to the size of the population.

Not only is the world population at a record level, the growth curve is one of never-ending increase. It took all of human history to reach the first 1 billion people in about the year 1800. However, life expectancy rose and infant mortality decreased after the Industrial Revolution as a result of new machines and medicines and improvements in health care, food distribution, and nutrition. The world population passed 2 billion in 1930, 3 billion in 1960, 4 billion in 1975, and 5 billion in 1987. It now stands at 5.6 billion, according to 1994 data.[35] Wars, diseases, and famine have had little effect on the general trend of population increase. As a simple graph of these numbers shows, the population is growing exponentially.

Readers of this text may be dismayed to learn this, since the implications of exponential growth are well known to engineers and geoscientists. Exponential growth cannot be sustained by finite resources, even resources as vast as the earth's. Moreover, the population is increasing at a time when there is a parallel growth in the expectations of people in less-developed countries. Satellite television shows them the conspicuous consumption of the Western world, and it is only human nature to want to share in this wealth of resources. The pressure to consume resources in even greater quantities, with the accompanying degradation of the environment, will reach disastrous proportions during the life span of the current generation of engineers and geoscientists. Obviously, we must improve the efficiency of resource use significantly if we are to avoid — or at least delay — a disaster.

Wildlife and plant populations do not increase in size indefinitely. Eventually, they encounter environmental resistance that places a limit on the size of the population.[36] For example, in animal populations, environmental resistance may take the form of food shortages as the population competes for dwindling prey. Less food leads to increased mortality, particularly infant mortality, until the birth rate, the death rate, and the food supply are in equilibrium. Every species in a given habitat has an equilibrium point for its population. Human population is a little different, since we can devise ways to improve the habitat and thus raise the equilibrium point. However, even for us there are limits since most of our essential resources are finite.

Many population forecasts have been made, using different assumptions. Peak world population is predicted to be reached between the years 2050 and 2100, when an equilibrium point, estimated to be from 10.6 billion[37] to 12 billion[38] people will be reached. The following quote summarizes the problem of population growth concisely, if bleakly.

> [T]he question of how many people the Earth can support is now at the top of the international agenda. In the next 35 years, even if the current trend of declining fertility rates continues, the United Nations forecasts that the Earth's population of 5.7 billion will balloon to nearly nine billion. Ninety-five percent of that growth will take place in the developing world. The upshot, say many experts, is that shantytowns like Nairobi's Mathare Valley and refugee-producing conflicts like the recent slaughter in Rwanda will proliferate — creating a 21st century of growing anarchy, warfare and disease in which masses of Third World migrants will be scrambling to get inside the protected citadel of the industrialized West. If those forecasts prove accurate, Canadians and other citizens of the developed world may face a stark choice: whether to open their borders to millions of new refugees, or to slam the door shut and turn their backs on the spreading misery.

> . . .

> While the steep rise in the world's population in the last half of the 20th century has brought calls for zero, or even negative, population growth, many conservative economists insist that there is no crisis over the Earth's ability to support the expected increase. Nicknamed "cornucopians," they argue that the international market will always find a substitute product or a new technology to circumvent shortages of particular resources. A case in point is copper: in the 1970s, some environmentalists predicted that the metal would be in short supply in the 1990s. Instead, there is a glut of copper and prices have plummeted because fibre-optic cable and plastic piping have replaced copper in many uses.

> As for crowded slums and food shortages in the developing world, the cornucopians point out that couples tend to have fewer children as their incomes rise. Economist Michael Walker of the Fraser Institute, a conservative Vancouver think-tank, says that the key is to increase the productivity of farmers ... by protecting property rights so that farmers can take out loans and invest in tools and crops.

> . . .

But while the general optimism of the cornucopians is comforting, it conflicts with the rough consensus emerging among most demographers, scientists and policy analysts involved in population and resource research. Their view is that a high percentage of the planet's peoples are doomed to live with poverty and violence unless population growth is dramatically reduced. That was the conclusion of a yearlong study by researchers at Cornell University's department of ecology and systematics. Interestingly, their report [released in February 1994] does not point to the depletion of nonrenewable resources like oil as the problem. Rather, they say, the Earth's biosphere can only produce enough renewable resources — food, fresh water and fish — to sustain [only] two billion people at a standard of living equal to that in Europe.

Another study by Cornell's David Pimentel, a professor of insect ecology and agricultural sciences, and Nobel-winning physicist Henry W. Kendall draws on statistics from the United Nations' Food and Agriculture Organization (FAO). Pimentel and Kendall state that even if the United Nations' population target of 7.8 billion were met, world food production would have to triple in the next 55 years for every inhabitant to have an adequate diet. That prospect is at best remote, they add, because less than half of the Earth's land is suitable for agriculture, and almost all of that is already exploited. Moreover, many of the benefits of the Green Revolution, which boosted crop yields with irrigation, fertilizer and pesticides, have already been realized — along with such unwelcome side-effects as nutrient depletion, pollution and water shortages.

Two long-term environmental problems, largely created by the industrialized countries, could also lower crop yields: increased ultraviolet radiation due to the thinning ozone layer and reduced precipitation because of global warming. Said Pimentel: "While the number of mouths to feed has increased, grain production has actually been declining since 1981."

According to yet another study [released in August 1994], by the Washington-based Worldwatch Institute, the solutions invoked by cornucopians are unlikely to stave off disaster. While Western countries helped avert large-scale famines in nations like India in the 1960s with Green Revolution aid programs, there is no new biotechnology or high-yield seed currently in development that will significantly boost world grain harvests. At the same time, marine biologists at the FAO report that all 17 of the world's major ocean fisheries are being fished at, or beyond, capacity. Nine of them are in decline or have been shut down — as in the case of Canada's Atlantic cod fishery.

Perhaps the most provocative research on the consequences of the looming gap between resources and population is being done by Thomas Homer-Dixon, head of the University of Toronto's Peace and Conflict Studies program. Homer-Dixon foresees a 21st century in which overpopulation, unequal distribution of wealth and environmental degradation combine to produce tribal warfare, mass migrations and the breakup of countries around the globe "with a speed, complexity and magnitude unprecedented in history." That bleak scenario may not be farfetched in a world where more than one billion people already go hungry and about two billion lack basics like running water or electricity.[39]

Clearly some action must be taken to avert these crises. Engineers and geoscientists are not in a position to alleviate the problems of overpopulation, or even to affect it significantly. What we can do is to ensure that we are not contributing further to the inefficient use of resources or the creation of environmental hazards. Ethical actions are always in our own interest when we take the long-term view.

EVALUATING RISKS TO SOCIETY

What is best for society? Who should take the benefits of new developments and the hazards that accompany them? The answers to these questions have usually been formulated using the utilitarian principle of creating the maximum good for the maximum number of people. In a report on risk management in the *IEEE Spectrum*, the following observation was made:

> In any applied technology that touches human lives, the decision to accept some level of risk as inevitable calls on subjective judgement about the worth of those lives. The classic, if callous, tradeoff is a cost/benefit analysis of the expense of installing safety systems versus the value of the lives they may save and the political effectiveness of the move. At times the monetary value of a human life is even assigned actuarially, up front, from insurance tables; sometimes it can only be inferred after the fact, from legal settlements for damages claimed.[40]

Therefore, as a general rule, the good of society is determined on a utilitarian basis, weighing the benefits against the disadvantages. When the benefits accrue to a large population, the risks are very small, and the potential damage is not life-threatening, then the good of society is served by encouraging the project to continue. This is true for the vast majority of engineering projects.

In projects where moderate risks exist, the safety of the public can usually be guaranteed simply by using established methods and accepted factors of safety. For a new, untested process, such as a chemical plant or nuclear facility, the potential for disaster must be rigidly controlled, and a "cradle-to-grave" systems approach is needed to ensure that all hazards are considered. The designers must foresee the problems of decommissioning and disposing of the plant, which may be 50 years in the future, as well as the immediate problems of design and construction. The operating hazards must also be considered and controlled. In these cases, sophisticated studies such as failure modes and effects analysis (FMEA), event tree analysis, and fault tree analysis would be required. The designers must examine every conceivable mode of failure, evaluate the probability of it occurring, and devise a remedy for combatting harmful results.

THE ETHICAL DILEMMA OF "WHISTLEBLOWING"

Each code of ethics discussed in this text requires engineers and geoscientists to consider their duty to society to be paramount. However, the codes also stipulate duties to clients, employers, colleagues, and employees. At what point does the duty to society exceed the duty to others, if the activities of others appear to be in serious conflict with the good of society? For example, the engineer is, on the one hand, obligated not to disclose confidential information concerning the client or employer. On the other hand, the engineer must report to the appropriate authority any situation that he or she believes may endanger the health or safety of the public. Failure to correct or report a hazardous situation is considered professional misconduct in every province and territory of Canada. Obviously, there may be occasions when these conflicting obligations create an ethical dilemma for the engineer.

Engineers who have concerns over unethical, unsafe, or illegal practices must first communicate these concerns to the people involved, whether they are clients, employers, colleagues, or employees. In fact, communication is generally all that is required to get most problems resolved. If needed, the provincial Association of Professional Engineers is available to provide advice and perhaps to mediate an exceptionally difficult case. It is therefore very rare for an engineer to resort to reporting to authorities to get action on a problem. This *whistleblowing*, as it is commonly called, is often a controversial act, and it is useful to define the term clearly. A good definition of a whistleblower is given in Connie Mucklestone's article "The Engineer as Public Defender":

> Whistleblowers are people (usually employees) who believe an organization is engaged in unsafe, unethical or illegal practices and go public with their charge, having tried with no success to have the situation corrected through internal channels.[41]

In a survey of 100 Ontario engineers at a seminar on whistleblowing in 1985, about 40 percent stated that they had seen practices during their engineering careers that justified reporting a client or employer, but few had done so. (One might wonder about the subjectivity of these statistics, since failure to report truly justifiable cases is ethically wrong.) Peter Osmond, a former Registrar of Professional Engineers Ontario, stated in 1990 that "the Association receives an average of one whistleblowing inquiry per month, but few are true whistleblowing situations. Most are really complaints of misconduct or incompetence.... Some are false alarms." In Osmond's first five years as Registrar, he dealt with four genuine whistleblowing cases.[42] We may therefore conclude that whistleblowing is occasionally necessary, but rare.

Recognizing that whistleblowing must exist as a last resort, some provincial Associations have made the reporting process more formal.[43] Two important points distinguish a whistleblower from a troublemaker: the motive of the engineer involved and the methods used to achieve the goal of protecting the public. These comments are summarized in the following quote:

> Engineers must act out of a sense of duty, with full knowledge of the effect of their actions, and accept responsibility for their judgement. For this reason any process which involves "leaking" information anonymously is discouraged. There is a basic difference between "leaking" information and "responsible disclosure." The former is essentially furtive, and selfish, with an apparent objective of revenge or embarrassment; the latter is open, personal, conducted with the interest of the public in mind and obviously requires that engineers *put their names on the action and sometimes their jobs on the line.*"[44] [Italics added.]

A whistleblower must also be aware that the process may involve public exposure and scrutiny and may place his or her career in jeopardy. Therefore, whistleblowing should not be done casually, unknowingly, or wantonly. The provincial Association should be contacted and its reporting process should be followed.

THE ENGINEER'S DUTY TO REPORT

Provincial Associations of Professional Engineers have frequently taken on the role of mediators or conciliators to help engineers who perceive their clients, colleagues, employers, or employees to be involved in unsafe, unethical, or illegal practices. The Association can serve a useful role in assisting the engineer to define the ethical issues involved, advising the engineer, communicating the concerns to the client or employer in an unbiased way, and generally helping to resolve the issue as informally as possible.

It is difficult to specify a procedure that will work in every case. Moreover, an engineer must take a much more aggressive approach if human life is at risk. And finally, one must decide whether it is the *situ-*

ation or conditions that must be reported, or the *individual*. Each case will be different.

For example, an engineer who has a serious concern over the ethical practice of a colleague might, as a last resort, contact the provincial Association. However, an urgent concern about an employer who put workers' lives at risk might be immediately referred to the police, particularly if a delay in acting might cause injury or death. Failure to take immediate action to protect human life would be considered professional misconduct. The excerpt that follows, taken from APEGGA's *Manual of Professional Practice under the Code of Ethics*, may give more guidance.

REPORTING UNPROFESSIONAL PRACTICE

Professional engineers, geologists and geophysicists shall advise the Registrar of any practice by a member of the Association that they believe to be contrary to this Code of Ethics.

[This rule is accompanied in the *Manual* by the following commentary:]

Through informal contact, normal working relationships, or special circumstances such as design reviews, one professional may develop the opinion that the work of another professional is deficient. The inadequacies may arise from unskilled practice and/or unprofessional conduct.

While it is not the role of the first professional to conduct a disciplinary investigation, he or she should be certain there is sufficient substance to warrant a serious allegation against a colleague. A professional should carefully consider the necessity and merits of disciplinary action for minor unprofessional inadequacies where protection of the public is not involved. But if it is decided to proceed, as a general rule, the first professional should discuss the situation with the second professional to clarify the facts and check for extenuating circumstances.

Ignoring unprofessional practices, either for expediency or sympathy, may indirectly endanger the public and certainly circumvents the responsibility of self-regulation that has been granted to the profession. Intentionally refraining from reporting substantive breaches of the Code of Ethics on the part of another member of APEGGA therefore constitutes unprofessional conduct.

If the immediate physical safety of the public is in jeopardy, speedy notification of the owner, operator or appropriate regulatory authorities is the immediate duty of the professional. So that a full investigation may either substantiate or dismiss the concern, notification to the Registrar is the professional's next duty. Prompt notification is necessary to prevent potential harm to the public through the continuation of unacceptable engineering, geological or geophysical practices. Professionals have a responsibility to be aware of hazards to society created by their profession, and also have a responsibility to report unethical practice so it may be dealt with through the disciplinary process.[45]

A similar idea is included in almost every code of ethics. Professional Engineers Ontario has defined the procedure for reporting even more clearly in a recent publication, *A Professional Engineer's Duty to Report — Responsible Disclosure of Conditions Affecting Public Safety*. An excerpt follows.

THE REPORTING PROCESS

Engineers are encouraged to raise their concerns internally with their employers or clients in an open and forthright manner before reporting the situation to PEO. Although there may be situations where this is not possible, engineers should first attempt to resolve problems themselves.

1. If resolution as above is not possible, engineers may report situations in writing or by telephone to the Office of the Registrar of PEO. In reporting the situation to PEO, engineers must be prepared to identify themselves and be prepared to stand openly behind their judgements if it becomes necessary.
2. The Office of the Registrar will expect the reporting party to provide the following information:
   a) the name of the engineer who is reporting the situation;
   b) the name(s) of the engineer's client/employer to whom the situation has been reported;
   c) a clear, detailed statement of the engineer's concerns, supported by evidence and the probable consequences if remedial action is not taken;
3. The Office of the Registrar will treat all information, including the reporting engineer's name, as confidential to the fullest extent possible.
4. The Office of the Registrar will confirm the factual nature of the situation and, where the reporting engineer has already contacted the client/employer, obtain an explanation of the situation from the client/employer's point of view.
5. Where the Office of the Registrar has reason to believe that a situation that may endanger the safety or welfare of the public does exist, the Office of the Registrar will take one or more of the following actions:
   a) report the situation to the appropriate municipal, provincial and/or federal authorities;
   b) where necessary, review the situation with one or more independent engineers, to obtain advice as to the potential danger to public safety or welfare and the remedial action to be taken;
   c) request the client/employer to take steps necessary to avoid danger to the public safety or welfare;
   d) take such other action as deemed appropriate under the circumstances;
   e) follow up on the action taken by all parties to confirm that the problem has been resolved.

6. Wherever possible, the Office of the Registrar shall maintain accurate records of all communications with the reporting engineer, any authorities involved and the client/employer.

In Summary: The Office of the Registrar will cooperate with any engineer who reports a situation that the engineer believes may endanger the safety or welfare of the public. Wherever possible, the confidentiality of reporting engineers and the information they disclose will be maintained. The Office of the Registrar will emphasize in all dealings with the engineer's client/employer and the public the engineer's duty to report under the Act and Regulations, and will provide the reporting engineer with an endorsement of the performance of his/her duty, provided that the Registrar has determined that the engineer has acted properly and in good faith.[46]

BEFORE USING THE REPORTING PROCESS — SOME HINTS

Before using either of the two reporting processes described in the previous section, an engineer would be well advised to reflect on the following three points:

- **Informal Resolution** It is extremely important that the engineer strive to resolve problems informally and internally in an open and professional manner. In the vast majority of cases, clear communication is all that is required. While the Association may give guidance and provide mediation, these steps represent a greater degree of formality. The engineer must ensure that an informal internal solution cannot be obtained before resorting to the Association. The engineer must assume the responsibility and consequences of any harm that results from a frivolous accusation.
- **Confidentiality** Even if it should be necessary to report an individual, the report should be made to the appropriate regulating body and not to the news media. The goal is to remedy a problem, not to embarrass individuals. Although full publicity may at some point be necessary, it is not the first step in reporting.
- **Retaliation** In extreme cases where it has been necessary to report an unethical, illegal, or unsafe act to public authorities, an employer may attempt to retaliate by firing the engineer. Engineers in such a position should know that their actions do not constitute just cause for firing, as explained in Chapter 8, and can file a lawsuit to recover lost wages and costs.

A DISSENTING VIEW OF THE ENGINEER'S DUTY TO SOCIETY

It appears to be clear and unequivocal that professional engineers have an obligation to the public. This is stated in the first or second clause of *every* code of ethics: engineers must consider their duty to society to be paramount or most important. However, one well-known expert on engineering ethics, Samuel Florman, spoke against this clause as a general guide because it does not have a precise meaning. His comments are quoted, in part, below.

If this appeal to conscience were to be followed literally, chaos would ensue. Ties of loyalty and discipline would dissolve, and organizations would shatter. Blowing the whistle on one's supervisors would become the norm, instead of a last and desperate resort. It is unthinkable that each engineer determine to his own satisfaction what criteria of safety, for example, should be observed in each problem he encounters. Any product can be made safer at greater cost, but absolute freedom from risk is an illusion. Thus, acceptable standards must be specifically established by code, by regulation, or by law, or where these do not exist, by management decision based upon standards of legal liability. Public-safety policies are determined by legislators, bureaucrats, judges, and juries, in response to facts presented by expert advisers. Many of our legal procedures seem disagreeable, particularly when lives are valued in dollars, but since an approximation of the public will does appear to prevail, I cannot think of a better way to proceed.

. . .

The regulations need not all be legislated, but they must be formally codified. If we are now discovering that there are tens of thousands of potentially dangerous substances in our midst, then they must be tested, the often-confusing results debated, and decisions made by democratically designated authorities — decisions that will be challenged and revised again and again.

. . .

This is an excruciatingly laborious business, but it cannot be avoided by appealing to the good instincts of engineers. If the multitude of new regulations and clumsy bureaucracies has made life difficult for corporate executives, the solution is not in promising to be good and eliminating the controls, but rather in consolidating the controls themselves and making them rational. The world's technological problems cannot even be formulated, much less solved, in terms of ethical rhetoric: especially in engineering, good intentions are a poor substitute for good sense, talent, and hard work.[47]

Florman's comments are thought-provoking and refreshing. However, it must be remembered that they were spoken in the U.S. context, where licensing regulations are somewhat different. Florman's recommendation for developing standards and regulations based on solid research would be useful to the practising engineer, particularly where dangerous chemicals are concerned. However, many well-known regulations and standards are already in print, yet some companies and individuals still do not follow them because of ignorance, inertia, or unethical attitudes. Developing more regulations will not change the attitudes of unethical people, and the professional engineer will still encounter cases where the public good must be weighed against the benefit of client and employer. Rather than discard the public safety clause in the code of ethics, as Florman suggests, there may be a need to provide more mediation between whistleblowers and their employers (as provincial Associations are now doing), and to provide protection against retaliation for engineers who, after exhausting all other routes of action, report unethical practices.

In spite of the apparent contradiction, both the code of ethics and Florman's suggestions are useful. The code of ethics is an ideal, while the regulations and standards proposed by Florman are a way of making that ideal more attainable.

TOPICS FOR STUDY AND DISCUSSION

1. Assume that you have graduated from university and have been working for three years as a plant design and maintenance engineer for a pulp and paper company in Northern Canada. The company is a wholly owned subsidiary of a large multinational conglomerate. When you received your P.Eng. licence, you were promoted to Chief Plant Engineer, and you work directly for the Plant Manager, François Bédard, who reports to the head office, which is not in Canada. The company employs about 150 people — most of the adult population of the nearby village — either directly as employees, or indirectly as woodcutters.

   In the course of your work, you have become aware that the plant effluent contains a very high concentration of a mercury compound that could be dangerous. In fact, since the plant has been discharging this material for 25 years, water in the nearby river is thoroughly unfit for drinking or swimming downstream from the plant. You suspect that a curious new illness in an Aboriginal village about 40 km downstream is really Minamata disease, the classic symptoms of which are loss of co-ordination, spastic muscle movement, and, eventually, death. You also suspect that the fish in the river are contaminated with the mercury and have spread the contamination to all the downstream lakes.

Remedying these problems would involve drastic changes to the plant and would cost at least one million dollars. At present, no one knows of your suspicions, except you and Bédard, with whom you have discussed the problem at length. Bédard is not an engineer, and has confided that the head office considers the plant only marginally profitable and an expenditure of this magnitude is simply not possible. The head office, he says, would close down the plant, causing massive unemployment in the area and probably forcing the workers to abandon their homes to seek work elsewhere. What should you do?

2. The greenhouse effect and ozone depletion are human-made problems, since the carbon dioxide and chlorofluorocarbon gases that cause the problems are mainly the result of human activity. It has been proposed that human-made solutions could reverse these problems. For example, it has been suggested that large quantities of propane released from aircraft at high levels would combine with the chlorine causing the depletion of the ozone, turning it into a harmless salt (HCL). Another plan proposes to reduce the amount of carbon dioxide in the air through photosynthesis, a chemical reaction in which water and carbon dioxide combine to form glucose and oxygen (photosynthesis is critically important in restoring oxygen to the atmosphere). Stimulating the growth of algae in the ocean would, through photosynthesis, simultaneously reduce greenhouse gases and increase the oxygen content of the atmosphere.

Using the information resources available to you, investigate and evaluate these proposals and any others you may discover. Prepare a brief summary that answers the following questions:

- Are these schemes chemically feasible?
- How much material would be required to stop the current trends?
- How much material would be required to reverse the trends and restore appropriate levels?
- How would the processes work in practice?
- How much would they cost?
- Who should pay this cost?
- Are there any foreseeable side effects?
- How could the proposals be tested before making a full commitment to apply them?

3. Population growth and the depletion of gas and oil reserves will put civilization on a collision course at some time in the future. Various sources put this date between 2020 and 2050, well within the professional lives of the engineers and geoscientists reading this text. Using the information resources available to you, investigate and evaluate these predictions. Prepare a brief summary of your evaluation of the problem and which alternative energy sources (solar, wind, wave, geothermal, fission, fusion, etc.) are the best bets for replacing gas and oil. Is it likely that we will be able to maintain our standard of living well into the future?

4. Between 40 000 and 50 000 people are killed every year in car accidents in North America, yet people apparently consider the benefits of the car to be worth the risk, in spite of the dreadful toll. Less than ten people have been killed in North America in nuclear reactor accidents, yet many people are frightened by nuclear power. Many people die every year in the process of producing food (even farming is dangerous), and thousands of coal miners have been killed in the twentieth century. This discrepancy between the perception of danger and the reality of the safety statistics occasionally influences public acceptance of engineering projects. Using the information resources available to you, answer the following questions:

   a. Examine the risks associated with the various energy sources (solar, wind, wave, geothermal, fission, fusion, etc.), and develop a fair method of comparing the risks and benefits of each. That is, find the statistics for the probability of injury or death per unit of energy produced. Compare this with automobile travel on the basis of risk per unit of energy consumed.

   b. Using the concepts contained in this chapter and in Chapter 6, state, in one or two pages, an ethical guideline for deciding when construction of a dangerous facility, such as a nuclear power plant, or production of a dangerous chemical, such as a pesticide, is morally justified. Include financial, engineering, or political arguments in your answer, as well as ethical concepts.

5. The North American lifestyle involves the largest consumption of energy and resources per capita in the world. Small changes in this lifestyle could reduce consumption significantly but are resisted by the general public for no apparent reason. The following is a brief list. Can you add to the list? How would you convince the general public to "do the right thing" in each of these cases?

   a. Although "blue box" recycling programs exist in many cities, some residents still insist on discarding bottles, cans, and plastics with garbage, thus increasing landfill needs.

b. Many homes have the cellar drainage sump or storm drains connected to the septic sewage system. Rainwater becomes polluted when mixed with the sewage, and the sewage treatment plant must process the otherwise clean rainwater along with the sewage. When there is a serious rain storm, the sewage plant may not be able to cope with the flow, and the overflow, polluted with sewage, is usually released into a stream or lake, fouling the environment.

c. Some car owners who change their own oil dump the used oil into a storm drain or septic sewer, rather than taking it to a gas station for recycling. In either case, a small amount of oil has a seriously harmful effect on the environment, and eventually reaches the local water table.

d. Many trailers and recreational vehicles have self-contained toilets that must be emptied on a regular basis. Some people who use these vehicles pollute the environment by dumping the toilet contents in parks, fields, or storm sewer systems, rather than in the septic systems that lead to sewage treatment plants.

6. Southern California is a desert.[48] It has a large population, has been using fossil water for years, and implements harsher water control programs every summer. Three possible solutions to this water shortage problem include

- buying water from areas with excess freshwater, such as B.C. or the Great Lakes area
- building seawater reclamation plants
- towing Antarctic icebergs north and melting them for their freshwater

Southern California also houses a large, active, and well-funded group of people who are hostile toward industry and engineering projects. They frequently contribute to organizations dedicated to stopping all industrial activity, such as Greenpeace, twice elected a governor (Jerry Brown) who favoured complete abandonment of all nuclear energy programs, and make movies in which engineers, scientists, and businesspeople are portrayed as evil individuals who are out to destroy the world for fun and profit (e.g., *China Syndrome, Dr. Strangelove, Medicine Man*).

a. Discuss the roles or responsibilities of engineers in implementing each of the three above-mentioned solutions to California's water shortage problem.

b. Name a movie made in the last twenty years in which engineers have been portrayed positively.

c. Why should engineers continue to undertake projects that will allow people in Southern California to maintain their swimming pools and household running water? Be specific and persuasive, but be logical.

7. Antarctica has never been opened to mining and oil exploration. Until recently, there was no point in doing so. However, it is believed that substantial resources exist there, and that expertise gained in Arctic environments has made Antarctic resource exploitation feasible. At the same time, known natural resources continue to be depleted, thus making recovery of Antarctic resources more economically attractive. However, Greenpeace is lobbying the world's governments to declare the whole continent a world park, off-limits to any human activity except controlled tourism and scientific research, for the next 30 to 50 years.

   a. Discuss the positive and negative aspects of Greenpeace's world park proposal. Evaluate each of these positive and negative aspects from the viewpoint of the following:

   - national governments that claim some Antarctic territory (Argentina, Chile, and Britain)
   - national governments that are not likely to claim Antarctic territory (most of the rest of the world)
   - an unemployed Chilean mining engineer
   - a Greenpeace member

   b. Who pays and who gains if the entire Antarctic continent is made off-limits to the resource development?

8. China's population of 1.2 billion people (as of 1994) exists on a land area that is slightly smaller than Canada, which has approximately 30 million people. Japan has a population of over 150 million people on a land area that is approximately the size of Newfoundland. Using the information resources available to you, investigate and answer the following questions, explaining the basis for each of your answers.

   a. China has implemented a severe "one-child" program to limit its birth rate. Couples who agree to have only one child receive free health care for the birth and for the child, whereas couples who do not agree must pay their own expenses. Under this policy, the birth rate declined to about 1.2 percent in 1981, but by 1991, it had risen to 1.4 percent. As well, there is some evidence that people are being forced to comply with the one-child policy. For example, more unwanted children are ending up in orphanages. Discuss the ethics of government policies that limit birth rates, using the ethical theories described in Chapter 6. Which should take precedence: personal freedom to bear children, or the duty-based concept that, in cases like this, everyone must share the responsibility and limit growth? How would you evaluate the greatest good for the greatest number? Is there a "virtue," mean, or compromise that can be identified in this case? Can you support the Chinese policy on an ethical basis?

b. Satellite television, recently available in many less-developed countries, shows the disparity in consumption between rich and poor countries. People throughout the world have begun to expect improved standards of living. Even if China is able to control population growth, it is likely to experience an increase in consumption patterns such as that now seen in South Korea and Singapore. What would be the effect on global warming if China consumed resources at the same rate as South Korea or Canada? How can the world meet the resource needs of China's existing population?

c. If you were a Japanese engineer, would you view population growth in your country as a desirable occurrence that will stimulate consumption of goods and services, thus causing economic growth and creating jobs for engineers, or as an undesirable occurrence that will make the Japanese islands too crowded for anyone to establish a high standard of living?

d. Should Canadian government policy encourage immigration? Should it encourage a high birth rate?

9. In the industrialized world, and especially in Western Europe and North America, considerable time and effort are spent on energy conservation programs aimed at getting people to drive less so that they will conserve oil. Canada has a potentially huge oil industry in tar sands recovery, and a large agricultural sector that will need new customers for grain when Russia and Ukraine eventually become stable and self-sufficient. One potential use for corn and canola is in the manufacture of synthetic lubricants and fuels. Canada also has considerable uranium and coal reserves. Using the information resources available to you, investigate and answer the following questions:

a. Do energy conservation programs in the Western countries that are aimed at getting people to drive less actually reduce worldwide consumption of oil?

b. Should Canada, through its tax policies, encourage conservation or faster consumption of foreign (Middle Eastern and South American) fossil fuels? What would be the effect of Canada totally eliminating all taxes on domestic fossil fuels? What would be the effect of Canada totally eliminating all taxes, including import tariffs, on all fossil fuels?

c. The developing countries are dismayed at suggestions that they should spend money on energy conservation and pollution control when developed countries were not subject to these controls until recently, and even now exhibit conspicuous overconsumption of resources. If you were an engineer in a developing country, what energy conservation and energy tax policies would you recommend that your country adopt?

NOTES

1. R. Howard, "Pollution," *The Canadian Encyclopedia*, 1st ed. (Toronto: McClelland & Stewart, 1985), 1448. Used by permission, McClelland & Stewart, Inc., The Canadian Publishers.

2. Commission for Environmental Cooperation (CEC), *Taking Stock: North American Pollutant Releases and Transfers* (Montreal: Communications and Public Outreach Department of the CEC Secretariat, July 1997).

3. Association of Professional Engineers, Geologists and Geophysicists of Alberta (APEGGA), *Environmental Practice: A Guideline* (Edmonton: APEGGA, 15 June 1994).

4. Professional Engineers Ontario (PEO), *Guideline to Professional Practice* (Toronto: PEO, 1988, revised 1996).

5. APEGGA, *Environmental Practice*, 6. Reprinted with permission of APEGGA.

6. R.W. Jackson and J.M. Jackson, *Environmental Science: The Natural Environment and Human Impact* (Harlow, Essex: Longman, 1996).

7. Ibid., 330.

8. Ibid., 334.

9. Ibid., 311.

10. R.E. Munn, "Air Pollution," *The Canadian Encyclopedia*, 32.

11. Jackson and Jackson, *Environmental Science*, 312.

12. Munn, "Air Pollution," 32.

13. Ibid.

14. Jackson and Jackson, *Environmental Science*, 315.

15. H.L. Ferguson, "Acid Rain," *The Canadian Encyclopedia*, 6.

16. Ibid.

17. Ibid.

18. D.E. Carr, *Death of Sweet Waters* (New York: Berkley, 1971), 41.

19. A.H.J. Dorcey, "Water Pollution," *The Canadian Encyclopedia*, 1923.

20. R.A. Smith, R.B. Alexander, and M.G. Wolman, "Water-Quality Trends in Nation's Rivers," *Science* (27 March 1987): 1607–15.

21. "Rescuing a Protein Factory," *Time* (23 July 1984): 84–85.

22. P.H. Abelson (editorial), "Carbon Dioxide Emissions," *Science* (25 November 1983).

23. Jackson and Jackson, *Environmental Science*, 318.

24. "The Politics of Climate," *EPRI Journal* (June 1988): 4–15.

25. M.F. Meier, "Contributions of Small Glaciers to Global Sea Level," *Science* (21 December 1984): 1418–21; *Carbon Dioxide and Climate: A Second Assessment* (Washington, DC: National Academy Press, 1982).

26. Jackson and Jackson, *Environmental Science*, 320.

27. M. Smith and M. Vincent, "Tanking a Killer Coolant," *Canadian Geographic* (September/October 1997): 40–44.

28. E. Titterton, "Nuclear Energy: An Overview," in H.D. Sharma, ed., *Energy Alternatives: Benefits and Risks* (Waterloo, ON: University of Waterloo, 1990), 146.

29. Jackson and Jackson, *Environmental Science*, 249.

30. Titterton, "Nuclear Energy," in Sharma, ed., *Energy Alternatives*, 146.

31. Jackson and Jackson, *Environmental Science*, 257.

32. S. Josey, "Why Hydro Failed," *The Toronto Star* (24 August 1997): 1.

33. R.D. Bott, "Nuclear Safety," *The Canadian Encyclopedia*, 1302.

34. "Letters: Chernobyl Public Health Effects," *Science* (2 October 1987): 10–11.

35. Jackson and Jackson, *Environmental Science*, 140.

36. Ibid.

37. "End of the Population Explosion?" *Discover* (July 1997): 14.

38. Jackson and Jackson, *Environmental Science*, 163.

39. P. Kaihla, C. Erasmus, J. Edlin, and B. Bethune, "Apocalypse When?: A United Nations Plan to Limit Global Population Growth Triggers an Acrid War of Words," *Maclean's* (5 September 1994): 22. Reprinted with permission.

40. T.E. Bell, "Managing Risk in Large Complex Systems," *IEEE Spectrum* 26, no. 6 (19 June 1989): 22.

41. C. Mucklestone, "The Engineer as Public Defender," *Engineering Dimensions* 11, no. 2 (March/April 1990): 29.

42. Ibid.

43. APEGGA, *Manual of Professional Practice under the Code of Ethics* (Edmonton: APEGGA, July 1990); PEO, *A Professional Engineer's Duty to Report: Responsible Disclosure of Conditions Affecting Public Safety* (Toronto: PEO, 1996).

44. PEO, *A Professional Engineer's Duty to Report*, 2. Reprinted with permission of PEO.

45. APEGGA, *Manual of Professional Practice under the Code of Ethics*, 4-28–4-29. Reprinted with permission of APEGGA.

46. PEO, *A Professional Engineer's Duty to Report*, 3–4. Reprinted with permission of PEO.

47. Samuel C. Florman, "Moral Blueprints," *Harper's Magazine* (October 1978). Copyright © 1978 by *Harper's Magazine*. All rights reserved. Reproduced from the October issue by special permission.

48. Problems 6 to 9 are adapted from exam papers by Dr. Jerry M. Whiting (formerly Professor Emeritus of Mining, Metallurgical and Petroleum Engineering, University of Alberta). Used with permission of Dr. Jerry M. Whiting.

CHAPTER ELEVEN

Engineering
Case Histories

As Canadians, we have many spectacular engineering achievements of which we can be proud, from the construction of the railway across Canada in 1885, to the design of the Avro Arrow in 1958, to the opening of the Confederation Bridge in 1997. In fact, we tend to take engineering success for granted when well-designed structures and devices work properly. However, when engineering structures fail, we are forced to focus our attention on the failure. We ask why it happened and how similar failures can be avoided in future. If the failure is particularly costly, in lives or money, an investigation panel or Royal Commission may be formed to examine it impartially and publicly. As a result, we often learn more from failures than from successes, and this chapter contains nine case histories that teach important lessons that were learned at great cost.

In reading the following case histories, it should be remembered that failure, itself, is not necessarily evidence of unethical or incompetent practice. Many engineering projects push the limits of knowledge. Novel or experimental projects always contain an element of risk, and even the most determined and ethical practitioners cannot guarantee success every time. Some failures must simply be accepted. Moreover, it should be emphasized that not every case of whistleblowing is justified; many cases have come to light in which the whistleblower was uninformed about the full situation and overreacted.

However, the nine engineering case histories that follow are historical summaries of avoidable incidents involving engineers and/or geoscientists, and were chosen because they involve an ethical aspect, such as negligence, incompetence, conflict of interest, corrupt practices, or the need to report such practices. Each case is fairly well known. Most of them concern events that took place in Canada, and each had an impact on the engineering profession to some degree. These case histories all ended in tragedy, but if we can learn how to avoid similar tragedies in future, these incidents will have served a useful purpose.

CASE HISTORY 11.1
THE QUEBEC BRIDGE DISASTERS

INTRODUCTION

The Quebec Bridge, opened officially in 1919, is the longest cantilever span in the world, with a centre-distance between supports of 549 m

▼ *A view of the pier and main truss pin of the completed Quebec Bridge. The sentry on duty during World War I shows the magnitude of the components.* Source: The Quebec Bridge over the St. Lawrence River near the City of Quebec: Report of the Government Board of Engineers, *Department of Railways and Canals Canada, 1919.*

(1800 ft). Although there are suspension bridges with longer spans, the massive size of the Quebec Bridge and the length of its span make it a very impressive structure. In fact, one must see it in person to fully comprehend the grandeur of the bridge and the achievement that it represents. However, the bridge is infamous for the harrowing accidents and great loss of life that occurred during its construction. *The Canadian Encyclopedia Plus* summarizes these tragic losses succinctly:

> **Quebec Bridge Disasters:** Construction on the Quebec Bridge, 11 km above Quebec City, officially began in 1900. On 29 Aug 1907, when the bridge was nearly finished, the southern cantilever span twisted and fell 46 m into the St Lawrence R. Seventy-five workmen, many of them Caughnawaga Indians, were killed in Canada's worst bridge disaster. An inquiry established that the accident had been caused by faulty design and inadequate engineering supervision. Work was resumed, but on 11 Sept 1916 a new centre span being hoisted into position fell into the river, killing 13 men. The bridge was completed in 1917 and the Prince of Wales (later Edward VIII) officially opened it 22 Aug 1919.[1]

HISTORY

The construction of a bridge over the St. Lawrence River was advocated by the residents of Quebec as early as 1852, and a site had been chosen at a narrowing of the river just upstream from the city. Designs were prepared, but no serious work was done until 1900. The success of the cantilevered Forth Bridge, built in 1890 in Scotland, was a factor in choosing the cantilever design, and there are some similarities in the general shape of the two bridges. The Forth Bridge, the first bridge built entirely of steel, has two spans of 521 m (1710 ft) each. They were the world's longest unsupported (cantilevered) bridge spans at that time; they remained so for 28 years until the Quebec Bridge was successfully completed.

At the time of the 1907 accident, four parties were directly involved in constructing the Quebec Bridge superstructure:

- the government of Canada, which had provided subsidies and a guarantee of bonds to
- the Quebec Bridge & Railway Company (known simply as the "Quebec Bridge Company"), which had responsibility for the complete structure, and had contracted with
- the Phoenix Bridge Company in Phoenixville, Pennsylvania, to design and construct the superstructure, and which, in turn, had subcontracted with the
- the Phoenix Iron Company, to fabricate the steel components

The Quebec Bridge Company employed a chief engineer, Edward Hoare, who was on site, and a consulting engineer, Theodore Cooper of New York, as well as many hundreds of erection and inspection staff. Cooper, the consulting engineer, was technically very competent: "In the extent of his experience and in reputation for integrity, professional judgement and acumen, Mr. Cooper had few equals on this continent...." In the early stages of the design, it was decided, at Cooper's insistence, that his decisions on technical matters would be final, and this authority was provided in written form as a government order-in-council.[2] However, although Cooper had ultimate design authority, he visited the Quebec site only when the supporting piers were being built and was never on site thereafter; he visited the Phoenix Iron Company shops only three times during the fabrication of the bridge components.[3]

Norman McLure was the Quebec Bridge Company's inspector of erection, appointed by Cooper with Hoare's agreement, and received instructions from both of them. He reported to Hoare mainly on "matters regarding monthly estimates, and to Cooper on matters of construction."[4]

The Phoenix Bridge Company's chief engineer was Mr. Deans, who was an experienced bridge builder but was more accurately described, after the accident, as its "chief business manager."[5] The design engineer was Mr. Szlapka, a German-educated engineer with 27 years of experience in designing many similar projects. Szlapka was responsible for generating the design details, and had the full confidence of Cooper.[6]

INITIAL CONSTRUCTION AND DISASTER

A competition for the design was held in 1898. Cooper reviewed the submitted plans and recommended the Phoenix Bridge Company's design, which showed a span of about 488 m (1600 ft) between the supporting piers. The contracts for detailed design and construction were signed, and work began in 1899. Cooper requested further investigation of the river bed to ascertain the best location for the supporting piers and, after considerable study, recommended that the piers be located closer to shore, thus increasing the unsupported span to 549 m (1800 ft). The work was slow at first, since there was some uncertainty about the financial status of the Quebec Bridge Company and its ability to pay its contractors. Government support was assured in 1903, and completion was urged before the Quebec Tercentennial in 1908.[7] The fabrication and erection of the superstructure proceeded fairly rapidly thereafter.

However, as the first span of the cantilever reached out over the water in 1907, unexpected deformation of some members was clearly evident, and the concerns were communicated to Cooper in New York. H. Petroski summarizes these fatal days concisely in his very readable book *Engineers of Dreams:*

▲ *The Quebec Bridge (Phoenix design) immediately before the collapse on 29 August 1907. Source:* The Quebec Bridge, *Department of Railways and Canals Canada, 1919.*

The south arm of the Quebec Bridge had been cantilevered out about six hundred feet over the St. Lawrence River by early August 1907, when it was discovered that the ends of pieces of steel which had been joined together were bent. Cooper was notified, by letter, by Norman R. McLure, a 1904 Princeton graduate who was "a technical man" in charge of inspecting the bridge work as it proceeded, who suggested some corrective measures. Cooper sent back a telegram rejecting the proposed procedure and asking how the bends had occurred. Over the next three weeks, in a series of letters back and forth among Cooper, chief engineer Deans, and McLure, Cooper repeatedly sought to understand how the steel had gotten bent, and rejected explanation after explanation put forth by his colleagues. Cooper alone seems to have been seriously concerned about the matter until the morning of August 27 when McLure reported that he had become aware of additional bending of other chords in the trusswork and, since "it looked like a serious matter," had the bends measured; he explained that erection of additional steel had been suspended until Cooper and the bridge company could evaluate the situation.

Yet, even as McLure went to New York to discuss the matter with Cooper, Hoare, as chief engineer of the Quebec Bridge Company, had authorized resumption of work on the great cantilever. As soon as McLure and Cooper had discussed the bent chords, Cooper wired Phoenixville: "Add no more load to bridge till after due consideration of facts." McLure had reported that work had already been suspended, and so contacting

Quebec more directly was not believed to be urgent, but when McLure went on to Phoenixville, he found that the construction had in fact been resumed. Some conflicting reports followed, thanks in part to a telegraph strike then in progress, as to whether Cooper's telegram was delivered and read in time for Phoenixville to alert Quebec.

In any event, the crucial telegram lay either undelivered or unread as the whistle blew to end the day's work at 5:30 P.M. on August 29, 1907. According to one report, ninety-two men were on the cantilever arm at that time, and when "a grinding sound" was heard, they turned to see what was happening. "The bridge is falling," came the cry, and the workmen rushed shoreward amid the sound of "snapping girders and cables booming like a crash of artillery." Only a few men reached safety; about seventy-five were crushed, trapped, or drowned in the water, surrounded by twisted steel. The death toll might also have included those on the steamer Glenmont, had it not just cleared the bridge when the first steel fell. Boats were lowered at once from the Glenmont to look for survivors, but there were none to be found in the water. Because of the depth of the river at the site, which allowed ocean liners to pass, and which had demanded so ambitious a bridge in the first place, the debris sank out of sight, and "a few floating timbers and the broken strands of the bridge toward the ... shore were the only signs that anything unusual had happened." The crash of the uncompleted bridge "was plainly heard in Quebec," and the event literally "shook the whole countryside so that the inhabitants rushed out of their houses, thinking that an earthquake had occurred." In the dark that evening, the groans of a few men trapped under the shoreward steel could be heard, but little could be done to help them until daylight. . . .[8]

THE REPORT OF THE ROYAL COMMISSION

Within hours of the accident, a Royal Commission was convened to determine the cause. The Commission's report is a thorough document containing lessons learned at great cost, but which have benefited structural engineers and bridge designers in Canada and around the world.[9] As G.H. Duggan later wrote:

> The report of the Royal Commission appointed to investigate the failure of the Phoenix[-designed] Bridge in 1907 is very comprehensive, and goes beyond the mere taking of evidence and the investigation of the faults of the bridge, as the Commission assembled most of the available data on other long span bridges, illustrated their important features, recorded the tests on large size compression members that had any bearing upon the work, and made a number of tests to supply some lacking experimental data of the behavior of large compression members under stress.[10]

▲ *The wreckage of the Quebec Bridge (Phoenix design) after the collapse on 29 August 1907. Source: The Quebec Bridge, Department of Railways and Canals Canada, 1919.*

The report concluded that Hoare's appointment as chief engineer of the Quebec Bridge Company was a mistake. Although he had a "reputation for integrity, good judgement and devotion to duty," he was not technically competent to control the work. The report stated: "[I]t must be recognized that in many cases, good executive ability is valued more highly or considered of more importance than special professional knowledge." Concerning Deans, chief engineer of the Phoenix Bridge Company, the report concluded that his "actions in the month of August, 1907, and his judgement ... were lacking in caution, and show a failure to appreciate emergencies that arose." However, the major cause of the bridge failure was attributed by the commission to errors in judgement on the part of Cooper and Szlapka.[11]

DESIGN AND COMMUNICATION DEFICIENCIES

The Commission report identified several serious deficiencies in the design and in the construction methods. The main deficiencies were that the design loads were underestimated, and very high unit stresses were known to exist in the compression members. The Commission also criticized the curvature and the "splicing and lacing" (joining and cross-bracing) of the bottom chords. As in any cantilever beam loaded downward at the end, the bottom chords are in compression, and the curvature and poor-quality connections reduced the ability of the chords to resist buckling. The compressive chords had been designed by Szlapka and examined and approved by Cooper.

In particular, the stresses were calculated by Szlapka based on an estimate of the total dead weight of the bridge made by Cooper at the start of the design process. However, as the detail design progressed and the precise shapes of the members were determined, the dead weight changed. The stresses should have been recalculated using more accurate estimates of the dead weight. This was not done. It is particularly important to note that the bridge span had originally been specified as 488 m (1600 ft), but Cooper recommended the placement of supporting piers that increased the span to 549 m (1800 ft). When the bridge span was increased, the dead weight increased significantly, but the increase was not included in the calculations. This point is explained more clearly by Petroski:

> In short, what Szlapka had done was to let stand an educated guess as to the weight of steel that the finished bridge would contain. Such guesses, guided by experience and judgment, are the only way to begin to design a new structure, for without information on the weight of the structure, the load that the members themselves must support cannot be fully known. When the loadings are assumed, the sizes of the various parts of the bridge can be calculated, and then their weight can be added up to check the original assumption. For an experienced engineer designing a conventional structure, a final calculation of weights only serves to confirm the educated guess, and so such a calculation may not even be made in any great detail. In the case of a bridge of new and unrealized proportions, however, there is little experience to provide guidance in guessing the weight accurately in the first place; a recalculation, or a series of iterated recalculations, is necessary to gain confidence in the design.... According to the findings of the commission, "the failure to make the necessary recomputations can be attributed in part to the pressure of work in the designing offices and to the confidence of Mr. Szlapka in the correctness of his assumed dead load concentrations. Mr. Cooper shared this confidence." Since Cooper was well known to have a "faculty of direct and unsparing criticism," his confidence in Szlapka's design work went unquestioned.

Just as Cooper had confidence in Szlapka's work, so did the resident engineer at the construction site have confidence in the work of them both. When a construction foreman expressed serious concern over the condition of the fatal member, the resident engineer thought the matter of little importance, telling the foreman, "Why, if you condemn that member, you condemn the whole bridge." After the collapse, it was reported that the resident engineer "had confidence in that failing chord because it was to him unbelievable that any mistake could have been made in the design and fabrication of the huge structure over which able engineers had toiled for so many years." . . .

The underestimation of the true weight of the bridge had actually come to Cooper's attention earlier in the design process, but only after considerable material had been fabricated and construction had begun. At this time, a recalculation of the stresses in the bridge led Cooper to consider that the error had meant that some stresses had been underestimated by 7–10 percent. All structures are designed with a certain margin of safety; he felt the error had reduced that margin to a small but acceptable limit, and so the work was allowed to proceed. In fact, some of the effects of the underestimated weights were, in the final analysis, of the order of 20 percent, and this was beyond the margin of error that the structure could tolerate.[12]

Other points were incidental to the main cause of the failure but might have prevented the tragedy or lessened its consequences. Cooper's poor health and age prevented him from visiting the construction site during the last two years of construction, and he was also criticized by Szlapka for making the bottom chords curved "for artistic reasons" and for failing to visit the Phoenixville plant where the bridge parts were fabricated. The Royal Commission report commented on Cooper's role, and on the design deficiencies and communication problems, as follows:

Mr. Cooper states that he greatly desired to build this bridge as his final work, and he gave it careful attention. His professional standing was so high that his appointment left no further anxiety about the outcome in the minds of all most closely concerned. As the event proved, his connection with the work produced in general a false feeling of security. His approval of any plan was considered by every one to be final, and he has accepted absolute responsibility for the two great engineering changes that were made during the progress of the work — the lengthening of the main span and the changes in the specification and the adopted unit stresses. In considering Mr. Cooper's part in this undertaking, it should be remembered that he was an elderly man, rapidly approaching seventy, and of such infirm health that he was only rarely permitted to leave New York.[13]

Cooper's distance from the construction site and his inability to travel created a communication problem that played a critical role in the days leading up to the disaster. Even in today's world, where cellular telephones, fax machines, electronic mail, and overnight courier service permit design work to be conducted off site, it is impossible to believe that the key consulting engineer, who was the authorized ultimate technical authority, would never visit the construction site.

ORGANIZATIONAL DEFICIENCIES

The Royal Commission also criticized the way in which the project was organized, both by the Quebec Bridge Company and by Cooper:

> Mr. Cooper assumed a position of great responsibility, and agreed to accept an inadequate salary for his services. No provision was made by the Quebec Bridge Company for a staff to assist him, nor is there any evidence to show that he asked for the appointment of such a staff. He endeavoured to maintain the necessary assistants out of his own salary, which was itself too small for his personal services, and he did a great deal of detail work which could have been satisfactorily done by a junior. The result of this was that he had no time to investigate the soundness of the data and theories which were being used in the designing, and consequently allowed fundamental errors to pass by him unchallenged. The detection and correction of these fundamental errors is a distinctive duty of the consulting engineer, and we are compelled to recognize that in undertaking to do his work without sufficient staff or sufficient remuneration both he and his employers are to blame, but it lay with himself to demand that these matters be remedied.[14]

Problems of this nature persist even today. Moreover, it appears that this lesson was not fully learned by the government's Board of Engineers. After the bridge reconstruction began (as described below), it was pointed out that the Board of Engineers took over two years and spent half a million dollars in preparing the specifications for the bridge, but then expected engineering companies to prepare detailed competitive bids in four months with no remuneration.[15]

RECONSTRUCTION AND SECOND ACCIDENT

In 1908, the government of Canada, recognizing that the bridge was a key link in the Transcontinental Railway, decided that the demolished bridge should be reconstructed. The government established a three-person Board of Engineers to prepare plans and specifications and to supervise the reconstruction. The duties and authority of the Board were clearly defined.

The Board reviewed the earlier plans and the report of the Royal Commission, and adopted a modified cantilever structure with a wider base between the side-trusses and with straight lower chords. Removing the twisted steel and debris from the 1907 disaster took two years, and new supporting piers, going down to bedrock, were built. Under the direction of the Board, the superstructure was designed, manufactured, and erected by the St. Lawrence Bridge Company, Ltd., of Montreal.

In the new design, the compressive chords were significantly larger than they were in the original design. As shown in the Table 11.1, the steel cross-sectional area of the rectangular chords of the original

Table 11.1 COMPARISON OF DESIGN SPECIFICATIONS

Bridge	Dimensions	Steel Cross-Sectional Area
Original design, Phoenix Bridge Co.	1.37 m (4′ 6.5″) high	0.543 m² (842 sq. inches)
Firth of Forth Bridge, 1890	3.66 m (12 ft) diameter circle	0.516 m² (800 sq. inches)
Final design, St. Lawrence Bridge Co.	2.21 m (7′ 3″) high	1.252 m² (1941 sq. inches)

Phoenix design was slightly greater than the area of the Forth Bridge, which had a circular cross-section. The circular cross-section gives a larger resistance (moment of inertia) against buckling, but more importantly, the circular sections do not require lattice-work, cross-braces, and other heavy links that are introduced to stiffen the rectangular plates.[16] These secondary members contribute greatly to the weight. Consequently, it is evident even from this data that the Phoenix bridge, with less-efficient compressive chords, must have been a very slender design. The use of circular chords was considered but deemed to be uneconomical. The circular chords of the Forth Bridge design could be built much more easily in Scotland, where the ship-building plants and artisans were accustomed to large structures and had the knowledge and machinery needed to fabricate the curved surfaces. Adequate facilities for a similar project of this magnitude were scarce in North America at that time. The final (St. Lawrence) bridge was designed to instil confidence in the structure: as shown in the table, its massive compressive chords are almost 2.5 times as heavy (per unit length) as those on the Forth bridge.

Regrettably, during the reconstruction, a second tragedy occurred with more loss of life. The original (Phoenix) erection plan was to construct the bridge entirely in place by building each cantilever out from the river bank, until they met at mid-span. However, for the second (St. Lawrence) design, the erection plan was to build the cantilevers only partway out from the shore, assemble the central part of the span on-shore, float it out, and raise it into position. On 11 September 1916, the weather and tides were suitable for floating the middle span to the bridge and raising it into place. All went smoothly, and by mid-morning, the span had been lifted about 7 m above the water. At about 11 A.M., a sharp crack was heard, and the centre span was seen to slide off its four corner-supports into the river. Thirteen men were killed, and fourteen more were injured to various degrees.

An investigation conducted by the St. Lawrence Bridge Company and the Board of Engineers found that the accident was unrelated to the de-

▲ *The instant of collapse of the centre span of the Quebec Bridge (St. Lawrence design) on 11 September 1916. Source:* The Quebec Bridge, *Department of Railways and Canals Canada, 1919.*

sign and was caused by a material failure in one of the four bearing castings that supported the central span temporarily while it was being transported and hoisted. The St. Lawrence Bridge Company assumed the responsibility for the failure, a second span was constructed, and the design of the support bearings was changed from a casting to a lead "cushion." The new middle span was successfully lifted into place, over a three-day period, in August 1917. The bridge was opened to traffic in 1918, and a formal ceremony attended by the Prince of Wales was held on 22 August 1919.

CONCLUSION

In the decade following the Quebec Bridge disasters, the first Acts to license professional engineers were put into law. The Ritual of the Calling of an Engineer (described in Appendix C) was instituted, and even today the chain and iron rings used in the ritual are rumoured to be made from the steel scrap that claimed the lives of so many men in the cold waters of the St. Lawrence. There are many lessons to be learned from Canada's worst bridge disasters, such as the importance of

- providing adequate capitalization for large-scale projects
- hiring capable and competent professionals
- defining clearly the duties, authority, and responsibility of professional personnel
- discussing design decisions and related technical problems openly and listening receptively
- reviewing details, especially in the iterative task of engineering design
- monitoring work on the site adequately
- ensuring that communication is rapid and accurate
- providing adequate support staff and remuneration for highly skilled professional people

▲ *A view of the completed bridge from the north shore. Source:* The Quebec Bridge, *Department of Railways and Canals Canada, 1919.*

Provincial regulation of engineering assists in achieving these goals. The Professional Engineer's stamp on engineering plans and specifications identifies unambiguously who is responsible for the accuracy of the documents and for the computations upon which they are based. These lessons were learned at great cost.

WHERE TO LEARN MORE

The Quebec Bridge over the St. Lawrence River near the City of Quebec: Report of the Government Board of Engineers, Department of Railways and Canals Canada, printed by order of the Governor-General in Council, 31 May 1919. This two-volume book describes the bridge design and construction in impressive detail. The book is a classic of project documentation, and its purpose was undoubtedly to restore public confidence in the safety of the final structure. It is available in most university libraries and is well worth reading, even almost a century later, by anyone interested in structural design.

CASE HISTORY 11.2
THE VANCOUVER SECOND NARROWS BRIDGE COLLAPSE

Although some failures involve design errors in the structure, many serious accidents occur because of design flaws in the temporary scaffolds or support structures erected during the construction. These temporary structures fail because they are rarely analyzed or designed as thoroughly as the permanent structures. The following case illustrates this point and is reprinted, with permission, from W.N. Marianos, Jr., "Vancouver Second Narrows Bridge Collapse."

BACKGROUND

On June 17, 1958, two spans of the Vancouver Second Narrows Bridge collapsed while under construction. The accident was caused by the collapse of a temporary tower supporting the partially completed bridge. Eighteen workers were killed in the collapse, which caused four million dollars in additional construction costs on the bridge.

The Second Narrows Bridge connects Vancouver, British Columbia, with its northern suburbs across Burrard Inlet, the city's harbor. The structure was built for the British Columbia Toll Highways and Bridges Authority. The main bridge, a steel cantilever truss structure, has a total length of over two thousand feet [610 m]. Unlike older, simpler, and shorter bridges whose spans or sections rest independently on their piers or abutments, those of a cantilever bridge run continuously over or extend beyond the piers. The main bridge has three spans: a 1100-foot-long [335 m] center span balanced by two side spans, one 465 feet [142 m] and one 466 feet [142 m] long. Four steel truss spans, each 276 feet [84 m] long, make up the northern approach to the main bridge. The structure was designed by Swan, Wooster and Partners, a Vancouver engineering firm. Dominion Bridge Company was the contractor for the construction of the steel spans. The foundations and bridge piers were constructed by Peter Kiewit Sons and Raymond International.

DETAILS OF THE COLLAPSE

By mid-June 1958, the approach spans were in place and erection of the northern side span of the main bridge was in progress. The length of the span required the use of two temporary supports for construction, since the side span would not be self-supporting until its full length was in place. Each temporary support, called a falsework bent, consisted of two columns, one under each side of the span. The columns were built on temporary piers in the harbor. These piers were supported by a group of foundation piles. The load from each column was distributed to the foundation piles by a grillage — a two-layer grid of steel beams. The lower set of beams sat on top of the foundation piles. The upper layer, a set of four beams set side by side, supported the column bases.

On June 17, the first side span was supported on a permanent concrete pier at one end, and was overhanging the first falsework bent, designated "bent N4," at the other. At 3:40 P.M. that afternoon, bent N4 collapsed, plunging the partially completed span into Burrard Inlet. The falling metalwork pulled the permanent pier it was resting on out of line, which caused the adjacent approach truss to collapse as well.

Immediately after the accident, the government of British Columbia appointed a royal commissioner, Sherman Lett, chief justice of the

▲ *Eighteen people were killed when failure of temporary construction supports caused the Vancouver Second Narrows Bridge spans to collapse on 17 June 1958. Source: UPI/Corbis-Bettmann.*

provincial supreme court, to determine the cause of the collapse. The commissioner selected five leading engineers to investigate and report on the matter: F.M. Masters and J.R. Giese of the United States; J.R.H. Otter and Ralph Freeman of Britain; and A.B. Sanderson of Canada. Materials testing and special investigations were conducted at the University of British Columbia and testing laboratories in Vancouver.

The commissioner's report concluded that the collapse was caused by failure of the four upper grillage beams. The webs (the vertical portion) of the beams buckled laterally, causing the collapse of the falsework bent columns.

Faulty design of the falsework or temporary columns led to the grillage failure. The commission discovered two major errors in the Dominion Bridge Company's grillage design calculations. The first mistake was in checking the grillage beam shear strength (the capacity of a beam to carry a load in its vertical plane; shear stress tends to tear a beam vertically, usually at supports or at points of concentrated load). The cross-sectional area of the entire beam was used in the calculation rather than just the areas carrying the load. This mistake would lead the grillage designer to believe the beam strength was about twice as much as it actually was.

A second calculation, which checked the need for web stiffeners, was also incorrect. Stiffeners are metal plates welded to beam webs to give them additional stiffness and resistance to buckling. The contractor's engineer had used the one-inch [2.5 cm] thickness of the beam flanges (the horizontal elements) rather than the actual 0.65-inch [1.6 cm] web thickness. This led to the erroneous conclusion that no stiffeners were needed.

A separate investigation by Dominion Bridge Company came to the same conclusion — that incorrect calculations led to a fatally inadequate grillage design. One of the errors in calculation was even discovered before the accident, but no corrective action was pursued. The two engineers responsible for the calculations were both killed in the collapse.

Wood blocks and plywood pads had been included in the grillage to provide some bracing of the beams. Laboratory tests indicated that these blocks were only marginally effective at best. Most of the wooden blocks were not even located at the most effective bracing points. The investigation performed at the University of British Columbia also indicated that the ability of the beam webs to resist buckling was not adequately predicted by the usual design formulas for column buckling.

In his report, the royal commissioner laid the blame for the collapse on the Dominion Bridge Company. The commission found the contractor negligent for "(a) failing properly to design and substantially construct false bent N4 for the loads which would come upon it …; (b) failing to submit to the engineers plans showing the falsework the contractor proposed to use in the erection …; and (c) leaving the design of the upper grillage of false bent N4 to a comparatively inexperienced engineer, and failing to provide for adequate or effective checking of the design and the calculations made in connection with the design."

The commissioner also found that a failure in the construction process had contributed to the accident. His report pointed out that the bridge design engineers, Swan, Wooster and Partners, had a responsibility to make sure the contractor submitted the falsework plans and calculations for their approval, as required by the project contract. The engineers certainly knew that the bridge was under construction, and they had prepared the section of the project specifications that required engineer's approval of the temporary falsework structures. Commissioner Lett concluded that "there was a lack of care on the part of the engineers in not requiring the contractor to submit plans of the falsework." Ironically, the satisfactory performance of Dominion Bridge on earlier projects may have contributed to the design engineer's laxness in pursuing the falsework plans and calculations for review.

The commissioner recommended that on future large bridge projects the consulting engineers recommend allowable stresses for temporary construction support structures, and that the contractor be required to submit all construction plans and calculations for approval prior to construction. The contractor, however, would always remain legally responsible for the adequacy of construction methods and temporary structures.

IMPACT

After the inquiry, construction resumed on the bridge. Two concrete bridge piers damaged in the collapse had to be rebuilt. This required the careful removal of two thousand cubic yards [1529 m³] reinforced concrete. The collapsed superstructure spans were salvaged, and some undamaged members were reused. Erection of the bridge continued according to the original plan, with the notable addition of careful checking and review of all construction calculations and plans. The additional time and materials required to reconstruct the damaged portions of the bridge added four million dollars to the original contract price of sixteen million dollars.

The editors of *Civil Engineering* magazine noted that the collapse "illustrates the ever-present risks that are inherent in construction, due to human error. The failure emphasizes the need for utilization of all possible checks on construction procedures."

Today, the leading bridge design firms continue to carefully review and check the contractor's construction plans and calculations. The collapse of the Vancouver Second Narrows Bridge was neither the first nor the last incident of mistaken temporary construction calculations leading to disastrous consequences. The accident vividly highlights the importance of independent checking of critical aspects of the construction process.[17]

WHERE TO LEARN MORE

"Faulty Grillage Felled Narrows Bridge in Vancouver, B.C.," *Engineering News-Record* (9 October 1958): 24.

A. Hrennikoff, "Lesson of Collapse of Vancouver 2nd Narrows Bridge," *Journal of the Structural Division, Proceedings of the American Society of Civil Engineers* (December 1959): 1–20.

F.M. Masters and J.R. Giese, "Findings on Second Narrows Bridge Collapse at Vancouver," *Civil Engineering* (February 1959): 60–63.

"What Happened at Vancouver? Probers Seek Cause of Bridge Collapse," *Engineering News-Record* (26 June 1958): 21–22.

CASE HISTORY 11.3
THE WESTRAY MINE DISASTER

INTRODUCTION

Mines are dangerous places, and coal mines are the most dangerous, since the rock is soft and the tunnels are deep, and unless the mine is properly maintained and ventilated, the coal dust and methane can be explosive. The first major Canadian mine disaster occurred in 1873, and thousands of lives have been lost in the years since then. In the Springhill, Nova Scotia, mines alone, 424 miners were killed in the period from 1881 to 1969, and in Canada's worst coal mine disaster at Hillcrest, Alberta, a total of 189 men were killed by an explosion in June 1914.[18] However, as the years passed, it was believed that the use of advanced ventilating, monitoring, and excavating methods had enhanced mine safety. This belief was shattered on 9 May 1992, when an explosion in the Westray mine in Plymouth (near Stellarton in Pictou County, Nova Scotia) killed 26 miners, and the later inquiry into the disaster revealed a "complex mosaic of actions, omissions, mistakes, incompetence, apathy, cynicism, stupidity, and neglect."[19]

DETAILS OF THE EXPLOSION

The Westray mine explosion occurred at 5:20 A.M. on the morning of Saturday, 9 May 1992, and the tremor was felt by most of the inhabitants of the small village of Plymouth. Within hours, mine rescue experts had assembled from neighbouring towns; with oxygen tanks on their backs, they descended into the battered mine. Rescue attempts were pointless, since it soon became clear that the explosion had caused instant death to all those below. The job of retrieving the dead was not totally successful. As this text goes to press, ten dead miners are still permanently entombed behind rock falls in the mine, much of which was flooded to prevent further explosions.

THE WESTRAY INQUIRY

Within days of the tragedy, anecdotal evidence of unsafe practices began to be widely reported. One miner described several infractions of the safety regulations, including the use of acetylene torches in areas where methane levels could be dangerous; a supervisor tampering with a methane level monitor to permit higher methane levels; and potentially explosive coal dust accumulating so thickly that it would prevent the machinery from operating.[20] The miners rarely complained about the safety infractions out of fear of retaliation or intimidation, since their managers appeared to put production ahead of safety.[21]

In the midst of bitter allegations and accusations, the provincial government appointed Justice K. Peter Richard to carry out a wide-ranging inquiry into how and why the 26 miners died. Shortly thereafter, the Royal Canadian Mounted Police started a criminal investigation, and in October 1992, the Nova Scotia Labour Department laid 52 noncriminal charges of unsafe practices against Curragh Resources, Inc., the company that owned the mine. These safety charges were later dropped to avoid jeopardizing the police investigation, which resulted in charges of manslaughter and criminal negligence being laid against Curragh Resources and two of its managers. These charges were later stayed, but on appeal, the Supreme Court of Canada upheld an order for a new trial (which has yet to take place as this text goes to press).

Throughout the court proceedings, the Westray Inquiry continued, and Justice Richard's final report, *The Westray Inquiry: A Predictable Path to Disaster*, was published in December 1997. Readers are urged to refer to the full report for a comprehensive discussion of the testimony, evidence, findings, and recommendations. The paragraphs that follow

▲ *The Westray coal mine explosion in 1992 killed 26 miners. In his report on the ill-fated mine, Justice Peter Richard blamed the coal company and the provincial government for the disaster. Source: Canapress/Andrew Vaughan.*

are a synopsis of the key facts exposed during the Inquiry, and are excerpted from the Executive Summary of Justice Richard's report.

PRELUDE TO THE TRAGEDY: HISTORY, DEVELOPMENT, AND OPERATION

The Westray mine is located at Plymouth, near Stellarton, in Pictou County, Nova Scotia. Westray was the only operating underground coal mine in Pictou County at the time of the explosion. The Pictou coalfield had been mined for some 200 years, and elements of the disaster rest in the nature of that coalfield with its thick and gassy seams. The Foord seam, which Westray was mining, has hosted at least eight mines. The Allan mine, the most productive and the one that lay just northwest of Westray's workings, finally closed in the 1950s, but during its 40-year lifetime, it experienced eight methane explosions.

The Westray project was controversial from the outset. Although various companies ... had been interested in the area with its low-sulphur coal, it was Curragh Resources Inc. that eventually put the pieces together, incorporated Westray Coal in November 1987, and some 16 months later began underground development. ...

The proposed mine developed amid opposition from the bureaucracy and unwavering support from the provincial government. As development proceeded, the mine was the subject of debate and criticism in the legislature and in the media. It also proceeded with an uncompromising and abusive Curragh negotiator, chief executive officer Clifford Frame, at the helm. For these reasons, it is not surprising that the negotiations for financial assistance between Curragh and government proved to be arduous and taxing. In the end, the strong and single-minded political backing for the project, by Donald Cameron in particular, prevailed. [Donald Cameron was, at that time, Conservative provincial Minister of Industry, Trade, and Technology, but was later to become Premier of Nova Scotia.] Westray received tremendous financial support from the public sector, which resulted in minimal equity investment by the company....

Before all the financing was in place, the underground work began. Early in 1989, Curragh's subcontractor, Canadian Mining Development (CMD), began driving the main access slopes. The Department of Natural Resources had approved Curragh's application for the mining lease in 1988, and, in January 1989, the department discovered that the tunnel alignment had been changed from the approved layout. CMD was to drive the two main slopes to the limits of the planned workings, and Westray would then take on the development of coal-producing sections off the mains. Meanwhile, several provincial government departments were engaged in continuing negotiations with Curragh. The Department of the Environment had a number of concerns about the effect of the development on the area. The Department of Labour expressed concern about training and certification, equipment approvals, plans for emergencies, and delays in setting up a workplace safety committee. The Department of Natural Resources was concerned that the new tunnel alignment would intersect major geological faults at oblique angles, resulting in extensive tunnel development through bad ground. Poor roof conditions in the earliest days of tunnel development gave credence to that concern....

Roof conditions emerged as a major problem in 1991. Westray took over development from CMD in early April 1991, at a much earlier stage of development than originally planned, and began using continuous mining machines to drive the mains. The company decided to scrap the original mine layout and to change direction so it could tap into the coal seam sooner. That change took development into the Southwest section of the mine. During the summer, development also continued down into the North mains, splitting the mine into two distinct sections, each with its own crews and supervisors. In the rush to reach saleable coal, workers without adequate coal mining experience were promoted to newly created supervisory positions. Workers were not trained by Westray in safe work methods or in recognizing dangerous roof conditions — despite a major roof collapse in August. Basic safety measures were ignored or performed inadequately. Stonedusting, for example, a critical and standard practice that renders coal dust non-explosive, was carried out sporadically by volunteers on overtime following their 12-hour shifts.

The official opening of the mine was on 11 September 1991. For that occasion, the mine was "spruced up" and stonedusted.

Four more roof falls were reported in September and October. The mine manager, Gerald Phillips, minimized the seriousness of roof problems, claiming that the falls were controlled and that they posed little threat to the miners or to production. To the contrary, realistic accounts of the miners' experiences revealed a series of near misses and increasing danger. There were approximately 160 employees at the site by October, a large majority of them working shifts underground. Management trivialized the concerns of workers, some of whom quit their jobs at the mine. Although the mine inspectors asked the company for roof support plans, as well as stonedusting plans, it repeatedly deferred supplying them. Westray is a stark example of an operation where production demands resulted in the violation of the basic and fundamental tenets of safe mining practice.

The first drive to unionize the workforce at Westray was officially begun on 2 October 1991 by local 26 of the United Mine Workers of America. The union was defeated by 20 votes in January 1992. In the spring of 1992, the United Steelworkers of America succeeded in its drive to unionize the workers, but certification was not granted until after the 9 May explosion.

The Southwest section was plagued with roof problems. The decision to drive into the Southwest section was proving a serious mistake. The levels of production and the quality of the coal were less than anticipated. Production remained behind schedule, and the company was not able to meet its commitments to supply coal. In late March 1992, the workforce was literally chased out of the Southwest 1 section by rapidly deteriorating ground conditions. In its determination to save equipment, the company put employees at extreme risk during the abandonment.

The Department of Natural Resources staff expressed concern about proximity to the old Allan mine workings, potential subsidence problems, and deviations from the approved mine plan. The department suggested that non-compliance could threaten the company's mining permit but inexplicably retreated from its position. Skeletal new plans submitted by the company were approved, and the department assisted the company in developing a surface mining operation to help meet its coal supply obligations. Federal and provincial money and expertise met most of the costs of technical studies for monitoring roof conditions and subsidence.

The regulatory framework in Nova Scotia requires that almost every person employed in underground coal mining hold a certificate of competency issued by an appointed provincial board of examiners. Section 11 of the Coal Mines Regulation Act (1989) sets out the education and work experience required for the various certificates. The administration of certification for mine rescue and for competency as a coal miner was delegated to the Department of Labour. In Nova Scotia, the company is responsible for training miners. The role of the Department of Labour is to ensure that the company complies with the Coal Mines Regulation Act and the Occupational Health and Safety Act.

It is clear that the company was derelict in carrying out its obligations for training. The testimony of the miners shows that training fell far short of need. Don Mitchell, mining consultant for the Department of Labour, concluded from his post-explosion investigation that the mine "had no program that was appropriate to the needs of that mine." And expert witness Dr. Malcolm McPherson referred to the inadequate training of mine workers as making an equally potent contribution to the propagation of a mine explosion as did the ventilation engineering deficiencies.

Quite simply, management did not instill a safety mentality in its workforce. Although it stressed safety in its employee handbook, the policy it laid out there was never promoted or enforced. Indeed, management ignored or encouraged a series of hazardous or illegal practices, including having the miners work 12-hour shifts, improperly storing fuel and refuelling vehicles underground, and using non-flameproof equipment underground in ways that violated conditions set by the Department of Labour — to mention only a few. Equipment fundamental to a safe mine operation — from the cap lamp to the environmental monitoring system — did not function properly.

It was equally clear that the Department of Labour was derelict in its duty to enforce the requirements of the two acts.

THE EXPLOSION: AN ANALYSIS OF UNDERGROUND CONDITIONS

... [V]entilation is the most crucial aspect of mine safety in an underground coal mine. Methane fires and explosions cannot happen if the gas is kept from accumulating in flammable and explosive concentrations. A coal mine can be quite "forgiving" with respect to other aspects of safety, as long as the ventilation system is properly planned, efficient, and conscientiously maintained. The other major requirement of coal mine safety is control of coal dust, through strict clean-up procedures and regular stonedusting.

The ventilation system of any underground mine is a network of interconnected passages, many of which are also used as transportation routes for personnel, vehicles, and the products of mining. Fresh air is drawn from the surface atmosphere. As the air passes through the underground passages, its quality deteriorates as a result of pollutants produced from the strata and from the effects of machines and mining procedures. The contaminated air is returned to the surface. A mine ventilation system has to deal with both gaseous and particulate pollutants. Methane is a dangerous pollutant present in coal. Although non-toxic, it is hazardous because of its flammability. It will explode in concentrations of between 5 and about 15 per cent by volume in air, and it reaches maximum explosiveness at about 9.6 per cent.

Methane is a natural component of coal, a by-product of the decomposition of the plant matter from which coal is formed. Methane is released as the coal-cutting machines break coal away from the face. As methane continues to emerge from the coal, it moves through fissures in the coal that remains after mining, and it can escape into the active roadways from abandoned or mined-out sections, depending on the effectiveness of the stoppings constructed at the entrances to abandoned sections. One of the principal functions of a ventilation system is to clear the methane at the working face of the mine and to exhaust it from the mine in non-explosive concentrations. It is clear that the Westray ventilation system was grossly inadequate for this task. It is also clear that the conditions in the mine were conducive to a coal-dust explosion.

The miners, faced with management pressure for production, undoubtedly indulged in many dangerous and foolhardy practices in the days immediately preceding 9 May 1992. In his various comments reported in the media following the explosion, Gerald Phillips blatantly blamed the miners for the explosion. In light of all the evidence of mismanagement, neglect, and incompetence at Westray, this simplistic explanation can only be regarded as a defensive ploy to deflect attention away from the real causative factors. Unfortunately, this explanation was picked up by former premier Donald Cameron. From all the evidence and the extensive analysis and studies by mining experts, however, it becomes abundantly clear that ventilation in the Westray mine was woefully deficient in almost every respect. The airflow was inadequate for the purpose of clearing methane from the working face during mining and preventing the layering of methane on the roof.

Therefore, I should like to put to rest the question raised by Cameron's testimony, as well as the statements of Phillips and Frame to the media: Had it not been for these unsafe practices attributed to the miners, would the explosion of 9 May have occurred? The answer must be yes, it would have. The consensus of the experts suggests strongly that Westray was an accident waiting to happen.

THE REGULATORS: DEPARTMENTAL AND MINISTERIAL RESPONSIBILITY

The Department of Natural Resources (the Department of Mines and Energy before September 1991) was charged with regulatory authority over the mine-planning approval process. As the testimony at the Inquiry unfolded, it became clear that the Department of Natural Resources had failed to carry out its statutory duties and responsibilities as they related to the Westray project. Natural Resources witnesses had mixed views on fundamental regulatory issues, such as whether the department was within its mandate to regulate for "safety," or whether its duty included monitoring Westray to ensure that it was operating in conformity with the approved mine plan.

The mandate of the department vis-à-vis the Department of Labour and the mine inspectorate was not formally defined in any way, and the changes affecting the departments over their history contributed to this lack of definition. Before 1986, both the mine engineering unit and the mine inspection unit were part of the Department of Mines and Energy, and their duties overlapped somewhat. When the inspectorate transferred to the occupational health and safety division of the Department of Labour in 1986, it lost its link to the engineering section. When the chief inspector left a short time later, the liaison between the two functions effectively ended. It is clear that the Department of Natural Resources, in spite of these changes, retained legislative responsibility to ensure, before permits are granted, that mining plans are not only efficient but safe.

In the view of the Department of Natural Resources, its responsibility for monitoring the Westray operation for compliance with the approved mine plan was limited to an annual review of plans submitted by the operator. Section 93 of the Mineral Resources Act (1990) is explicit: the permit holder "shall conduct mining operations in conformity with the approved mining plan." The Department of Natural Resources was ill-advised in approving the Westray mine proposal in the form submitted. The department did not insist that the company submit sufficient information to support its application. Furthermore, it did not insist that the company submit any changes to approved plans. Consequently, for a critical period, the department was not aware that Westray was working an unapproved section of the mine. The department's explanation was that such day-to-day monitoring was the responsibility of the Department of Labour. What it did not explain was why the department failed to shut down a company that was undeniably in violation of the Mineral Resources Act — an action that fell squarely within its own mandate. The evidence of the public servants of the Department of Natural Resources is replete with examples of neglect of duties, submissiveness to Westray management, and just plain apathy.

The Department of Labour shares with the Department of Natural Resources the responsibility for failure to coordinate the several aspects of mine regulation. The Department of Labour was responsible for regulating occupational health and safety at the mine, and as such was the body most responsible for the exercise of regulatory authority respecting safe mining at Westray. What is clear from the testimony of Labour witnesses at the Inquiry is that the department did not discharge its duties with competence or diligence, and thereby failed to carry out its mandated responsibilities to the workers at Westray and to the people of Nova Scotia.

The Report enumerates in detail the many ways in which Westray Coal violated the regulations governing mine operations. The Department of Labour's mine inspectorate should have detected these violations and ensured compliance. To give just one example, despite the company's repeated violations of the Coal Mines Regulation Act in the matters of clearing coal dust from the working sections of the mine and applying stonedust to render the coal dust inert, the mine inspectorate did not use the means at its disposal to ensure compliance. It was not until 29 April 1992 that inspector Albert McLean gave oral orders, followed up by written orders, to Westray underground manager Roger Parry and mine manager Gerald Phillips to clean up and treat the coal dust immediately and to produce the stonedusting and dust sampling plans that had been promised in September 1991. McLean failed to follow up on his orders during his visit to the mine on 6 May 1992.

The Report also examines the involvement of politicians in the development of the Westray project and their very active support of a project that would mean jobs in Pictou County. The three provincial politicians most involved with the Westray project were John Buchanan, Donald Cameron, and Leroy Legere. Cameron had the most prominent and enduring role in the project, serving as minister of industry, trade, and technology from April 1988 until he succeeded Buchanan as premier in February 1991, a position he held until late spring 1993. Legere was appointed minister of labour in February 1991. It became clear in the course of the Inquiry that Buchanan, Cameron, and Legere had disparate understandings of their roles as ministers of the crown. The fact that they had such an imperfect understanding of the nature of their responsibilities suggests that a formal clarification of constitutional responsibilities is required....

... Clearly, the aim of mining legislation should be the protection of the miner in the mining environment. Coal mining is inherently hazardous, and safety regulations must protect the miner in a way that is consistent with the economic viability of the undertaking. This goal has been expressed in terms of safe mine production. "Attitude," which may be the most significant single factor in attaining safe mine production, cannot of course be legislated. It must, however, be cultivated within an organization, whether it be a mining company, a union, or a government agency charged with enforcement of safety legislation.

THE AFTERMATH: RESCUE EFFORTS AND THE INQUIRY

[It is essential to] comment on the selfless bravery shown by the rescue teams in the days following the explosion. The conditions in the mine were terrifying. The force of the explosion resulted in severe instability within the roof and walls of the mine. Rock falls, of varying degrees of intensity, were almost continuous. Signs of the devastation were rampant, as were signs of impending danger. The poisonous, unbreathable atmosphere and the actively "working" ground surrounding the mine openings, with the attendant grinding and cracking, were extremely stressful. Yet these men, miners trained in mine rescue, each wearing his personal life-support system, went unquestioningly into that perilous environment with the hope of finding some of their comrades alive. The rescuers came from mainland Nova Scotia, Cape Breton, and New Brunswick. We can only be thankful for this valiant display of concern for fellow workers....[22]

CONCLUSION

During the Westray Inquiry, many witnesses observed that the terms of the Coal Mines Regulation Act were outdated and inadequate. In his final report, Justice Richard stated:

> As outdated and archaic as the present act is, it is painfully clear that this disaster would not have occurred if there had been compliance with the act.
>
> - If the "floor, roof and sides of the road and the working places" had been systematically cleared so as to prevent the accumulation of coal dust;
> - If the "floor, road and sides of every road" had been treated with stone-dust so that the resulting mixture would contain no more than 35 per cent combustible matter (adjusted downward to allow for the presence of methane); and
> - If the mine had been "thoroughly ventilated and furnished with an adequate supply of pure air to dilute and render harmless inflammable and noxious gases," then . . .

the 9 May 1992 explosion could not have happened, and 26 miners would not have been killed. Compliance with these sections of the Coal Mines Regulation Act was the clear duty of Westray management, from the chief executive officer to the first-line supervisor. To ensure that this duty was undertaken and fulfilled by management was the legislated duty of the inspectorate of the Department of Labour. Management failed, the inspectorate failed, and the mine blew up.[23]

CASE HISTORY 11.4
THE LODGEPOLE WELL BLOWOUT

INTRODUCTION

Residents of the Drayton Valley area of Alberta will long remember the autumn of 1982. At 2:30 P.M., on 17 October 1982, the Amoco Lodgepole oil well being drilled near Drayton Valley encountered "sour gas" (gas laden with hydrogen sulphide) and blew out of control. During the next two months, residents living within a 20- or 30-km radius of the well were twice exposed to the rotten egg smell of hydrogen sulphide (H_2S) and the threat of H_2S poisoning while specialists fought to regain control of the well. The first H_2S exposure period lasted sixteen days and the second period twelve days. During attempts to cap the blowout, two employees were overcome by H_2S and died, and the well was twice engulfed in flames. About 28 people were voluntarily relocated to avoid the H_2S, and several residences were ordered evacuated during particularly heavy H_2S concentrations on 29 October and 17–24 November. Even people at great distances from the well were subjected to noxious and unpleasant odours depending on the prevailing winds. The well was not successfully capped until 23 December 1982. In January 1983, a Lodgepole Blowout Inquiry Panel was convened to investigate the causes of the blowout, the actions taken to prevent it and to regain control, the hazard to human health, the impact on the environment, and what should be done to avoid future blowouts at wells in Alberta. The panel issued a comprehensive report in December 1984.[24]

EVENTS LEADING UP TO THE BLOWOUT

The Amoco Lodgepole oil well, known officially as Amoco Dome Brazeau River 13-12-48-12, is located about 140 km west of Edmonton (or about 40 km west of Drayton Valley). The well gets its name from the nearest village, the small hamlet of Lodgepole, which is situated about halfway between the well and Drayton Valley. The Amoco Canada Petroleum Company Ltd. obtained a licence to drill the Lodgepole oil well from the Alberta Energy Resources Conservation Board (now superseded by the Alberta Energy and Utilities Board), and the well was "spudded" (started) in August 1982. Drilling proceeded to a depth of about 3000 m without problems. An intermediate casing was then installed, and the drilling crew began coring operations to examine the strata prior to drilling into the oil-bearing formation. Two cores were obtained without apparent problems. On 16 October, the crew was obtaining a third core when they recognized that fluid was entering the well from the oil- and gas-bearing formation.

The drill crew stopped the coring operations to deal with this problem, known as a "kick," since the entry of reservoir fluids into the wellbore forces the drilling mud out of the well. For the next sixteen hours, the crew fought to regain control of the well, but finally the drill pipe "hydraulicked" up the hole, the kelly hose was severed, and the well was out of control. The intense pressure caused a continuous, uncontrolled flow of mud and sour gas into the atmosphere. The exact flow rate is unknown, but during the inquiry, it was estimated at 1.4 million m^3 of gas per day. Later tests indicated that the flow could have been even greater.[25]

EMERGENCY MEASURES

Amoco immediately implemented its Major Wellsite Incident Response Plan, and key Amoco personnel were notified of the blowout. People and equipment were dispatched to the site, including safety personnel, paramedics, breathing apparatus, ambulances, helicopters, and firefighting equipment. Hydrogen sulphide monitoring equipment was ordered, both for on-site and off-site monitoring. The company immediately hired specialists to cap the well, and special tools and equipment were ordered.

Over the next two months, several plans for capping the well were conceived but were unsuccessful. On 1 November, a failed attempt resulted in a fire that engulfed the well. A new control plan was developed, and on 16 November, the fire was extinguished with explosives preparatory to implementing the plan. Two days later, while attempting to execute the plan, an accident occurred that resulted in the deaths of two employees who were overcome by H_2S. On 25 November, the well was again on fire. It was later determined that this fire probably resulted from an undetected underground muskeg fire that had been smouldering for some time. Amoco decided to try to cap the well while it was still on fire; however, the well specialists declined to attempt this procedure, which had seldom been successful for other blowouts. On 1 December, new well-capping specialists were hired, and by 23 December, they had installed a blowout preventer (BOP) over the stub of the intermediate casing. Lines were then connected to flare off the gas and pump mud into the well. Over the next five days, 96 m³ of mud were pumped into the well, the pressure was stabilized, and the crisis ended.

WHAT WENT WRONG AT THE LODGEPOLE WELL?

The basic reason why the blowout occurred is that Amoco personnel were unsuccessful in controlling the hydrostatic pressure in the well. This is a critically important and delicate balancing procedure, and the control strategy, usually called the well control plan, sets out the basic principles and procedures that must be followed to ensure that a well will not blow out during drilling, completion, or production operations. The well control plan is rarely a single document, but is the sum total of all drilling program documents, special instruction bulletins (including those posted at the site), company procedure manuals, and other books, manuals, and written and verbal instructions that guide drilling procedures.

To understand the importance of the control plan and how it applies to the critical balancing of hydrostatic pressure in the well, the following explanatory note is reproduced from the inquiry report, with permission.

> The drilling fluid (mud) system has a dominant position in the general well control plan for any well. The plan requires that the hydrostatic pressure in the wellbore be greater than the formation pressure. Hydrostatic pressure depends on the height of the column of drilling mud and the density of the mud. A reduction in either or both of these will reduce the pressure that results from the column or head of drilling mud. If the hydrostatic pressure is too low, a state of underbalance exists and fluids from the formation, such as gas, may flow into the wellbore. Unless this flow is properly controlled, a blowout will result. On the other hand, if the hydrostatic pressure is too great and a state of excessive overbalance exists, the drilling mud may flow into the formation. This is referred to as "lost circulation" and will result in a loss of hydrostatic pressure which can also lead to a blowout.

... In developing a drilling program, an operator must therefore consider formation pressures encountered at other wells in the general vicinity of the well being planned for and select a drilling mud density which will ensure a modest overbalance. To ensure that neither a state of underbalance nor of excessive overbalance develops as the well is drilled, close attention must be given to:

(a) mud density,
(b) any contamination of the drilling mud that will change its effective density, such as by drill cuttings (increase) or by air introduced during tripping (decrease),
(c) maintaining a full mud column,
(d) the rate of lowering the drill pipe,
(e) the rate of hoisting the drill pipe, and
(f) pumping rate and pressure.

Should the hydrostatic pressure from the mud column prove to be insufficient, and as a consequence, formation fluids such as gas enter the wellbore, a kick condition would exist. Procedures have been developed such that those fluids may be controlled within the wellbore using the BOP [blowout preventer] system mounted on the well casing. The gas is flared at the surface and control operations are continued until the flow is progressively restricted and stopped. If the kick has occurred and efforts to control and contain the in-flowing formation fluids have failed, an uncontrolled flow or blowout results and re-establishing control may be both technically difficult and dangerous.

The general well control plan must include the procedures to be used to circulate out the kick, and it must also ensure that proper equipment will be available should a kick occur. This includes the BOP system but additionally, casing and drill pipe design and selection are important components of the plan.

In order to carry out the general well control plan, the detailed drilling plan must provide for the integrity of the drilling fluid system throughout the drilling operation regardless of the circumstances encountered. The operator should have on site, at critical times, experts in geology and mud properties. The operator must also ensure that the drilling and well equipment, particularly the BOP, is functioning properly. Finally, for the general well control plan to be implemented effectively, on-site supervisors and the drilling crew must be properly trained, regularly briefed, and always prepared to act promptly in carrying out prescribed kick control procedures.[26]

EVALUATING AMOCO'S ACTIONS
Amoco assisted the Inquiry Panel by providing complete documentation on its drilling plan, drilling mud program, rig, and well equipment, and on the qualifications of well-site crew and supervisors. It also provided a detailed chronology of events leading to the blowout

and expert witnesses to explain the events surrounding the blowout. A particularly important point was the density of the drilling mud used. Obtaining the right mud density is a key part of the balancing act: it requires a knowledge of the formation pressures and careful monitoring of the hydrostatic pressure. The problem is explained in the inquiry report as follows:

[I]t is necessary to use a mud density which is neither too low, thus allowing an influx from the formation, nor too high, which could result in lost circulation. This means there is a range of mud density within which operations must take place. The closer one is to the upper or lower limit, the more careful drilling procedures must be. If the mud density is within the range but towards the "high limit," care must be taken to avoid lost circulation. For example, the mud volume must be carefully monitored and the crew must be ready to add lost circulation material. If the mud is closer to the "low limit," drilling must proceed slowly, the potential for a kick must be carefully monitored, and plans to quickly weight up the mud must be in place.

... In developing its drilling plan, Amoco reviewed information concerning formation pressures encountered at a large number of other wells in nearby areas. These indicated that the pressures in the formation of interest at the [Lodgepole] well would ordinarily be around 33 000 to 35 000 kPa. In isolated cases, pressures as low as 22 430 kPa and as high as 46 540 kPa had been reported. Amoco decided to design the drilling mud density to meet a pressure of 33 600 kPa with provision for increasing the density if higher pressure was encountered.

Amoco also indicated that its normal practice was to design mud density to provide a mud column overbalance of some 1500 kPa above the expected formation pressures. During drilling or coring operations, mud pump pressures would add further to this, resulting in an overbalance which should avoid the possibility of excess fluid head and lost circulation but at the same time prevent influx of reservoir fluid.

Calculations by the Panel indicate that, at the predicted Nisku reef depth of 3035 m, the planned mud density of 1150 kg/m³ would result in an overbalance of some 630 kPa relative to the expected pressure of 33 600 kPa ...

... The range of appropriate mud densities varies for each situation, and at the [Lodgepole] well, because of the high reservoir pressure with very good permeability, [the range] was likely relatively narrow. The planned pressure overbalance of 630 kPa plus or minus the effects of operations such as pumping or pulling out of the hole, was less than the 1500 kPa normally used by Amoco....

... In summary, the Panel concludes that the planned mud density for the [Lodgepole] well was on the low side and therefore extra care should have been specified during the critical period of drilling into the Nisku zone. Although substantial seepage losses were reported and these might have been interpreted by on-site personnel as an indication that the mud was on the heavy side, an analysis of the situation indicates that the reported losses were likely due to errors. This is an indication that the drilling practices being used were less than satisfactory.[27]

ASSIGNING RESPONSIBILITY

Although the Lodgepole Inquiry Panel concluded that "no single element in the chain of events was the sole cause of the blowout," the panel's examination of the events led it to conclude that the initial kick occurred primarily because "drilling practices during the taking of cores were deficient" and, when combined with the marginally adequate mud density being used, this permitted the entry of reservoir fluids into the wellbore.[28]

Amoco expected the Lodgepole well to find "sweet" oil, but the company's control plan definitely recognized and accounted for the possibility of encountering sour gas. The Panel accepted this testimony, and did not believe that ". . . the expectation of sweet oil played a direct role in the cause of the blowout. However, it may have influenced planning for the well and may have led to less caution in the drilling operations than might have been the case if the well was being drilled specifically for sour gas."[29]

The Inquiry Panel also concluded that the kick was likely not controlled because the drilling crew did not immediately recognize the problem and therefore did not immediately apply and maintain standard kick-control procedures. As contributing factors, the Panel noted that several pieces of vital equipment did not function properly, and supplies of mixed drilling mud were not adequate during the kick-control operations. As the Panel wrote in its report:

> The unexpected entry of reservoir fluids into the wellbore was probably due to a combination of Amoco not adhering to sound drilling practices and only marginally adequate mud density. If the degasser had operated effectively, the initial kick might have been circulated out of the system, and subsequent kicks would likely have been avoided. Even with the failure of the degasser, control might have been maintained if there had been sufficient and properly weighted mud on hand to pump into the well. Additionally, if the casing pressure instruments had been operational from the outset, the crew might have recognized the kick at an earlier stage and implemented standard kick-control procedures when fluid influx was still relatively small. If the decision had been made to use hydrogen-sulphide(H_2S)-resistant pipe for the full drill string, the pipe might not have parted and the succession of kicks might still have been successfully circulated out. And finally, if the travelling block hook latch had not failed, it may have been possible to retain control of the well by "top kill" methods.[30]

The Panel was satisfied that Amoco applied reasonable judgement in selecting the type of drilling rig, the degasser, and the type of drill pipe even though, in retrospect, other choices might have been better. However, the Inquiry Panel concluded that

Amoco's actions were deficient with respect to:

(a) drilling practices during coring operations (cores No. 2 and 3),
(b) implementation of standard kick control procedures,
(c) ensuring adequate mixed drilling mud was available at all times, and
(d) maintaining equipment in satisfactory operating condition (casing pressure instruments).

It appears to the Panel that the fundamental problem was that Amoco did not apply the necessary degree of caution while carrying out operations in the critical zone. Amoco did not appear to be sufficiently aware of the potential for problems that could occur when coring into the Porous zone and thus the need to be fully prepared in the event of a fluid influx. Consequently, when a kick developed, there were delays in responding to it. Then, when equipment problems occurred and supplies of mixed mud were inadequate, Amoco was forced into further delays of precious time in implementing kick-control actions.[31]

In a second phase of the inquiry, the Inquiry Panel made a series of recommendations to reduce the possibility of future blowouts.[32]

CONCLUSION
Oil drilling is a demanding, uncertain, and dangerous job. The Lodgepole blowout, with two tragic deaths and millions of dollars in financial losses, illustrates this point. There are other losses that are harder to put a price on, such as the threat of H_2S poisoning, the disruption of life, and the inconvenience caused to almost all the residents within a 30-km radius of the well. It is also difficult to evaluate the magnitude of loss suffered by Amoco and its technical staff as a result of the negative publicity resulting from the blowout. Safe, standard procedures on drill sites are essential if these dangers are to be avoided.

CASE HISTORY 11.5
THE BRE-X MINING FRAUD

INTRODUCTION
In the spring of 1997, Bre-X Minerals Limited, a mining company in Calgary, was the focus of a spectacular mining fraud. The Bre-X story is an intriguing but scandalous saga of a gold discovery that was reported-

ly richer than any other ever found; in fact, the discoverer received an award for his achievement. However, after a mysterious death, the fraud began to unravel, and the still-evolving scandal has ruined the lives of almost everyone involved. Most seriously, the fraud caused financial calamity for many thousands of investors, from pension funds to individual investors, some of whom staked their life savings on geologists' reports. An independent investigating team from Strathcona Mineral Services later stated: "The magnitude of tampering with core samples ... is of a scale and over a period of time and with a precision that, to our knowledge, is without precedent in the history of mining anywhere in the world."[33]

For our purposes, the story begins in May 1993 when David Walsh, the founder, chairman, and CEO of Bre-X Minerals Limited, announced the discovery of a gold deposit in Busang, Indonesia. One site, drilled previously by an Australian company, was reported to contain an estimated one million ounces of recoverable gold.[34] This modest estimate was to escalate as the months passed, generating an investment frenzy that pushed Bre-X stock prices from pennies in March 1994 to the equivalent of over $200 per share in September 1996.[35]

After the revelation of the fraud in early 1997, the stock value dropped sharply and soon became worthless. The resulting panic in the sale of mining stocks and precious-metal mutual funds immediately eroded public confidence in gold in general and in Canadian mines in particular. Estimates of the total loss to investors as a result of the Bre-X fraud run as high as $6 billion.

The Bre-X scandal had a serious harmful effect on the Canadian mining industry, and even junior mining companies that have no link to Bre-X had trouble raising the necessary capital to develop their discoveries.

SEQUENCE OF EVENTS

The roles played by various Bre-X geological staff are not completely clear as this text goes to press, although they may be clarified when the many pending lawsuits come to trial. John Felderhof, a 1962 geology graduate of Dalhousie University, is frequently referred to as the chief geologist[36] of Bre-X Minerals, although he has since declared that his role was that of an administrator or commercial manager. Michael de Guzman, a geologist from the Philippines, was "Bre-X's No. 2 geologist." De Guzman was running four Bre-X camps in Indonesia, so much of the work at the Busang site was reportedly supervised by a fellow Filipino geologist, Cesar Puspos.[37] Walsh was actively involved in raising funds in Calgary, Felderhof was in Jakarta, and de Guzman and Puspos were in Busang. In addition, about twenty others worked as geologists or project managers for Bre-X in Indonesia.

A reputable Australian drilling company was hired to drill core samples to evaluate the gold content of the Busang site. In March 1996, Bre-

X reported estimates of 30 million ounces of gold at the Busang site, which was soon raised to 70 million ounces, with a potential of 100 million ounces. In early 1997, Felderhof increased the "official" reserve estimate at Busang to 200 million ounces of gold.[38]

However, the golden glow began to tarnish in January 1997 when a storage building containing the core samples at the Busang site burned down, allegedly destroying the records of the drilling results. World attention was drawn to the unravelling fraud on 19 March 1997, when de Guzman committed suicide in a spectacular jump from a helicopter, leaving a suicide note stating that the cause of his suicide was poor health. A body, later confirmed to be his, was recovered from the Indonesian jungle. At the time, de Guzman had been en route to a meeting with a geological team to discuss the discrepancies in test results. Additional test holes had been drilled next to the Bre-X drill-holes by Freeport-McMoRan Copper & Gold, Inc., a company in partnership with Bre-X, and the results were quite different from the glowing results quoted by Bre-X.

The key events in the "discovery" of the gold and the uncovering of the fraud are summarized concisely as follows:

De Guzman worked on the [Busang] site with a team of Filipino technicians who processed the samples — two-meter-long cylinders of rock — which were then sent to an independent laboratory in Samarinda, the nearest big town, for assaying. (For a single evaluation commissioned by Bre-X last January [1997], more than 16 000 samples were submitted.) Felderhof visited the mine monthly. In public announcements, the chief geologist steadily increased his estimate of the lode's size: from 2.5 million ounces to 30 million, to 70 million to an eventual 200 million, which would be worth nearly $70 billion at current prices.

As a relatively small operator, Bre-X lacked the capital and expertise to exploit the discovery by itself, so it took on Freeport as a partner — or rather, the relationship was forced upon Bre-X by the Indonesian government. After Freeport raised questions about the extent of the find in March, Bre-X commissioned an independent firm, Strathcona Mineral Services of Toronto, to make a thorough analysis of the rock taken from Busang. Strathcona found that the samples had been "salted" — the industry term for adding minerals like gold where none exists. Salting is not especially difficult to detect if it is suspected. Feasibility tests carried out on Bre-X samples by several companies from as early as 1995 showed that the gold found in Busang samples was not the type that comes out of the ground, but rather the kind that is found in rivers — and that Dayak tribesmen had been finding in small amounts for years. Since the tests were never made public and the companies that carried them out say they weren't hunting for fraud, Bre-X never got caught on that point, or on several other breaches of normal sampling methods. Core-sample cylinders are normally sliced up the middle, with one part kept intact and the other crushed, marked with the precise location of its extraction, and sent to a laboratory for evaluation. Bre-X never maintained the uncrushed half for later verification. So it was impossible to say with precision where each sample had come from.

Analysts say the pattern of salting at Busang was remarkably skilled: samples have to have varying quantities of gold to be consistent with the pattern of how the metal is distributed in an underground seam. "This was not just a matter of tapping the salt shaker," says Freeport-McMoRan's [CEO Jim Bob] Moffett. "The results had to give a very specific three-dimensional picture of a plausible deposit. The whole picture had to make sense. It had to be very well-planned and well-executed."

The chief candidate for that role is de Guzman, who had the knowledge: he frequently gave speeches to geological societies on his theory about gold in earthquake zones. . . .[39]

AFTERMATH

In the aftermath of the fraud discovery, Bre-X hired an investigative team, Forensic Investigative Associates (FIA), to perform an independent audit. The FIA report, published in October 1997, exonerated Bre-X's senior staff and stated that FIA had "reasonable and probable grounds" to conclude that de Guzman and others at the Busang site were responsible for the ore salting.[40]

Many Bre-X employees had profited personally by selling shares that they had purchased with stock options. The FIA report estimated that de Guzman received $4.5 million in stock sales, Puspos $2.2 million, and Walsh about $36 million.[41] Felderhof reportedly sold about $30 million of his shares.

In March 1997, just before the fraud was discovered, Felderhof was named Prospector of the Year at the annual meeting of the Prospectors and Developers Association of Canada in honour of the Busang discovery, which was believed at that time to be the world's largest single gold deposit. A few months later, he agreed to return the award. He was also asked to resign from Bre-X, and as this text goes to press, he resides in the Cayman Islands.[42]

Bre-X has been delisted from the stock exchange, and its shares are essentially worthless. Lawsuits are being prepared by many investors, both corporations and individuals, who believe that the corporation and the individuals who controlled its geological activities should have shown greater diligence in controlling the assay samples and verifying the gold estimates. The lawsuits and related problems will persist for years, perhaps decades.

CONCLUSION

The Bre-X scandal is a case of skilled geological fraud, apparently perpetrated by de Guzman. Were he alive, de Guzman would face criminal charges for fraud, as well as discipline for professional misconduct. Although he was working for a Canadian company, the criminal activities occurred in a foreign country, so the laws of that country would have to be followed (although Canadian authorities could act if any of his work was actually conducted in Canada).

In the face of such deliberate fraud, it may seem trivial to refer to the code of ethics, since it is clear that the individuals involved were not concerned about breaking the code. Nevertheless, these examples demonstrate the classical results of ignoring the code of ethics: the perpetrator has come to a bad end, but in the process, thousands of people have suffered serious financial harm, and an entire industry has been held up to contempt and ridicule. Many lawsuits have begun, and there will undoubtedly be more ruined lives before the scandal ends.

The vital importance of clear and unambiguous duties and titles is evident, after the fact. The chief geologist was reportedly Felderhof. He certainly seemed to consider himself qualified for this title when he made estimates of the gold content in Busang and when he accepted the Prospector of the Year award, although he later claimed to be merely an administrator. In any case, the chief geologist — whether Felderhof or some other person — had a responsibility to show due diligence in safeguarding the core samples and ensuring that the gold assay was properly done, that the gold content, based on the samples, was accurately calculated, and that double-checks were made to confirm the results. As mentioned above, core samples are usually split before testing, and half of each sample is retained for further confirmation, if needed. However, it is clear in the Bre-X case that none of this was done. Security was loose, and all of the Busang samples were crushed and tested under the control of a single individual, permitting the salting to proceed undetected for many months.

Although stock promoters who unwittingly encouraged the investment of billions of dollars in a fraud may need to question the standards of due diligence in their own profession as well, the Bre-X case emphasizes once again the critical dependence of the mining and resource industries on professional geologists with high ethical standards.

CASE HISTORY 11.6
THE *CHALLENGER* SPACE SHUTTLE EXPLOSION

INTRODUCTION

On 28 January 1986, the U.S. National Aeronautics and Space Administration (NASA) launched the space shuttle *Challenger* at Cape Canaveral, Florida. The launch had been delayed by bad weather, and since part of the cargo was a satellite that required a specific "launch window," there was some concern about the delay. It had been rather cold overnight, and since icicles that had formed on the shuttle might break off during the launch and damage the tiled heat shield, the launch was delayed even further for inspections.

At 11:38 A.M., the rockets were finally ignited. At first, the shuttle rose according to the flight plan; however, at 59 seconds into the flight a plume of flame was evident near the booster rockets. At 64 seconds, the flame had burned a hole in the booster, and at 72 seconds, the booster's strut detached from the external tank. At 73 seconds into the flight, the loosened booster hit *Challenger*'s right wing and then hit the fuel tank. The tank exploded. The shuttle was at an altitude of 14 600 m and travelling at about Mach 2 when the explosion occurred. Although the explosion may have killed some of the seven crew members, later evidence showed that the crew module, which separated from the rocket during the explosion, made a 2-minute, 45-second free fall, hitting the ocean at a speed in excess of 320 km/h, killing the remaining crew members. Fragments of the shuttle continued to rain down on the rescue team for about an hour after the explosion.[43]

The *Challenger* explosion represented the first deaths of U.S. astronauts during a mission (although there had been three U.S. deaths in a ground test for the first *Apollo* mission, and three Soviet deaths when parachutes failed to deploy at the end of the first *Soyuz* mission). The *Challenger* launch, however, had an immediate and personal impact on people around the world who watched it on live television. The disaster was also a serious setback for the U.S. space program.

INVESTIGATION

U.S. President Reagan convened a commission to investigate the *Challenger* explosion. The investigation involved over 6000 people, and the resulting 256-page report was issued in June 1986.[44] After exhaustive deliberation, it was finally decided that the accident was caused by hot gases blowing past one of the seals in the rocket boosters. It was concluded that the seal had been unable to do its job properly because of the unusually low temperature in Florida on the day of the launch — about –8°C. Claims were made that the management of Morton Thiokol, the manufacturer of the boosters, had had engineering information that cast doubt on the seals, but had decided to go ahead with the launch anyway.

After an earlier launch, a postflight examination of the solid rocket motors (these are recovered from the ocean after each flight) showed that hot gases had blown past the primary seal in one of the joints, although the gases had been stopped by a secondary seal. Each joint was provided with two seals, because the motor sections could distort under pressure; if the primary seal failed, then the secondary seal would maintain joint integrity — or so it was thought. Attempts had been made to improve the situation, but without much result.

It is not generally known that this distortion of the joint was revealed by testing as far back as 1978, eight years before *Challenger* blew up. Furthermore, the joint was labelled unacceptable by NASA engineers, and a redesign was demanded. Morton Thiokol argued that the joint design was derived from that used in many successful flights using the Titan rocket and had proven to be very reliable. Furthermore, the new design was better than Titan's because it used two O-rings for sealing instead of one. This was perceived as conservative, with safety as the rationale.

In following this sorry tale, it is worth bearing in mind that the proven reliability of the Titan system dominated the thoughts of almost everyone and finally proved to be the source of the self-mesmerization that was at the core of the problem. One of the witnesses before the presidential commission put it this way: "All of the people in the program ... felt that this solid rocket motor ... was probably one of the least worrisome things in the program."

But there were a couple of important differences from the Titan design. First, the *Challenger* joint was more flexible than the one in the Titan, so more distortion took place. Second, when the sections of the rocket were put together, there was a gap leading from the O-rings right up into the combustion area. Zinc chromate putty was placed into this gap to protect the O-rings from the hot gases. After a number of tests in which the joints apparently were successful, the viewpoint gradually became "After all, it's worked for years in the Titan, so isn't it the safest thing to do?" Everything might still have come out all right if it had not been for another step taken to ensure safety: before each flight, the joints had to be tested under pressure to make sure the O-rings were sealed properly. But the people conducting the tests wanted to be sure they were really testing the O-rings and not just the putty, so they put on enough pressure to blow a hole through the putty. There you have it: the putty was essential in protecting the O-rings from the flames, but to test the rings, a hole was put in the putty that would lead directly from the flames to the O-rings. Even after this fact became known, it was still believed that if the first ring did not hold, the second one would. Besides, there was the comforting image of Titan's reliability.

O-RING DESIGN

Our primary source of detailed information about the O-rings is Roger M. Boisjoly, who was an engineer with Morton Thiokol at the time and who was directly responsible for the seals in question. Here is a summary of Boisjoly's version of the events.[45]

Following a shuttle launch a year earlier, Boisjoly had noticed, in a postflight examination of the boosters, that hot combustion gases had apparently blown past the primary seal in one of the booster joints, although the gases had been stopped by the secondary seal. The ambient temperature for that launch had been in the 18 to 22°F (–8 to –6°C) range for several days prior to launch and was 60 to 65°F (16 to 18°C) at launch time. It was calculated that the seal itself was at a temperature of about 53°F (12°C).

Laboratory simulations of the conditions for the previous launch were set up by Morton Thiokol, but were not conclusive. At 50°F (10°C), if the seals were compressed 0.040-inch (0.1 cm) and then separated 0.030-inch (0.07 cm), there was loss of seal contact for more than ten minutes. (The 0.030-inch [0.07 cm] separation was to simulate the effect of booster-section expansion under internal pressure.) If the seals were compressed 0.040-inch (0.1 cm) and then separated only 0.010-inch (0.03 cm), the seals successfully maintained contact. In another test, the seals maintained their integrity down to 30°F (–1°C), but this test fixture was rigid, and did not simulate the expansion under pressure.

Boisjoly wrote memos and reports to document his concern about the seals. He recommended the formation of a special team to work on the sealing problem, to consist of three engineers and four technicians. Approximately six months before the accident occurred, in a memo to Morton Thiokol's vice-president of engineering, he said:

> It is my honest and very real fear that if we do not take immediate action to dedicate a team to solve the problem, with the field joint having the number one priority, then we stand in jeopardy of losing a flight along with all the launch pad facilities.[46]

The night before the disaster, a teleconference was held between Morton Thiokol and NASA personnel. The overnight low temperature in Florida was predicted to be 18°F (–8°C). Boisjoly presented his calculations, which showed a predicted seal-ring temperature of 27 to 29°F (–3 to –2°C) under those conditions. He urgently recommended that no launch take place unless the temperature reached 53°F (12°C) or higher. Management, placing its reliance on the laboratory test that showed that the seal had retained its integrity down to 30°F (–1°C) (but without providing for distortion) and on its belief that if the primary seal did not hold then the secondary one would, ordered the launch to proceed. The next day, *Challenger* blew up.

Boisjoly was appointed to the failure investigation team and was placed in charge of redesigning the seal. During subsequent hearings before a presidential commission, he freely volunteered information and handed over packets of his internal company correspondence, contrary to instructions from Morton Thiokol's attorney. Boisjoly says that his actions incurred strong company disapproval and that he was subsequently isolated from any contact with NASA and from the main redesign activity of the seals. Feeling that the environment at Morton Thiokol had turned hostile, he resigned in July 1986. The company gave him six months of extended sick leave and after that placed him on two years of long-term disability at 60 percent of his salary.

Boisjoly then filed a $1 billion lawsuit against Morton Thiokol and a $10 million lawsuit against NASA for his lost salary and ruined career. He places the blame for the accident on a subtle shift in philosophy in the space shuttle program. In prior days, Boisjoly said, it was necessary to prove that conditions were safe before a launch was given the go-ahead. In the case of *Challenger*, it was necessary to prove that conditions were not safe before a launch would be cancelled.

CONCLUSION

In reflecting on the *Challenger* explosion, engineers can learn two lessons. First, in an enormously large organization, it is easy for decision making to fall through the cracks; in fact, this was the principal conclusion drawn by the members of the presidential commission, and they directed their harshest criticisms at NASA's administrative structure. Second, it is very easy for engineers to fall into the comforting belief that they are following a conservative course (the Morton Thiokol engineers apparently thought so), when in fact they are deviating into dangerous territory. Disasters are easy to create — safety comes hard.[47]

CASE HISTORY 11.7
THE DC-10 PASSENGER AIRCRAFT DISASTER

INTRODUCTION

The design of the cargo door latches on the DC-10 passenger aircraft gives us a prime example of a case in which danger was clearly perceived and someone should have taken corrective action but did not. In this case, the whistle was *not* blown, and a tragedy resulted.

The DC-10 is a three-engine wide-body jet airliner that is slightly smaller than the Boeing 747. It was first certified to carry passengers in 1971. In the DC-10, as in almost all airliners, the pressurized space in the fuselage below the passenger cabin is the cargo compartment. In 1972, a cargo door blew out of a DC-10 over Windsor, Ontario, and the explosive decompression of the cargo compartment caused part of the cabin floor to collapse, opened a large hole in the bottom of the fuselage, and, most seriously, severed most of the hydraulic control lines. In the DC-10, the control systems are routed through the cabin floor (instead of through the ceiling, as in the 747 aircraft), so when the floor collapsed and the hydraulic lines were cut, almost all control of the ailerons and rudder was lost. Miraculously, the pilot was able to steer and land the aircraft safely by manipulating the engine throttles.

The Windsor incident was a portent of the tragedy to come. Two years later, on 3 March 1974, nine minutes after takeoff from Paris, DC-10 flight THY 981 lost its cargo door. The people on this flight were not so lucky. The resulting decompression of the cargo compartment again caused the cabin floor to collapse; control of the ailerons and rudder was lost, and the plane crashed, killing all 346 people aboard.[48] At that time, the accident was the worst air disaster involving a commercial airliner. (There have been at least three commercial airline crashes of similar magnitude since that time.) A recent television documentary about the crash showed that, almost 25 years later, the Ermonenville Forest near Paris is still strewn with hundreds of small fragments of the doomed airliner.

DETAILS OF THE CARGO DOOR DESIGN

In 1968, McDonnell Douglas, the manufacturer of the DC-10, gave a subcontract to the Convair Division of General Dynamics to perform the detail design of the fuselage of the DC-10, including its cargo doors. (Such arrangements are common in the aircraft industry, whereby one manufacturer subcontracts portions of the design to others.) Initially, the specifications required the use of hydraulic actuators to drive the cargo latches, but later Douglas told Convair that electric actuators were to be used instead, since they were lighter. The distinction between hydraulic actuators and electric actuators is important to the case. If, for some reason, hydraulic latches failed to seat properly, a fairly moderate degree of internal fuselage pressure would force them open. Such a sequence of events, if it occurred, would take place at a fairly low altitude and pressure differential, and the resulting decompression would not be disastrous — the aircraft could land safely. But if electric latches did not fully seat, the cargo doors would probably be forced open at a time when a much higher pressure differential had developed, so that the resulting decompression would be catastrophic.

HISTORY OF SIMILAR FAILURES

In the year prior to the certification of the DC-10 for service, there had been five instances in which cargo doors on DC-8s and DC-9s had blown open during flight. Because the DC-8s and DC-9s were equipped with hydraulic actuators, the accidental openings occurred under moderate pressure differentials, and the planes landed safely. During the design of the DC-10 fuselage, Douglas asked Convair to prepare a failure mode and effects analysis (FMEA) for the cargo door system. In the FMEA, Convair pointed out that little reliance could be placed on the use of warning lights to indicate the presence of an improperly latched cargo door, because the warning lights themselves were subject to malfunction. Even less reliance could be placed upon the alertness of ground crews to check such things as fully closed latches, because any such procedure was too much subject to human error. Beyond this, the FMEA described a number of scenarios that could prove hazardous to life. One of these involved a failure of the latch to seat properly, leading to an explosive decompression of the cargo compartment, collapse of the cabin floor, and loss of control of the aircraft.

In 1970, during the ground tests of the first DC-10, the aircraft was being subjected to pressurization tests when a cargo door suddenly blew open. The accident was blamed on the failure of a mechanic to close the door properly. But even before the accident, Douglas had decided that further safeguards were needed to prevent the possibility of such an accident. A hand-operated locking handle was provided, together with a small door near the locking handle. The door was supposed to stand open until the locking handle was operated. A member of the ground crew would look through the door to be sure that the locking pins had fully seated, then the door would be closed. However, there was no necessary connection between the two. If the locking handle failed to seat properly, the door could still be closed. Thus, the DC-10 system depended on a member of the ground crew understanding and following the correct steps.

INADEQUATE REMEDIAL ACTION

After the cargo door blew out on the DC-10 over Windsor, Ontario, the United States Federal Aviation Administration (FAA) was officially in the picture. (Although the incident occurred in Canadian airspace, the airline was American.) FAA personnel prepared a draft of an "airworthiness directive" that would have ordered McDonnell Douglas to take certain actions before the DC-10 could resume operation. But the airworthiness directive was not issued. Instead, an informal agreement (referred to later as the "gentlemen's agreement") was worked out between Douglas and the head of the FAA, detailing certain steps that Douglas would take to correct the situation. (These steps apparently included the provision of small inspection windows or "peepholes," deeper latch engagement, and stiffer latch linkages.) The ill-fated DC-10 that crashed near Paris was on the production line when the modifications were ordered, but for some unknown reason did not receive all of the modifications, even though the inspectors responsible for the aircraft certified that all modifications were done. A further relevant factor is that the cargo door latching instructions printed on the aircraft were in English. The ground crew member who was responsible for the cargo door closure knew several languages, but English was not one of them. The FAA investigators later learned that even cargo handlers in the United States were unaware of the purpose of the peepholes, and assumed they were there for the convenience of the mechanics.[49]

Shortly after the Windsor blowout, Convair's director of product engineering wrote an internal memo to his superiors declaring that the safety of the cargo door latching system had been progressively degraded since the inception of the program. After the first blowout (the one that occurred during ground testing), Convair had discussed possible corrective action with Douglas, including the possibility of providing "blowout panels" in the cabin floor. If the cargo area suddenly lost pressure, these panels would blow out, but the cabin floor would not collapse. Instead of that alternative, however, Douglas had opted to install the peepholes that would allow visual observation of the manual latching system. The memo also suggested strengthening the cabin floor so that it could resist a sudden decompression, but this would add 1400 kg in mass — a serious matter for an aircraft. The memo contained this paragraph:

> My only criticism of Douglas in this regard is that once this inherent weakness was demonstrated by the July 1970 test failure [the ground test], they did not take immediate steps to correct it. It seems to me inevitable that, in the twenty years ahead of us, DC-10 cargo doors will come open and I would expect this to usually result in the loss of the airplane.

The Convair senior management's rationale, in response to the ideas raised in the memo, was essentially as follows:

- Since Convair had not raised any objections to Douglas's design philosophy at the beginning of the program, Convair, in effect, had agreed that the proper design philosophy was to design a safe cargo door latching system in lieu of designing a stronger floor or providing blowout panels.
- A design philosophy involving a safe latching system would satisfy FAA safety requirements.
- Douglas had unilaterally redesigned the cargo door latch system and had previously rejected the proposal of blowout panels.
- Convair management had been informally advised that Douglas was making corrections to the latching system and was reconsidering the provision of blowout panels.

As a result of the above, Convair management made no formal communication to Douglas concerning the safety of the cargo door. The company's justification was that the arguments advanced in the memo were already well known to Douglas, and it was not likely that any additional actions would be taken beyond those that Convair understood were already taking place. In addition, an adversarial relationship between Douglas and Convair had developed, in which Douglas was seeking to shift the costs of redesign to Convair. It was feared that if Convair now questioned a design philosophy with which it had originally concurred, Douglas would use this as a further pretext to shift all the costs of redesign to Convair.[50] An opportunity to prevent the flight THY 981 disaster was lost.

CONCLUSION

Probably the saddest feature of this case is that the FAA failed so badly in its role of safety watchdog. The FAA certified the cargo door design as airworthy, in spite of the failure during the ground test. Then, after the near-disaster over Windsor, the FAA merely advised the airlines to follow the manufacturer's service bulletin, rather than issuing a directive that would have required all DC-10s to be retrofitted.[51]

This case underscores the serious responsibility for safety borne by engineers and the importance of developing proper design philosophies. These strategies, usually followed by a failure modes and effects analysis, are the basis of safe design. However, clear communication and clear delineation of responsibilities are critically important in implementing any design philosophy. If someone in the design team, in management, or in the FAA had insisted on greater safety, there were certainly many opportunities to provide a safe locking system. For example, when the inspection door was provided to go with the manual latching system, the design engineers could also have provided an absolute interlock that would have prevented the door from being closed unless the locking pins were fully seated. A safety system that depends on human beings to execute a series of actions, such as making visual checks to see if latches are fully seated, is almost guaranteed to fail at some time. In aircraft, and in any structure where many lives are at stake, such systems must be fail-safe; that is, safety must be designed into the system hardware itself.

CASE HISTORY 11.8
TOXIC POLLUTION: LOVE CANAL, MINAMATA, BHOPAL, SUDBURY

The improper disposal of toxic or environmentally harmful waste, whether through deliberate secret dumping or through naïve or incompetent operation of a complex processing plant, is professional misconduct. This dangerous ethical problem occurs around the world and is not unique to any particular country, culture, or political system. The following brief history reviews four well-known cases of harmful waste being disposed of improperly in four different countries. The cases are so well known that the names of the areas — Love Canal, Minamata, Bhopal, and Sudbury — have become synonymous with pollution.

LOVE CANAL, NEW YORK — DIOXIN

The Love Canal is named after one of the early residents of New York state. The canal was originally excavated for boats and barges, but since it was never fully completed for navigational purposes, the term "canal" is a misnomer. In the years between 1942 and 1953, the Olin Corporation and the Hooker Chemical Corporation saw it as a convenient hole for burying waste chemicals. Over 18 000 tons of chemical waste, including dioxins, were buried, and eventually the "canal" was again a flat piece of land. Only the people who had buried the chemicals knew what lay beneath the soil.

In 1953, the Hooker Chemical Company donated the land to the local board of education but said nothing about the chemicals buried there. Over time, the land became a residential area, and homes, playgrounds, and a school were built on it. In 1976, after several seasons of heavy rains, people began to notice the terrible smell of chemicals. Homes reeked unpleasantly, children complained of chemical burns, and pets died or became sick. These problems were minor compared with the miscarriages, birth defects, and cases of cancer that appeared in the Love Canal area more and more frequently as years passed. Residents soon demanded some action, on the basis that these problems were occurring at much higher rates than should have been expected. In 1978, a government study of the area exposed some remarkable — and scary — statistics: over 80 different chemicals were detected in the air, some of which were carcinogenic (cancer-causing); the chemical pollution in the air was 250 to 5000 times safe levels; there was an unusually high (almost 30 percent) rate of miscarriages; and of 17 pregnant women in the Love Canal area in 1978, only 2 gave birth to normal children.[52]

New York state authorities recognized the serious health threat posed by the buried chemicals and moved a few hundred families out of the area. The school was closed and surrounded by barbed wire. The area over the buried chemicals became a ghost town, with derelict houses, empty streets, and "No Trespassing" signs. Neighbouring residents who lived only a small distance from the chemicals were concerned about their health but faced a dilemma: they wanted to move away from the danger, but since the term "Love Canal" had become a synonym for hazardous waste, their homes were worthless. To move, they would likely have to abandon their homes. In 1980, residents' demands forced the government to carry out further testing. The tests showed high levels of genetic damage among the neighbouring residents. The area was declared a disaster area by the U.S. president, and another 710 families were relocated. Many of the abandoned homes were demolished, and the chemical wastes were excavated for treatment and proper disposal. The total cost of the cleanup was estimated at $250 million.[53]

The ethical issues in this case focus on the actions of the Hooker Chemical Corporation, which buried much of the chemical waste and then donated the land to the board of education. The company did not warn the board about the chemical contamination, but it is clear that the company was aware of the problem and concerned about it: the agreement to transfer the land contained several clauses to protect

Hooker against any future claims for liability. While the company took credit for its public generosity, it kept the danger of the buried chemicals secret. Even when the area was being developed for residential use and (as documents later showed) the company was aware that children had suffered chemical burns from playing there, Hooker maintained its secret. When the extent of the disaster became public, chemical industry spokespeople ridiculed the residents as hypochondriacs.[54] This toxic secret caused tragedy for many families who risked their life savings in worthless homes, whose children were born deformed, and who, even now, may live in fear of contracting cancer.

The only benefit to come from the Love Canal case has been the heightened awareness of the importance of ethical conduct and the need for environmental regulations and guidelines. In the years following the Love Canal revelations, the U.S. Environmental Protection Agency (EPA) discovered that there were between 32 000 and 50 000 other toxic waste dumps scattered across the United States, of which possibly 1200 to 2000 might pose significant risks to the public.[55] Moreover, only about 10 percent of the toxic waste generated each year was being disposed of properly. The resulting furor over the callous disregard for public welfare at Love Canal eventually led to more stringent laws and more severe punitive measures for improper disposal of waste.

Unfortunately, the public awakening to the danger of improper and unethical dumping was too slow in coming. Lois Gibbs, the resident who took the leadership role in drawing public attention to the environmental disaster at Love Canal, said recently that the areas in which dioxin was being released into the environment were "canary communities," and was dismayed that it took thirteen years for scientists to verify the "barefoot epidemiology" of the housewives who discovered that dioxin caused birth defects in their children.[56]

MINAMATA BAY, JAPAN — MERCURY POISONING

Since 1953, thousands of residents of the Minamata Bay area of Japan have fallen ill as a result of organic mercury poisoning. Mercury is the familiar liquid metal (also called "quicksilver") commonly found in thermometers and barometers. Mercury, whether as a pure liquid element or in compound form, can cause serious renal and neurological dysfunction. Subtle cases may mimic amyotrophic lateral sclerosis (ALS), but serious poisoning is evidenced by clumsiness, a stumbling "drunken" gait, serious mental or behavioural problems, and partial or complete loss of speech, taste, and hearing.

The Chisso company, a nitrogen fertilizer company in Minamata, the main city on Minamata Bay, first began producing acetaldehyde in 1932. Mercury was needed as a catalyst in this process and for other chemicals that the company would produce later, such as vinyl chloride. The mercury was used in liquid (or organic) form, but during the production process, a portion of the mercury was lost — washed into Minamata Bay with the waste water. In the bay, microbes acted on the mercury and converted it into an organic (methyl or carbon-based) mercury compound. The organic mercury was absorbed by shellfish; since mercury is not excreted by mammals, it becomes more concentrated as it moves up the food chain. Over time, the concentrations are sufficiently large that the toxic effect becomes apparent. The first humans to be affected were fishermen and their families, who had a diet rich in fish, including shellfish. Since family cats ate fish, they were also observed to exhibit curious behaviour. The seriousness of the problem was first recognized by physicians and health officials in the Minamata area about 1956, although cases were later traced back to 1953, about twenty years after the first mercury began to wash into the bay.[57]

The medical director of the hospital associated with the Chisso company became sufficiently concerned about the problem in 1956 that he began a series of tests on cats. He identified the manufacturing plant effluent as the cause of the problem, but when he reported his results to his superiors at Chisso, he was ordered to stop his tests and forbidden to report his findings to the local health authorities.[58]

It was not until 1959 that medical authorities requested help from the Kumamoto medical school and an investigation was begun. By 1962, they were certain that the problem was caused by organic mercury and that the source of the mercury was the Chisso effluent.[59] Initially it was estimated that about 2900 people had contracted "Minamata disease," as the debilitating neurological syndrome has come to be known. However, a 1995 estimate places the number between 7000 and 8000 people.[60]

The government indicted Chisso, but it took years for the various cases to work their way through the legal system. The first decision was made in 1970, and some compensation was also awarded by the government. However, as of 1995, some 100 of the victims were still pressing lawsuits for compensation.

The ethical issues in this case are clear. The Chisso company's actions — stopping the medical tests, suppressing knowledge of the problem, and continuing to permit mercury to be dumped in Minamata Bay — were unethical and inexcusable. The only mitigating factor is the fact that the damaging effects of mercury were not well known in 1932. The engineering knowledge, testing facilities, and environmental standards of today are, of course, far superior to anything they might have had in those days. However, the company knowingly inflicted personal tragedy on thousands of unwitting people.

An outbreak of Minamata disease occurred in Grassy Narrows, Ontario, in 1970. Members of two Ojibwa bands living near the Wabigoon River began to show the distressing and debilitating symptoms of mercury poisoning, and the source of the pollution was traced to the Reed Paper company in Dryden, just upstream from the Ojibwa reserves.

The provincial government ordered Reed to reduce its mercury usage, and the pollution was eventually eliminated. Although a settlement was eventually paid to the two bands as compensation, the economic and social effects on the bands were tragic.[61]

BHOPAL, INDIA — METHYL ISOCYANATE

In the early morning hours of 3 December 1984, a poisonous cloud of methyl isocyanate gas escaped from the Union Carbide plant in Bhopal, India. It killed thousands of people up to 6 km away, some sleeping in their beds. It was likely the worst industrial accident in human history: the estimates of the number of casualties are about 3000 to 12 000 dead, about 30 000 with permanent injuries, 20 000 with temporary injuries, and 150 000 with minor injuries. The sheer numbers of casualties are almost inconceivable to most North Americans, who have never had a personal connection with a tragedy of this proportion and for whom the memories of war casualties are now getting dim. However, even these massive numbers are disputed by victims' rights organizations, who say the real numbers are higher.[62]

The Union Carbide plant was established in 1969 as a mixing factory for pesticides. Methyl isocyanate, used in large quantities in the production process, is a highly volatile and toxic substance estimated to be ten times as deadly as the phosgene gas (or "mustard gas") used during the gas attacks of World War I. Methyl isocyanate reacts vigorously with many common substances and must be maintained at very low temperatures to prevent uncontrolled reactions from occurring. The precise cause of the disaster is not known, but the most probable cause has been traced to an employee who closed a valve on a piping system so that a filter connected to the pipe could be washed. A metal disc should have been inserted to guarantee that the valve could not leak, but this was not done. During the washing process, the water leaked past the closed valve, entering piping that was connected to the methyl isocyanate holding tank. The water reacted with the methyl isocyanate, generating immense heat. The pressure in the tank increased dramatically and pushed past pressure relief valves into the atmosphere. Safety measures were either inadequate or did not work. Over a period of 90 minutes, about 40 tonnes of methyl isocyanate and other reaction products were released into the atmosphere. Since the vapour was heavier than air, it filled low-lying areas, entered the houses through openings, and asphyxiated thousands of sleeping residents.

An inquiry followed the disaster. The construction, operation, and maintenance of the Bhopal plant were examined, as well as management decisions that permitted such a potentially dangerous plant to operate in such an unsafe manner, in an urban area, with no suitable

▲ *The Union Carbide factory in the background looms over relatives and friends carrying a victim's body to cremation. Source: Canapress/Associated Press (AP).*

emergency plan. The Indian government charged the company management with negligence, brought murder charges against its chief executive, and demanded $3.3 billion (U.S.) to settle victims' claims. However, in 1989, the Indian Supreme Court announced a settlement of all claims for $470 million (U.S.) conditional on dropping the criminal charges.[63] However, shortly after the settlement was announced, a new Indian government disallowed the claim and sought to reinstate the criminal charges.[64] The Indian government is presently supporting the survivors of the Bhopal disaster while the litigation continues.

SUDBURY, ONTARIO — SULPHUR DIOXIDE

Canada is the world's second-largest producer of nickel, and the mines in the Sudbury region of Ontario are the main source of this metal. Nickel was first discovered in Sudbury in 1856, but it was not until the Canadian Pacific Railway reached Sudbury in 1883 that the full size of the ore body was recognized.[65] The nickel is in the form of sulphide ore, and it cannot be directly converted into metallic form; it must first be smelted — burned to remove the sulphur and convert the ore to nickel oxide, which can be reduced to pure nickel. In the early 1900s, the first conversion step was typically done in huge, open "roasts," where layers of timber were interspersed with layers of ore. The roasts burned around the clock and emitted a toxic cloud of sulphur dioxide.[66] The ecological aspects were not given much thought a century ago when the Sudbury area was sparsely populated, and few people realized that the sulphur dioxide, when dissolved in water, became acid rain. In 1928, the federal government became aware of the problem and banned the use of open roasts.[67] However, enclosed smelting is still done.

The problem was improved by the erection of larger smokestacks, which spread the pollution over a greater area, thus reducing the concentration. The largest superstack, at the Copper Cliff mine in Sudbury, was built in 1972 and is about 380 m high.

The environmental effects of acid rain on the Sudbury region are immense: in the areas surrounding the smelters, trees are stunted and sparse, lakes are devoid of fish, and bird life is nonexistent or composed only of very hardy species. In 1978, the Regional Municipality of Sudbury began an ambitious program to restore 10 000 hectares of barren land. By experiment, it was discovered that a combination of agricultural lime to neutralize the acidity, fertilizer, and a seed mixture of grass and legumes would grow into a healthy grass cover in most locations. Two years after the grass grew, crews returned to plant trees and shrubs. Fifteen different tree species were planted, and, by 1995, many of these trees were more than 3 m tall. Although the acidity is still very high, the levels of heavy metals have been reduced; populations of insects, birds, and small mammals have increased; and successful tree growth has averaged 70 percent across all species. About 3000 hectares have so far been reclaimed.[68] As a result of its extensive environmental damage, Sudbury has ironically become one of the most closely studied ecological areas in the world.

CONCLUSION

These four incidents show that environmental damage can occur anywhere, and that it can be sudden, as in Bhopal; or it can be slow but insidious, as in Minamata and the Love Canal; or it can take decades, as in Sudbury. These environmental disasters are not, of course, of equal magnitude, nor do they contain equal evidence of unethical action.

Thousands of deaths and injuries were directly connected to the Bhopal incident — the world's worst environmental disaster — and to the Minamata incident. Many illnesses, miscarriages, and cancer cases were related to the Love Canal, but no such human tragedies have been directly related to the Sudbury sulphur dioxide emissions, although the pollution is evident in the lakes and vegetation downwind of the smelters. Each of these incidents, to varying degrees, involves ignorance, carelessness, or incompetence, and most of them involve an arrogant lack of ethical action. Industrial processes that involve chemicals carry a great potential for disaster and require a particularly high level of competence, diligence, and ethical decision making from their managers. As these incidents show, the consequences of negligence can be terrifying.

CASE HISTORY 11.9
NUCLEAR SAFETY: THREE MILE ISLAND AND CHERNOBYL

INTRODUCTION
One of the major sources of public concern over engineering and technology involves generating electricity by nuclear fission, even though U.S. and British studies show that power generation by coal is 250 times more hazardous, and oil is 180 times more hazardous than nuclear power. Only natural gas poses fewer hazards to workers and the general population.[69] There were a few serious radiation releases in the early, experimental days of Canadian and U.S. nuclear development, and a few deaths in North America have been attributed to power reactor accidents. However, the U.S. reactor accident at Three Mile Island in 1979 and a later accident at Chernobyl in the Soviet Union in 1986 involved partial or complete meltdown of the nuclear reactor cores, and were therefore much more serious than any other accidents before or since. The fact that meltdowns — the most unthinkable nuclear accident — could occur shocked the general public and created fear, uncertainty, and dislike of nuclear power that has lasted for decades. The only benefit to come out of these tragedies may, in fact, be the heightened caution and greater attention to safety in operating these massive power plants.

THREE MILE ISLAND
Three Mile Island is located on the Susquehanna River in southern Pennsylvania. Construction on its two-unit nuclear power plant began in the late 1960s, and the second unit (unit TMI-2) was completed, tested, and put on-line in December 1978. At the peak of production, the two units were designed to produce 880 megawatts of electrical power. The TMI-2 unit experienced many minor problems during its commissioning and early operation, and it had been in operation for

only three months when it was the source of North America's worst nuclear accident.[70]

The problem began at 4:00 A.M. on 28 March 1979 during a routine maintenance operation — changing water in a pipe system — and lasted about five hours. During the maintenance operation, air was accidentally admitted into the pipes, causing a blockage in one of the main feedwater lines. This caused the TMI-2 reactor to trip, and within eight seconds it had shut itself down. That would have been the end of the event except for the fact that a pressure relief valve stuck open, allowing radioactive water to escape from the system for more than two hours — leaving the reactor core partially uncooled — before the problem was correctly diagnosed and fixed.[71] Before it was all over, more than a third of the reactor core had melted and fallen to the bottom of the reactor vessel.[72]

Three Mile Island was the worst possible nuclear accident come true: some fuel rods in the core had melted. In the scenario for a meltdown, the molten mass is believed capable of burning its way through the bottom of the reactor vessel, then through the thick concrete lining at the bottom of the containment shell, to finally penetrate the ground-water table. In a cynical version of this event, the molten mass continues to eat its way through the interior of the earth until it emerges in China, thus giving it the name "the China Syndrome." However, it is more likely that, had the containment been breached, a conventional explosion would likely have occurred that would have released massive amounts of radioactive materials into the air, with the subsequent contamination of several eastern U.S. states. The accident was frightening in how close it came to the worst possible consequences.

Fortunately for the areas neighbouring Three Mile Island, except for the fact that the core did indeed melt, none of the rest of the scenario took place. The molten mass did not even penetrate the shell of the reactor vessel, but instead acted as an insulating layer as it cooled. The amount of radiation that escaped into the environment turned out to be negligible. One unexpected result was a buildup of steam and hydrogen in the containment dome. The steam was expected, but the nuclear designers had not known that the radiation would cause the steam to dissociate into hydrogen and oxygen. The risk of a hydrogen explosion was very great, and removal of the hydrogen was one of the first priorities of the clean-up. Through a combination of filtered release to the atmosphere and chemical action, the hydrogen was safely removed from the containment dome, thus avoiding an explosion.[73]

Although experts had predicted that an enormous amount of radio-iodine would be released in gaseous form, very little actually escaped. Instead, it was converted into an iodine that readily dissolved in water and thus remained inside the reactor containment.[74] Virtually all the other radioactive materials also remained inside the containment shell, except for inert gases like krypton and xenon, which pose little hazard. It is believed that the person who received the worst exposure from the accident was a man working on a nearby island in the Susquehanna River who received an estimated exposure of 40 millirems, which is relatively safe. (Natural background radiation causes an annual average dose of about 100 millirems for each person in North America.)[75]

The Three Mile Island accident may not have been a catastrophe for the public, but it was for the utility that owned it. The billion-dollar TMI-2 unit was a total loss. Furthermore, just the obligatory clean-up of the radioactive materials took eleven years and cost nearly $1 billion. The total TMI-2 reactor costs were therefore almost $2 billion, including construction and clean-up, and the unit never generated any significant electricity. It is small wonder that American utilities immediately began a retreat from nuclear power and, in some cases, abandoned plants that were nearly finished, incurring billions of dollars in costs of cancelled projects and conversion of plants to coal. About nuclear plants, one U.S. utilities executive said, "They are just too expensive for a company like us to construct any more." Another said, "Don't build nuclear plants in America. You subject yourself to financial risk and public abuse."[76] These observations followed: although 17 nuclear plants that were under construction in the United States in 1979 have since been opened, no new nuclear plants have been ordered, and 59 planned reactors were cancelled.[77]

However, as serious as it was, the accident at Three Mile Island pales in comparison with the tragedy that was to occur at Chernobyl seven years later.

CHERNOBYL

The Chernobyl nuclear power plant is located about 100 km north of Kiev, the capital of Ukraine. The plant consists of four reactors constructed between 1977 and 1983, and at full capacity, the plant generated 4000 megawatts of electricity. The reactor design was based on a graphite-moderated nuclear reaction.

At 1:23 A.M. on 26 April 1986, the No. 4 nuclear reactor in Chernobyl exploded, releasing huge, sinister clouds of radioactive plutonium, cesium, and uranium dioxide into the atmosphere. It was the worst nuclear accident in history involving a nuclear generating plant. The Ukraine was then part of the USSR, and regrettably, the Soviet authorities were slow to issue a warning or to release any details surrounding the accident. The first information about the accident was, in fact, released by Swedish experts, who had noticed the nuclear fallout over Scandinavian countries, almost two days later. The details of the story contain a useful lesson for engineers and for those who manage the large electrical generating plants that are essential to industrialized society.

The Chernobyl accident frightened almost everyone in the northern hemisphere, and for good reason. Not only did the core melt, but the graphite blocks used for moderation caught fire and burned fiercely for days, spreading measurable amounts of radiation throughout Europe. (Canadian nuclear power plants use heavy water, not graphite, for moderation.)

The accident occurred during a supposedly routine low-power test. To prevent the reactor from automatically shutting itself off, the automatic trip safety system was disconnected. Much publicity has been given to the improper procedures used by the reactor operators that contributed to the accident, but the real problem lay with the design. At low power levels, the Chernobyl-type reactor is inherently unstable, because it has what is called a "positive void coefficient." In broad terms, this means that, as the available water in the core decreases (i.e., when steam bubbles form), the reactivity *increases*. In an American light water reactor (LWR), just the opposite happens. As steam bubbles form, or if the water decreases for any other reason, the reactivity decreases. Thus, an LWR is inherently stable. (Almost all commercial power reactors in the United States are LWRs.) The Canadian CANDU reactor is even safer than the American LWR design; the presence of heavy water is needed for the fission to proceed. Loss of heavy water causes the fission process to stop automatically.[78]

At Chernobyl, the operators apparently thought they could manage the reactivity of the plant by manipulating the controls. But the instability of the reactor at low power was so extreme that there was no chance of doing so. When the accident began, the plant was being run at about 6 percent of its full power rating. As the operators started the planned test, the power level began to rise. Because of the positive void coefficient, the rise became exponential. Within 2.5 seconds, the power level had gone from 6 percent to 120 percent of full load. In just 1.5 seconds more, it was 100 times the normal full load. The reactor exploded.[79]

About 7 million curies of iodine-131 and 2.4 million curies of cesium-137 were released. The smoke plume carried these high into the atmosphere and around the world. By contrast, about 15 curies of iodine-131 and essentially no cesium were released by the Three Mile Island accident.[80] Thirty-one people died in the Chernobyl accident: 29 firefighters and two reactor operators. Of the latter two, one was on top of the reactor when it exploded and was killed instantly; the other ran into the reactor building trying to find out what had happened. About 200 people were exposed to high levels of radiation and developed acute radiation sickness. Some of them will die prematurely as a result of their exposure.

At first, the release of radioiodine at Chernobyl dominated the concern of public health officials because of the way in which iodine enters the food chain. But iodine-131 has only an eight-day half-life, and it later became apparent that the radiocesium was more threatening than the iodine, partly because its half-life is 30 years. One year after the accident, a report by the U.S. Department of Energy (DOE) estimated that the number of cancer deaths among the exposed population worldwide might be 14 000 to 39 000 people more than would normally occur. The "normal" number of cancer deaths among the same population is expected to be 630 million, so the estimated increase in deaths because of Chernobyl lies between 0.002 and 0.006 percent. The authors of the DOE report warned that, because of the uncertainty in the statistical methods being used, they could not rule out the possibility that the future cancer increase might actually be zero. But even if the highest number of estimated future deaths occurs, the authors of the report said it will be impossible ever to identify them as such among the "normal" cancer deaths.[81] A further problem, according to the report, is that our current knowledge of radiation cancer risk is based on high doses usually delivered at high rates. However, there seems to be substantial amelioration of latent health effects for those exposed to low doses, based on animal experiments. Thus, the authors say, the actual health effects of Chernobyl might turn out to be as little as one-tenth as severe as those given in their report.[82]

New secrets about the Chernobyl disaster were revealed by Grigori Medvedev in his book *The Truth about Chernobyl*,[83] first published in the Soviet Union in 1989 and translated into English in 1991. The following brief review and summary of the book was written by Joseph Schull of *Maclean's* magazine, and is reproduced with permission.

Medvedev brings expertise and bitter experience to the tale. He worked as deputy chief engineer at Chernobyl in the 1970s, at the time of its construction. Later, he watched in frustration as the project fell into the hands of administrators who won prestigious and well-paid positions despite their lack of experience in the nuclear field. By 1986, Medvedev was deputy director of the Soviet energy ministry's department of nuclear power-plant construction in Moscow. He returned to Chernobyl in the days following the disaster to investigate its causes. His research has yielded a meticulous reconstruction of events on a minute-by-minute basis laced together with first-person accounts from people involved in the plant's operation and management.

The first truth that emerges in Medvedev's book is that the Soviet nuclear industry was run by incompetents from top to bottom: officials in charge of the construction and management of nuclear power stations simply had no training in the field, while their underlings at Chernobyl were no better prepared. Meanwhile, secrecy surrounded the industry and fostered utter ignorance about its potential dangers. Information about previous nuclear mishaps, including the 1979 accident at Three Mile Island, was reserved for high-placed officials unable to draw the appropriate lessons. A state bureaucracy that acknowledged successes but not setbacks was equally damaging. Attention to safety implied the possibility of accidents, and that could only mean that errors might be committed — a possibility that nearly everyone, from minister to technician, wanted to deny. Failure was not in the Soviet vocabulary.

The disease that led to the explosion continued through its aftermath, according to Medvedev. Several costly hours were wasted as the plant's managers denied incontrovertible evidence that the reactor had exploded, passing on to Moscow the reassuring myth of a minor explosion in an emergency water tank. A day and a half passed before the nearby town of Pripyat was finally evacuated; Chernobyl itself did not follow until May 5. The accident merely triggered a chain reaction of human errors.

The coverup continues even now. Soviet authorities have admitted to only 31 deaths in the immediate aftermath and have kept secret the numbers who have died since then. But Vladimir Chernousenko, the scientific director now in charge of the 32-km exclusion zone surrounding the Chernobyl power station, recently estimated that fatal casualties to date number between 7,000 and 10,000....[84]

CONCLUSION

The Three Mile Island and Chernobyl accidents warn us of the immense potential energy of nuclear plants and also of dams, reservoirs, power distribution systems, and other huge structures. The possibility of disaster may be infinitesimally small, but it is not zero. Controlling this energy requires continual alertness, and such responsibility must never be treated casually. These tragic accidents also expose the danger of making design decisions for political or military reasons, and the danger of secrecy and information management in engineering. Competence; clear, quick communication; and high professional and ethical standards are obviously essential in the design and operation of such facilities.

NOTES

1. H.A. Halliday, *The Canadian Encyclopedia Plus* (Toronto: McClelland & Stewart, 1995). Used by permission, McClelland & Stewart, Inc., The Canadian Publishers.
2. Canada, Royal Commission, *Quebec Bridge Inquiry Report*, Sessional Paper No. 154, 7–8 Edward VII (Ottawa, 1908), 49.
3. Ibid., 50.
4. Ibid., 50.
5. Ibid., 51.
6. Ibid., 52.
7. Henry Petroski, *Engineers of Dreams: Great Bridge Builders and the Spanning of America* (New York: Vintage Books, 1995), 102.

8. Petroski, *Engineers of Dreams*, 104–105. Copyright © 1995 by Henry Petroski. Reprinted by permission of Alfred A. Knopf Inc.

9. Canada, Royal Commission, *Quebec Bridge Inquiry Report*.

10. G.H. Duggan, *The Quebec Bridge*, bound monograph prepared originally as an illustrated lecture for the Canadian Society of Civil Engineers, 10 January 1918.

11. Ibid.

12. Petroski, *Engineers of Dreams*, 108–109. Copyright © 1995 by Henry Petroski. Reprinted by permission of Alfred A. Knopf Inc.

13. Canada, Royal Commission, *Quebec Bridge Inquiry Report*, 49.

14. Ibid., 50.

15. Petroski, *Engineers of Dreams*, 115.

16. Duggan, *The Quebec Bridge*.

17. W.N. Marianos, Jr., "Vancouver Second Narrows Bridge Collapse," from N. Schlager, ed., *When Technology Fails: Significant Technological Disasters, Accidents, and Failures of the Twentieth Century* (Detroit: Gale Research, 1994), 191–95. Reprinted by permission of the publisher.

18. H.A. Halliday and J. Joegg, "Mining Disasters," *The Canadian Encyclopedia Plus* (Toronto: McClelland & Stewart, 1996).

19. Justice K. Peter Richard, "Executive Summary," *The Westray Story: A Predictable Path to Disaster*, report of the Westray Mine Public Inquiry, published on the authority of the Lieutenant Governor in Council, Province of Nova Scotia (1 December 1997). This report is also available on the Internet at http://www.gov.ns.ca/legi/inquiry/westray.

20. MacIssac, "Miners Testify at Westray," *Maclean's* (29 January 1996). See also *The Canadian Encyclopedia Plus*.

21. MacIssac, "Miners Testify at Westray."

22. Justice Richard, "Executive Summary."

23. Ibid.

24. *Lodgepole Blowout Inquiry: Phase 1 — Decision Report*, report to the Lieutenant Governor in Council with respect to an inquiry held into the Blowout of the Well: Amoco Dome Brazeau River 13-12-48-12 (Calgary: Energy Resources Conservation Board, December 1984).

25. Ibid., 7-17.

26. Ibid., 5-2. Reprinted with permission of Alberta Energy and Utilities Board.

27. Ibid., 5-5. Reprinted with permission of Alberta Energy and Utilities Board.

28. Ibid., 1-1.

29. Ibid., 5-30.

30. Ibid., 1-2. Reprinted with permission of Alberta Energy and Utilities Board.

31. Ibid., 1-2. Reprinted with permission of Alberta Energy and Utilities Board.

32. *Lodgepole Blowout Inquiry: Phase 2 Report — Sour Gas Well Blowouts in Alberta; Their Causes, and Actions Required to Minimize Their Future Occurrence* (bound as Appendix 5 of *Lodgepole Blowout Inquiry: Phase 1 — Decision Report*) (Calgary: Energy Resources Conservation Board, April 1984).

33. A. Willis and D. Goold, "Bre-X: The One-Man Scam," *The Globe and Mail* (22 July 1997): 1.

34. J. Stackhouse, P. Waldie, and J. McFarland (with files from C. Donnelly), "Bre-X: The Untold Story," *The Globe and Mail* (3 May 1997): B1.
35. A. Spaeth, "The Scam of the Century," *Time*, Canadian ed. (3 May 1997): 34.
36. J. Wells, "The Bre-X Bust," *Maclean's* (7 April 1997): 50.
37. Stackhouse, Waldie, and McFarland, "Bre-X: The Untold Story," B1.
38. Wells, "The Bre-X Bust," 50.
39. Spaeth, "The Scam of the Century," 34. © 1997 Time Inc. Reprinted by permission.
40. P. Waldie, "De Guzman Led Tampering at Gold Site," *The Globe and Mail* (8 October 1997): 1.
41. Ibid.
42. Spaeth, "The Scam of the Century," 34.
43. L.C. Bruno, "Challenger Explosion," from N. Schlager, ed., *When Technology Fails: Significant Technological Disasters, Accidents, and Failures of the Twentieth Century* (Detroit: Gale Research, 1994), 613.
44. *Report of the Presidential Commission on the Space Shuttle Challenger Accident* (in compliance with Executive Order 12546 of 3 February 1986), Washington, DC, 1986. Accessible through http://www.ksc.nasa.gov/shuttle/missions/51-1/docs/rogers-commission.
45. R.M. Boisjoly, "Ethical Decisions: Morton Thiokol and the Space Shuttle Challenger Disaster." Paper presented at the ASME Winter Annual Meeting, Boston, 13–18 December 1987; S. Jaeger, "Failures in Ethics Doomed Challenger, Says Engineer," *Engineering Times* (September 1987): 12.
46. R.M. Boisjoly, *Interoffice Memo to R.K. Lund*, Vice-President, Engineering, Wasatch Division, Morton Thiokol, Inc., 31 July 1985, supporting documentation for Internet article, *Roger Boisjoly on the Challenger Disaster*, accessible through the Ethics Center for Engineering & Science, Case Western Reserve University, http://www.cwru.ed/affil/wwwethics/moral.html.
47. J.D. Kemper, *Introduction to the Engineering Profession*, 2nd ed. (Fort Worth: Saunders College Publishing, 1993), 139.
48. R.J. Baum, ed., *Ethical Problems in Engineering*, 2nd ed., vol. 2: *Cases* (Tory, NY: Center for the Study of the Human Dimensions of Science and Technology, Rensselaer Polytechnic Institute, 1980), 175–95; see also J.H. Schaub and K. Pavlovic, eds., *Engineering Professionalism and Ethics* (New York: John Wiley & Sons, 1983), 388–401.
49. R.J. Serling, "Turkish Airlines DC-10 Crash," from Schlager, ed., *When Technology Fails*, 55.
50. Baum, *Ethical Problems in Engineering*.
51. Serling, "Turkish Airlines DC-10 Crash," 56.
52. L.G. Regenstein, "Love Canal Toxic Waste Contamination: Niagara Falls, New York," in Schlager, *When Technology Fails*, 354–60.
53. Ibid., 357.
54. Ibid., 358.
55. Ibid., 358.

56. Alliance for Environmental Technology, "Activists Call for Ban on All Dioxin Sources," *AET News Splash*, 1, no. 5 (September/October 1994). Also available at http://www.aet.org/news/splash/nssept.html.

57. J. Larson, "Mercury Poisoning: Minamata Bay, Japan," in Schlager, *When Technology Fails*, 367–71.

58. Ibid., 370.

59. Ibid., 370.

60. "Approval of Plan for Minamata Disease Victims," *Environmental Newsletter*, No. 56. © 1996 the Tokio Marine and Fire Insurance Co., Ltd., October 1995. Accessible through http://www.tokiomarine.co.jp/e0300/html/EnvOct95-2.html#A.

61. M. Bray, "Grassy Narrows," *The Canadian Encyclopedia Plus*.

62. L. Ingals, "Toxic Vapour Leak: Bhopal India," in Schlager, *When Technology Fails*, 403–10.

63. "Damages for a Deadly Cloud," *Time* (27 February 1989): 53.

64. "Haunted by a Gas Cloud," *Time* (5 February 1990): 53.

65. B. Sutherland, "Nickel," *The Canadian Encyclopedia Plus*.

66. L.C. Ritchie, "Ecological Disaster: Sudbury Ontario," in Schlager, *When Technology Fails*, 340–44.

67. Ibid., 341.

68. International Council for Local Environmental Initiatives, *Region of Sudbury, Canada: Land Reclamation*, ICLEI Project Summary Series, Project Summary #22. Available from http://iclei.org/iclei.htm.

69. R.D. Bott, "Nuclear Safety," *The Canadian Encyclopedia*, 1st ed. (Toronto: Hurtig, 1985), 1302.

70. D.E. Newton, "Three Mile Island Accident," in Schlager, *When Technology Fails*, 510.

71. "Prelude: The Accident at Three Mile Island," *EPRI Journal* (June 1980): 7–13.

72. W. Booth, "Postmortem on Three Mile Island," *Science* (4 December 1987): 1342–45.

73. Newton, "Three Mile Island Accident," 513.

74. C. Norman, "Assessing the Effects of a Nuclear Accident," *Science* (5 April 1985): 31–33.

75. "Assessment: The Impact and Influence of TMI," *EPRI Journal* (June 1980): 25–33.

76. "Pulling the Nuclear Plug," *Time* (13 February 1984): 34–41.

77. Newton, "Three Mile Island Accident," 513.

78. J.A.L. Robertson, "Nuclear Power Plants," *The Canadian Encyclopedia*, 1300.

79. J.F. Ahearne, "Nuclear Power after Chernobyl," *Science* (8 May 1987): 673–79.

80. A.P. Malinauskas, et al., "Calamity at Chernobyl," *Mechanical Engineering* (February 1987): 50–53.

81. E. Marshall, "Recalculating the Cost of Chernobyl," *Science* (8 May 1987): 658–59.

82. M. Goldman, "Chernobyl: A Radiobiological Perspective," *Science* (30 October 1987): 622–23.

83. G. Medvedev, *The Truth about Chernobyl*, trans. E. Rossiter (New York: HarperCollins, 1991).

84. J. Schull, "A Fatal Coverup: The Deadly Lies of Chernobyl Come to Light," *Maclean's* (13 May 1991): 65. Reprinted with permission.

CHAPTER TWELVE

Product Safety, Quality, and Liability

Although engineers certainly want to design safe, high-quality products and the vast majority of engineered products are indeed safe, engineers and manufacturers occasionally find themselves defendants in lawsuits for damages caused or allegedly caused by unsafe products. The ethical issue is a complex one: how safe is "safe"? Should a product that is not completely safe ever be produced? In this chapter, we will adopt the pragmatic definition that safe products are those that do not result in lawsuits for damages. The good news is that engineers who are basically ethical and competent build safety into their products and rarely need to worry about lawsuits. Safe products protect the public, are rarely challenged in the courts, and are easily defended.

The purpose of this chapter is to introduce some of the basic concepts of product safety, quality management, and legal liability. The emphasis is on professional methods of protecting the public and helping engineers to design safe products. The chapter concludes with some specific advice to engineers and manufacturers for improving product safety and an introduction to ISO 9000 standards, which require all members of a manufacturing corporation (not just the engineers) to make a commitment to product safety and quality. Although the examples in this chapter apply mainly to manufactured goods, the term "product" also includes devices, machines, structures, facilities, resources, and so on. Note that this chapter is a basic review of product safety and legal liability and does not constitute legal advice; consult your lawyer if you have specific questions about your legal liability.

BASIS FOR LEGAL LIABILITY

If a person suffers injury or loss as a result of an unsafe or defective product, then there are at least three ways in which a design engineer, the manufacturer of the product, or both may be held legally liable. Liability may result if it can be proved that the defect was caused by the engineer's negligence (including incompetence or carelessness), or if the defect constitutes a breach of the product warranty, or (in the United States) if the manufacturer cannot mount an adequate defence against "strict liability." Each of these topics is discussed in this chapter.

Generally, a lawsuit will be brought against the design engineer only in the case of alleged negligence. Most lawsuits involving product liability are brought against the manufacturers and sellers of the products, and are usually based on breach of warranty or on strict liability. However, some of these lawsuits have been enormously costly, so safety is a good investment. When an engineer makes a product safer, this simultaneously protects the public from harm and protects the manufacturer from financial loss.

The concept of *strict liability* applies mainly in the United States, but most Canadian engineers and manufacturers must be aware of it, since the North American Free Trade Agreement (NAFTA) permits a freer flow of their products across the border into the United States. Moreover, since Canadian and U.S. law are both based on the British concept of established precedents, legal decisions made in one country may, over time, be applied in the other country. Many of the legal decisions quoted in this chapter are U.S. decisions, since strict liability is usually more stringent than negligence or warranty law. To emphasize the importance of safety in design, this chapter quotes a few past court decisions on product safety. The courts are not totally reliable guides, of course, because their decisions are not always consistent. One court may go one way, another court may go a different way. A case won at the trial court level may be reversed on appeal, and the law itself is changing as years pass.

NEGLIGENCE AND LIABILITY

Engineers, as professionals, are liable for negligent, incompetent, or careless acts that result in damages to others, including the employer. Engineers are not expected to be infallible, and perfection is not required, but engineers are required to use reasonable care, established practices, and well-tested engineering principles. Most provincial Acts define negligence as an act or omission that "constitutes a failure to maintain the standards that a reasonable and prudent practitioner would maintain, in the circumstances."[1] Obviously, the precise meaning of this definition depends on the accepted standards of the engineering discipline concerned.

If the product design is carried out under contract, then negligence, incompetence, or carelessness may be a basis for a lawsuit over breach of contract. Obviously, this risk applies mainly to engineers in private practice. For example, if an engineer agrees to carry out specified design, analysis, or inspection services and then performs the work incorrectly or is unable to complete the contract, the engineer may be liable for breach of contract. This form of liability can easily be avoided by making more careful contractual commitments. That is, the engineer must not undertake work for which he or she is not competent and must seek help when a problem arises that is outside his or her area of competence. Obviously, it is common sense, good law, and good engineering to consider all the reasonable ways in which an agreement may go wrong, to foresee the damages that may result, and to include clauses in the contract that specify bonuses for good results, payments for damages, and/or limits to liability.

However, even when there is no contract, negligence, incompetence, or carelessness may result in legal liability based on tort law. The word "tort" means injury or damage. For a claim to be successful under tort law, the plaintiff must prove, beyond a "balance of probabilities," that the facts of the case satisfy three basic tort principles: that the defendant had a duty of care to the plaintiff, that the defendant breached that duty, and that the loss or damage was a direct result of the defendant's breach of duty. A duty of care may exist between two people who have no contract and, in fact, may never have met. The most common example occurs on our highways: drivers have a duty of care to follow the provincial traffic rules and to avoid injury or damage to other drivers. If a collision occurs, one driver may be found negligent in this duty of care and therefore liable for the damages suffered by the other driver. Provincial law requires drivers to carry liability insurance to guarantee that the innocent victim of a motor vehicle accident will be reimbursed for damages.

Many court cases have been based on allegations of negligence on the part of an engineer. For example,

A house in England was built on a garbage dump, and should have had deep foundations to avoid settling. The foundations were supposed to be inspected by the municipal inspector, but the inspector approved the foundations without inspecting them. Over a period of time, the foundations settled, and could not carry the weight of the building, which collapsed. The municipal authority who employed the inspector was held liable to a later purchaser of the house. In his 1972 judgement, the judge said: "... in the case of a professional man who gives advice on the safety of buildings, or machines, or material, his duty is to all those who may suffer injury, in case his advice is bad."[2]

Obviously, liability insurance is a wise investment — and in most provinces, is a compulsory requirement — for engineers in private practice. Liability insurance is also essential for manufacturers that employ engineers since, under the concept of *vicarious liability*, the engineer's employer is liable for any loss or damage that results from a tort caused by an employee. Although an employee engineer may be shielded financially by this concept, a negligent engineer is always subject to the disciplinary action of the provincial Association, as discussed later in this text.

PRODUCT WARRANTIES

Warranties and guarantees are promises that a manufacturer makes about a product. The term *warranty* is usually applied to goods and products, while the term *guarantee* is usually applied to services or agreements. If the manufacturer's product fails to meet the terms in the manufacturer's warranty, then obviously the manufacturer may be liable for resulting damages. Warranties take two forms: express warranties and implied warranties.

Express warranties are promises that the product has a certain quality and/or it will perform for a certain period of time. Express warranties are familiar to all of us from the warranty cards that accompany new appliances and promise that they will perform satisfactorily for a specified period of time. Express warranties need not be in writing provided that a court can be convinced at a later time that an oral warranty was indeed made. In such cases, some reasonable allowance for "puffery" is necessary. For example:

- In a Florida case, the court decided that no warranty regarding the sale of a ladder had been made when the salesperson said that the ladder was "strong," would "last a lifetime," and would "never break." The ladder did break, but the court said that the statements, although exaggerated, did not constitute a warranty.
- In contrast, in an Oklahoma case involving the sale of a used car, the court held that the salesperson's statements did constitute a warranty. The salesperson had said, "This is a car we know — in A-1 shape." Because the statement presumably was based on expert knowledge, the court said it was an express warranty.[3]

Implied warranties are unstated promises that exist as a matter of common sense. Unlike the express warranty, which is essentially a part of the contract of sale, the very fact that a product is offered for sale implies that it is of average quality and will function. A washing machine should be capable of getting clothes clean; a refrigerator should be capable of keeping food cold. In Canada and the United States, consumer protection laws require goods to be of acceptable average quality, fit for the ordinary purposes for which such goods are used, and conforming to the statements made on the label. In the United States, the implied warranty applies even to the packages that contain the product. For example:

- In a North Carolina case, a bottle of carbonated beverage exploded in a customer's hand as she was carrying it to the check-out counter in a grocery store. The customer won her suit because one would not normally expect this to happen.[4]
- However, in an Oklahoma case, a person was opening a box bound with steel straps. One of the straps, when cut, flew up and injured the person's eye. The court said there was no liability because one should expect such a thing to happen when a strap is cut.

STRICT LIABILITY

In the case of negligence and in the case of warranties, it is necessary to prove that a certain standard of conduct was required, the accused party did not conform to this standard, and the failure to conform caused injury to someone. But in the case of strict liability, which covers product defects and consumer safety, the focus is on the product itself, and no questions of negligence arise. D.L. Marston, in his text *Law for Professional Engineers*, states:

> In products liability cases in the United States, a manufacturer may be strictly liable for any damage that results from the use of his product, even though the manufacturer was not negligent in producing it. Canadian products-liability law has not yet adopted this "strict liability" concept, but the law appears to be developing in that direction.[5]

The Canadian requirement for care was set by a 1932 case, in which the judgement stated in part:

> A manufacturer of products which he sells in such a form as to show that he intends them to reach the ultimate customer in the form in which they left him, and with no reasonable possibility of intermediate examination, and with the knowledge that the absence of reasonable care in the preparation or putting up of the products will result in injury to the consumer's life or property, owes a duty of care to the consumer to take that reasonable care.[6]

In the United States, the American Law Institute has published the following two-part rule, which is somewhat different from the Canadian judgement regarding strict tort liability (as mentioned earlier, a tort is a wrongful civil act committed by one person against another, other than a breach of contract).

1. One who sells any product in a defective condition, unreasonably dangerous to the user or consumer or to his property, is subject to liability for physical harm thereby caused to the ultimate user or consumer, or to his property, if
   (a) the seller is engaged in the business of selling such a product, and
   (b) it is expected to and does reach the user or consumer without substantial change in the condition in which it is sold.
2. The rule stated in Subsection (1) applies although
   (a) the seller has exercised all possible care in the preparation and sale of his product, and
   (b) the user or consumer has not bought the product from or entered into any contractual relation with the seller.[7]

A number of things should be noted in comparing the above definitions. The key difference is that, in the U.S. definition, rule 2(a) specifically states that the rule applies even though the seller has taken all possible care; in other words, it applies even if no negligence can be shown. Second, the U.S. definition applies to sellers, but since manufacturers are necessarily sellers, it also applies to manufacturers. Both definitions use words like "reasonable," "defective," and "unreasonably dangerous," which are difficult concepts to pin down in court. What seems to one person to be a defect may not appear so to another, and people also rarely agree on whether something is reasonable or unreasonable. For example, knives are dangerous, but are they unreasonably so? A car moving at a legal speed on a highway is certainly dangerous if it hits something, but is it unreasonably dangerous?

A number of other things should be noted in the U.S. statement of strict liability. Rule 1(b) says that the product must reach the consumer without substantial change ("substantial" is another word whose meaning must often be settled in court). Rule 2(b) says that the user who is injured does not have to be the original buyer; there could have been many intervening owners. Obviously, in this last case, the product is not expected to remain safe for an infinite number of years: wear and tear are factors in a court's determination as to whether something is "reasonably" safe.

UNREASONABLE DANGER

Any attempt to define the word "unreasonable" leads to more definitions. For example, the American Law Institute attempts to define "unreasonably dangerous" as "dangerous to an extent beyond that which would be contemplated by the ordinary consumer ... with the ordinary knowledge common to the community."[8] Now we are left with the job of finding out what "ordinary" means instead of "unreasonable." But a judge in Massachusetts left no doubt as to what he thought "unreasonable" meant in what might be called "the fish chowder case." In this case, the plaintiff ate some fish chowder, and a bone became lodged in her throat. It took two operations to remove the bone. The plaintiff brought suit and lost. The court said that consumers "should be prepared to cope with hazards of fish bones, the occasional presence of which in chowders is to be anticipated." The court added that its opinion was based in part upon its unwillingness to overturn 'age-old recipes.'"[9]

The fish chowder case gives a definition of what "unreasonable" means, at least with regard to fish bones in chowder. The reader is left with the uncomfortable feeling, however, that a different judge, perhaps with less respect for "age-old recipes," might have reached a different decision. Nevertheless, courts do tend to show a reluctance to overturn customs that are long established. They are not likely to rule, for example, that all knives must be equipped with cumbersome safety guards or that all cars must be operated at speeds of less than 30 km per hour, even though such rules might prevent many injuries and deaths. In some instances, courts have even balanced safety issues with economic benefits, as will be seen later. A couple of additional cases help to show where some courts have drawn the line between reasonable and unreasonable.

- In a Georgia case, a child was injured while riding after dark on a bicycle that was not equipped with a headlight or reflector. The child's father attempted to collect damages from the seller of the bicycle but lost. The court ruled that the father was aware of the danger and that the seller had no duty to protect against obvious, common dangers.
- On the other hand, in a South Carolina case, a child was injured by lawn mower blades, and the plaintiff did collect. The court agreed that the danger was obvious, but said that the seller of the lawn mower nevertheless had a duty to improve its safety. In this case, the court was swayed by the gravity of the danger and by the relative ease with which the danger could have been reduced, either through better design or through warnings.

For design engineers, it is especially important to note that the viewpoint of U.S. courts appears to be undergoing a significant change away from judging cases on the basis of what an "ordinarily prudent person" might do and toward judging on the basis of what an "occasionally careless person" might do.[10]

CONCEALED DANGERS

Courts are likely to be especially critical when it comes to concealed dangers.

- In a Florida case, the manufacturer of a folding aluminum lawn chair was held liable in a case in which a person lost a finger. If the chair was not fully unfolded, and if the user sat in it and simultaneously had a finger in the danger spot, the further unfolding of the chair under pressure would neatly amputate the finger.[11]
- In a somewhat more complicated case, one person died and another was severely injured by accidental overdoses of radiation from a machine used to treat cancer. The problem here was in the software. The operator initially set the machine for too high an energy level. She recognized her error and entered the instructions to reset for a lower level, but she did this so quickly that the operating system for the machine was still engaged in setting itself up for the higher level. The result was that the machine ignored a part of her input, and the patient received 2.5 times as much radiation as intended.[12] There was no reason why the operator should have been aware of such a possible outcome, so this was an error on the part of the software designer.
- Another case involved a tractor that had a steering wheel made of rubber and fibre (other manufacturers used metal or wood). The plaintiff, in using the tractor, occasionally supported himself on the wheel while turning. On one such occasion, the wheel broke and the driver fell under the tractor and was run over. The manufacturer was held liable because tests showed that the steering wheel would break under this kind of load, and the court said the manufacturer should reasonably have expected such things to occur.[13]
- A somewhat similar case involved a tractor operator who slipped on a step while dismounting, injuring his back. It was shown that the operator could not see the step in dismounting, that mud had been thrown up by the tractor treads and collected on the step, and that the step had no anti-skid material on it. The tractor manufacturer lost. More importantly for design engineers, it was revealed during the trial that the tractor manufacturer had violated its own design guidelines, which required, among other things, that steps and ladders be designed to minimize the accumulation of mud and debris and that steps be provided with anti-skid material.[14]

OBVIOUS DANGERS

For the manufacturer of a knife, there is no duty to protect against the sharp edge because the danger is obvious. However, just because a danger is obvious does not mean that the designer is relieved of responsibility.

- In a California case involving an earth-moving machine, a person standing behind the machine was run over and killed. The machine had a blind spot in the rear caused by the presence of a large engine box, and there were no rear-view mirrors that gave a view of the blind spot. The manufacturer was held liable, the court saying that even though the danger was obvious, it was unreasonable.
- In an Illinois case, a 7-year-old girl playing near a lawn mower slipped and fell as it approached her. There was a 20-cm gap under the mower, and she slid into it, losing her leg. The manufacturer in this case was not held liable on the grounds that no defect existed and that the danger was obvious. However, this case has been heavily criticized in subsequent decisions, because it offers too much encouragement to manufacturers to make their products attractive on the basis of a low price.[15]

The doctrine of strict liability has evolved in the United States over many years. This is very apparent if one examines how the law has been applied to cars during recent decades. In the 1960s, plaintiffs often lost cases involving injury from an auto accident. The doctrine in such cases essentially said that the driver suffered injury because of an unintended use; the intended purpose of a car does not include collision with other objects. Today, however, the doctrine has moved far beyond that. Any use of a car creates a certain risk of accidents, and manufacturers are now expected to design cars in such a way as to protect occupants from aggravated harm.[16] As a result, features such as collapsing steering columns, padded interiors, fuel tanks protected from rupture, and stronger roofs to protect occupants during a rollover, all unknown in the 1960s, are now standard automotive components.

FORESEEABILITY — ANTICIPATING DANGER

It sometimes seems as if the designer of a product is expected to foresee everything and to detect every manner in which a product might be used or misused. As unfair as it may seem, this is almost precisely the case, particularly in the United States. Lawyers and courts may be allowed to use perfect 20/20 hindsight to pass judgement on what engineers did or did not do, but engineers have no such privilege. By the very nature of their calling, they must look ahead, anticipate danger, and be right most of the time. One student of product liability has said: "Many engineers and designers assume their products are safe if they meet all regulations and standards and if moving parts are protected with a guard when the product leaves the factory. In fact, a review of litigation resulting from product failure shows that accidents are caused less often by mechanical failure than by the designer's failure to consider how the product would be used."[17] Here are a few cases that illustrate the point of foreseeability:

- Doors, designed to contain glass panels, were packed in bundles for shipment without the glass in them. They were surrounded by cardboard, so the bundles were essentially large cavities covered only with thin corrugated cardboard. A worker walked across one of these bundles as it lay in a warehouse, fell through the cardboard, and injured himself. At the trial court level, the worker won. The court said the situation was dangerous and that a prudent manufacturer would have protected against the danger by warnings or otherwise. On appeal, the decision was reversed. The court said the manufacturer was not obliged to protect against the danger unless it was aware of the practice of walking on packages of doors. However, on appeal to the U.S. Supreme Court, the decision was reversed again. The Supreme Court said that the dangerous condition was foreseeable and that it could have been corrected with little effort.[18]
- A coal-mining machine was 9.4 m long and travelled through the mine on crawler tracks with the operator moving alongside, on hands and knees if necessary. It was typical of such a machine that, if it encountered an irregularity in the mine floor, it might suddenly swerve to one side. In one incident, a machine did swerve and crushed the operator against the mine wall, killing him. In subsequent investigations, it was not possible to prove that this was the precise cause of the accident, but it became apparent that the placement of the machine's controls contributed to the death of the worker. It was also shown that, in machines made by other manufacturers, two sets of controls were provided so that the operator could choose whichever set would be safer to use.[19]
- Another case involved a printer-slotter machine. It was frequently necessary to separate the machine into two freestanding halves with a 76-cm passage between them to gain access to the inside of the machine to change dies. When this occurred, half of the machine was "dead," whereas the other half continued to be supplied with power so that an automatic roller-washing operation could continue while the dies were being changed. The purpose, of course, was to minimize change-over time; otherwise, the machine would have to be shut down completely, lengthening the time required for change-over. During such a change-over, a worker walked between the two halves of the machine. He was carrying a rag, and the rag was caught by the moving rollers on the "live" half of the machine. His arm was drawn into the rollers and subsequently had to be amputated. He won his court case on the grounds that the machine should have been equipped with a safety ("interlock") switch that would cut off all power to both halves of the machines whenever it was opened.[20] This

seems like a simple case merely involving the provision of an interlock switch — something engineers deal with frequently. But some
legal scholars have pointed out that it is not as simple as it seems.
Given the pay incentive scheme being used by the employer, there
was a strong desire on the part of the operator to minimize change-
over time. Even if an interlock switch had been provided, there
would have been a strong incentive for the operator to disable the
switch in order to cut down the change-over time. (Machine operators are notorious for disabling safety devices that are intended for
their protection.)

Although U.S. law on strict liability appears to be unfairly biased
against manufacturers, one more case will be cited in which a trade-off
between safety and economics governed the outcome. In this case, the
court was mindful of the fact that some products (such as cars), although admittedly dangerous, still provide benefits that society wants.

• The case in question involved serious injury to the plaintiff by a
huge earth-moving machine. The operator of the machine saw the
plaintiff standing in front of the machine but could not avoid the accident because the brakes were filled with mud and he could not stop
quickly enough, nor could he swerve the machine fast enough. At
the trial court level, the jury's verdict was in favour of the plaintiff,
but the case was reversed on appeal. The court said the manufacturer was "not required to expend exorbitant sums of money in research
to devise a sophisticated braking system which would price its product completely out of the market." As for the steering system, the
court said it "was reliable and durable. It was the standard steering
system used in the heavy construction industry and no better turning device available was disclosed by the evidence."[21] However, design engineers should note that there was strong minority dissent in
this case, and a different court might have disagreed entirely.

ADVICE TO DESIGN ENGINEERS

Many texts and papers have discussed methods that engineers and
manufacturers could follow to make products safer and to avoid product
liability lawsuits.[22] The following is a summary of key ideas from a variety of sources that should be useful to design engineers:

FORMAL DESIGN REVIEWS

Formal design reviews are probably the best way to ensure safe products and avoid legal liability. Design reviews are formal meetings held

at critical points in the design process that include representatives from the various parts of the corporation: design, manufacturing, sales, management, and so on. It might be added that formal design reviews are required in the ISO 9000 Quality Management process, discussed later in this chapter.

Typically, three design reviews are needed for a large project: conceptual, feasibility, and final. The *conceptual* design review should be held very early in the design sequence. (In the design process explained in Chapter 6, the conceptual review would be held at the end of step 2, after gathering information.) The various design criteria, including the design constraints and goals that the final product must satisfy, should be stated clearly, and a permanent record of these criteria should be made to guide design decisions. The *feasibility* design review should be held after a variety of alternative configurations have been suggested and a preliminary analysis has been made of their feasibility (a first pass through steps 2 and 3 in the design sequence). The *final* design review should be held after the design has been fully analyzed and optimized (step 5 in the design sequence). The goal is to ensure that all of the design criteria have been satisfied.

If including a safety feature creates a severe cost penalty, then a formal cost–benefit analysis should be carried out and discussed at the design review meeting, and a written record of the recommendation must be thoroughly understood and confirmed at the highest levels of the organization.

CODES AND STANDARDS

Be aware of, and adhere to, established government, industry, and company standards. In particular, establish and apply your own company's design standards unless a thorough written analysis indicates that the standards should be changed. In court, admitting that your company deviated from its own standards for no reason is very damaging testimony. Standards are unfortunately a chaotic and confusing issue, since they are set by engineering societies, government agencies, industry groups, and internal company procedures. Some standards are enforced by government law or regulations: the National Building Code and the certification of elevators, escalators, vehicles, and aircraft are obvious examples. Some standards are developed by engineering societies: the Society of Automotive Engineers (SAE) standards for vehicle design and the American Society of Mechanical Engineers (ASME) pressure vessel code are used universally, even though they are not legally enforced in some jurisdictions. Some of the best standards are set by the industries themselves, such as the American Gear Manufacturers Association (AGMA), which has the definitive standard for gear strength in North America, although a metric ISO standard is in draft form. Finding the standards may take some searching; however, regardless of the product, a relevant standard almost always exists, and these codes and standards must be found and applied.

STATE-OF-THE-ART DESIGN METHODS

The state of the art once applied to established and accepted procedures but is gradually coming to mean the new, "cutting-edge" analysis techniques and other methods of ensuring safety. In a lawsuit, a judge will condemn negligent practices if they are shown to be the minimum practices and standards of a negligent industry. Where safety is concerned, industries must use up-to-date design, analysis, and manufacturing techniques and recent standards. When industry standards represent minimum requirements, compliance with the standards becomes a weak defence. As a general rule, engineers and manufacturers must make choices that lean toward safety; this is particularly true if the safety has little or no cost penalty and merely requires foresight on the part of the design engineer.

FORMAL HAZARD ANALYSIS

Hazard analysis is a formal review of a product or process that involves four steps. The engineer examines each aspect of the proposed design and (1) identifies the hazards that may be created and (2) tries to prevent or eliminate the need for creating the hazard. If the hazard cannot be eliminated, then (3) it should be treated as a signal that emanates from a source and follows some path to a receiver (the user of the design), where the hazard inflicts some damage. There are three locations where action can be taken to shield the hazard and prevent damage: at the source, along the path, or at the receiver. Finally, if the above steps prove to be unsuccessful and the product is unsafe, then (4) some remedial action is essential: recall the unsafe product, notify people of danger, or assist the injured as appropriate.[23]

FORMAL FAILURE ANALYSES

Hazard analysis looks at the risks that exist during the normal operation of a product, but it is also important to examine whether failure of a component of a system may result in disastrous results for the system as a whole. Formal failure analysis techniques are complex and are typically applied only to large systems, such as electrical power plants, aircraft, control systems, and so on. The two best-known methods are failure modes and effects analysis (FMEA) and fault tree analysis (FTA).

Failure modes and effects analysis (FMEA) is a bottom-to-top process. For instance, in a chemical plant with many processing steps, the "bottom" might be an individual piece of equipment, such as a distilling column, a cooler, or a pump. The steps to be followed when carrying out an FMEA for such a system are to list each of the components in the system, make a list of all of the ways (modes) in which each component could conceivably fail, and then determine the effect of the component failure on the system as a whole. This is an orderly way to determine which components might cause a disastrous failure of the system. It is important, then, for the engineer to describe how the component failure mode can be prevented (or at least detected) so that the system failure can be avoided.[24]

Unlike FMEA, fault tree analysis (FTA) is a "top-to-bottom" process. The first step in FTA is to envision a disastrous system failure, such as a collapse of a cooling tower, or the leak of a toxic substance, that must be avoided. This is called the "top event." The next step in the analysis is to determine any (and all) events which could cause the top event. These "contributor" events are then analyzed, in turn, to determine the events upon which they depend. The process is continued until events are reached which are primary (or basic) fault events — usually human errors or failures of equipment. The result is a tree-like structure, which gives the method its name. Probabilities can then be assigned to the basic events, and by simple mathematics, the probability of the top event can be determined. The engineer must then decide whether this probability is low enough to be an acceptable risk. If not, then the system must be redesigned to eliminate this failure mode or, at least, to reduce its probability to an acceptable level.[25]

DESIGN RECORDS

In case a product that you have designed is alleged to be unsafe or defective, it is critically important to have a record of the design decisions, up-to-date drawings or CAD files, and minutes of the design review meetings. The most convincing defence, in some cases, is to show that the danger was properly evaluated and shown to be within acceptable limits. Good engineering records will not prevent product liability lawsuits, but will be extremely useful in convincing a judge that prudent and reasonable care was taken to produce a safe product.

ADVICE TO MANUFACTURERS

Manufacturers are also under pressure to ensure that their products are safe. C.O. Smith has listed several actions that manufacturers could follow to improve product safety.[26] These suggestions are explained below.

QUALITY ASSURANCE AND TESTING

Quality assurance is an obvious first requirement for the manufacturer. In the past decade, certification of the manufacturer under ISO 9000 standards has become the best guarantee of good quality management, as described later in this chapter.

INSTRUCTION, WARNING, AND DANGER SIGNS

When hazards cannot be completely eliminated and a final design still poses some risk of injury or damage, then appropriate instructions or warnings must be given to the user. The Society of Automotive Engineers (SAE) publishes a recommended practice for such signs. The SAE distinguishes between instructional or educational signs, which should be headed with the words "IMPORTANT," "ATTENTION," or "NOTICE," and the warning signs "CAUTION," "WARNING," and "DANGER," which should be used only when personal safety is at risk. These indicate progressively increasing risks, as follows:

- CAUTION is used to warn of risks that might result from unsafe practices.
- WARNING denotes a specific potential hazard.
- DANGER indicates a serious hazard to personal safety, near the sign.

The SAE recommends that the caution and warning signs be yellow and black and the danger sign red and white for easier identification. Since people understand symbols or pictorial warnings easier than written warnings, it is advisable to include an explanatory symbol in the sign.[27]

INSTRUCTION MANUALS

Any product that is dangerous or that could be used dangerously should usually be accompanied by a service or instruction manual. This manual should give basic operating and servicing information in simple, clear writing. Safe procedures must be encouraged, and all caution, warning, or danger signs on the product should be explained. Particular emphasis should be placed on hidden dangers or potentially hazardous operating procedures, and any hazard that cannot be indicated by a warning sign should be fully explained. Alert customers to any dangers that might result from using unauthorized parts, and explain appropriate safety features and procedures for assembly/disassembly and troubleshooting.

WARRANTIES, DISCLAIMERS, AND OTHER PUBLISHED MATERIAL

Ideally, all information published about a product should be reviewed by the design engineer and by a products liability lawyer who is familiar with the product. Warranties and disclaimers must be written in clear, simple, and easily understood language and must accompany the product. Warranties may help sell a safe product, but disclaimers, no matter how well written, are a weak defence for an unsafe product. All advertising, promotion, and sales literature must be carefully screened, since an implied warranty may be created by the wording on labels, instruction pamphlets, sales literature, advertising, and so on, even though no warranty is intended. There must be no exaggeration: the product must be able to safely do what the advertising says it will do.

CONSUMER COMPLAINTS

Complaints from consumers can be a useful early-warning system. Each complaint should be investigated quickly, carefully, and completely. An efficient system for investigating consumer complaints permits problems to be corrected on the production line or a recall to be instituted before large numbers of people are exposed to an unsafe product.

CUSTOMER RECORDS

A mistake may occasionally occur that results in an unsafe product. In this situation, rapid remedial action — such as a warning bulletin or a product recall — may be required. This requires a record of customers who own the defective or unsafe product. Also, if only one specific model is identified as unsafe, then time and money can be saved by contacting only the owners of that particular model. Clearly, good customer records are essential to safe and efficient operation.

OBTAINING CANADIAN STANDARDS

One of the key methods of improving product quality is to follow the accepted standards. Although standards exist for almost every type of product, finding them is sometimes difficult. An excellent starting point in such a search is the Standards Council of Canada (SCC), a federal Crown corporation with a mandate to promote standardization in Canada. The following description of the role of SCC is excerpted from SCC publications, with permission.[28]

Standards: Standards are publications that establish accepted practices, technical requirements, and terminologies for diverse fields of human endeavour. There are standards for everything from the simplest screw to the most complex telecommunications equipment. Standardization — the development and application of standards — brings people together to pursue better, safer, and more efficient methods and products. Standardization is a means of improving on what was done correctly before, and correcting what was not.

The Standards Council of Canada: The Standards Council of Canada (SCC) is a federal Crown corporation whose mandate is to promote efficient and effective voluntary standardization in Canada in order to advance the national economy; support sustainable development; benefit the health, safety, and welfare of workers and the public; assist and protect consumers; facilitate domestic and international trade; and further international cooperation in relation to standardization. Located in Ottawa, the SCC has a staff of approximately 70, and a governing Council of 15 members. The SCC's activities are carried out within the context of the National Standards System, a federation of organizations providing standardization services to the Canadian public. The SCC is manager of the National Standards System.

The National Standards System: Members of the National Standards System write standards, certify conformance of products or services to standards, test products, or register the quality systems of companies. These services are fundamental to the well-being of Canadians. Therefore, the Standards Council requires that members of the National Standards System satisfy certain established criteria. Members of the system must be stable, impartial, and technically competent. Where applicable, they must follow internationally accepted procedures. The Standards Council operates accreditation programs which ensure that members of the System meet all the necessary criteria for the services they wish to provide. The National Standards System promotes better standardization within Canada, and makes Canada more competitive internationally.

Who's Who in the NSS? The Standards Council of Canada accredits four types of organizations:

1. Standards-developing organizations coordinate the work of committees of volunteers that write standards;
2. Certification organizations certify that products or services comply with the requirements of standards, and allow their mark to be used as an indication of compliance;
3. Testing and calibration laboratories perform tests and measurements indicated by standards;
4. Quality registration organizations register quality systems of companies to quality standards like the ISO 9000 series.

International Standardization: The Standards Council coordinates the contribution of Canadians to the two most prominent international standards-writing forums — the International Organization for Standardization (ISO) and the International Electrotechnical Commission (IEC). ISO and IEC standards are well-respected around the world, and are often adopted by countries for inclusion in national rules and regulations. Through committees administered by the Standards Council, thousands of Canadians contribute to the standards published by these two bodies in such important fields as information technology, the environment, health technology, quality, and commodities. In addition to international standards-writing forums, the Standards Council participates in international efforts to bring uniformity to certification, testing, and calibration.

National Standards of Canada: The Standards Council of Canada may approve a standard, published by an accredited standards developing organization, as a *National Standard of Canada*. To obtain this designation, a standard must conform to certain published criteria. The standard must be developed by a balanced committee representing producers, consumers, and other relevant interests. It must undergo a public review process and be available in both official languages. It must not be framed in such a way that it will act as a restraint to trade. Further, the standard should be consistent with or incorporate appropriate international standards as well as pertinent national standards.

Standards Information and Sales: The Standards Council of Canada operates an information and sales service geared to helping Canadians compete in the global marketplace. The information service provides individuals and companies with the latest technical, safety, and quality requirements of standards and regulations around the world — including such important markets as the European Community, the United States, Asia, and Canada.

The Standards Council's sales centre stocks foreign and international standards as well as some National Standards of Canada. The sales centre greatly simplifies the purchase of non-Canadian standards by eliminating the need to deal with foreign distributors and currencies.

Finding Out More: More information on the programs and services of the Standards Council of Canada is available from the Standards Council of Canada website at http://www.scc.ca or from:
Standards Council of Canada
Communications Division
Suite 1200 – 45 O'Connor Street
Ottawa, Ontario K1P 6N7

ISO 9000 AND ISO 14000 STANDARDS

A standard exists for almost every manufactured object (although it may take some digging to find it), and standards have a major impact on maintaining and improving our quality of life. In order to make standards more widely available and to "standardize the standards" from country to country, the International Organization for Standards (ISO) was founded. By 1996, it had over 100 member countries participating in standards activities, including Canada, Britain, and the United States.

Whenever a new ISO standard is proposed, a technical committee is formed with experts taken from the various member countries. Each nation that participates in a technical committee typically sets up an advisory group composed of experts from that nation, which generates a national consensus opinion of the proposed standard for that country. The proposed standard must pass through three draft stages in which differing opinions and alternative wording are proposed by members of the technical committee before it is voted on by the member countries. If the final draft standard receives a two-thirds positive vote, it becomes an ISO standard and is translated into the three official ISO languages: English, French, and Russian. Each country may take a further step and adopt the ISO standard as a national standard, and may publish the standard in the language of that country.

Although ISO standards for many different products have been in use for many years, two recent standards have generated a great deal of interest among engineers, engineering managers, and manufacturers: ISO 9000 and ISO 14000. Each of these standards is described briefly below.

ISO 9000 — QUALITY MANAGEMENT AND QUALITY ASSURANCE STANDARDS

The ISO 9000 standard is very different from most ISO standards for products, because it is a standard for effective management of a manufacturing corporation in order to maximize the quality of the manufactured products. The standard is very effective and has been widely adopted. Since 1987, when the first version of ISO 9000 was released, over 100 000 corporations have obtained ISO 9000 certification (as of 1996), and all 13 000 first-tier suppliers to the "big three" automakers (Chrysler, Ford, and General Motors) were required to adopt an extended form of ISO 9000 before the end of 1997. Although some of the suppliers required an extension to fulfil registration requirements, it is expected that every supplier to the automotive industry will eventually have to be certified to ISO 9000 standards in order to do business.[29] It is estimated that the investment in ISO 9000 certification is typically repaid in three years through increased productivity and reduced scrap.[30]

The ISO 9000 standard is very comprehensive and requires a corporation to examine virtually every aspect of its management, design, purchasing, inspection, testing, handling, storage, packaging, preservation, delivery, and documentation functions. Improving the quality of these functions permits the manufacturing process to be evaluated effectively and shows where quality improvements are required. A key part of the ISO 9000 process involves the development of a *Quality Manual* that documents the four key aspects of the certification process. The manual documents the

- quality policies for every aspect of the corporation's operations
- quality assurance procedures, which involve twenty clauses in the ISO 9000 standard
- quality process procedures (or practices or instructions), which include all of the company's production processes
- quality proof: a repository for all of the forms, records, and other documentation that give objective evidence, or proof, that the quality system is operating properly

An important aspect of the ISO 9000 quality management process is that it is arranged to permit internal and external audits, much like a financial control system for which audits are an established, accepted, and routine procedure. A customer who purchases the company's products may be invited to examine the company's operations to ensure that the quality management system complies with ISO 9000 standards. The audit is typically carried out by independent quality auditors, or "registrars," to ensure impartiality. These audits should occur every six months or so, with a complete recertification audit carried out every third year.[31]

ISO 9000 certification is a long and detailed process. Every aspect of a company's operation must be examined in detail, and fifteen-step processes, which may take more than a year to implement, are common. However, ISO 9000 is clearly becoming the accepted world standard for quality management, so many companies may need it in order to survive in business.

ISO 14000 — ENVIRONMENTAL MANAGEMENT SYSTEMS

As companies around the world register to the ISO 9000 standard for quality management, many of these companies are also adopting another new ISO standard, ISO 14000, for environmental management. The ISO 14000 standard was developed using the international consensus procedure, as for all ISO standards, and fits well with the ISO 9000 standard. In fact, companies that have registered to ISO 9000 standard should find that ISO 14000 certification is very similar.

The ISO 14000 process requires the company to examine every function of its operations with the goal of identifying activities with a significant environmental impact, and committing the company to preventing pollution in all of its many forms. The standard does not set acceptable environmental levels; this is left to regulatory agencies. However, the standard does require that these environmental levels be determined and followed. Monitoring and measurement are, of course, essential, and procedures for corrective action and emergency response are also required. This may require setting procedures and performance criteria, defining and assigning responsibilities, providing training, and ensuring adequate communication. Although ISO 14000 does not require the writing of an environmental management system manual, most companies would probably want to develop such a document.

ISO 14000 was released very recently (1996), and very few companies have certified as this text goes to press. However, major companies such as Ford, Toyota, Sony, and several others have already made a commitment to implement this new standard.[32] It appears that ISO 14000 will follow ISO 9000 as a standard that companies must adopt if they want to compete in the future.

THE FUTURE OF PRODUCT SAFETY AND QUALITY MANAGEMENT

There is no question that the emphasis on quality standards has led to significant improvements in our quality of life. However, there is some concern about the increasing incidence in the United States of lawsuits based on strict liability, and the possibility that this philosophy may also become commonplace in Canada. Although the concept of strict liability serves to compensate victims who have suffered grievous harm and discourages bad design, there are excesses in the system. The lawsuits are usually brought against those with "deep pockets" who can pay large damage awards, and not necessarily against the offending engineers or manufacturers. Moreover, the fear of lawsuits has stifled innovation by many small companies, and only large corporations that can afford protection against immense judgements can accept the legal risks associated with new products. These large manufacturers obviously pass these costs on to consumers if they can, so the public ends up paying. The philosophy of strict liability has not yet become dominant in Canada, and in the interests of innovation and local prosperity, this trend should be resisted.

In the areas of quality, safety, and liability, many subjective ideas must be quantified. Although we might all agree that an unsafe product should never be placed on the market, we must remember that even the word "unsafe" requires definition. Indeed, there are some products, such as land mines, that are so dangerous that they should be banned altogether. However, motor vehicles kill 40 000 or more people a year in the United States, meaning that cars are probably the most dangerous things with which we commonly come in contact. However, the personal convenience (and, for some, the luxury) of owning a car is so great that no one would seriously suggest removing cars from the market just because they are so dangerous. The obligation to protect the public from harm is the first statement in most codes of ethics, and engineers must apply their efforts to ensure that quality management, risk assessment, and risk–benefit analysis are properly applied to eliminate or reduce danger to the public.

The following case illustrates some of the concepts discussed in this chapter.

CASE HISTORY 12.1
SAFETY OF MOTORCYCLE HELMETS

Product quality is not the sole responsibility of the engineer. In 1996, the Bell motorcycle helmet, arguably one of the safest designed, was found liable for injuries caused to a Toronto teenager about ten years earlier. A summary of the judgement, reproduced from *The Globe and Mail*, follows.

An Ontario jury found a helmet manufacturer partly liable for catastrophic head injuries that a teen-aged motorcyclist suffered 10 years ago when his helmet flew off in a crash. In holding Bell Helmets Inc. 25 percent liable for Steven Michael Thomas's injuries, jurors found that the California-based manufacturer neglected to advise users of a simple test to ensure that its helmets stay on.

Although the parties are awaiting a final judgment, lawyers for the plaintiff said yesterday that Bell Helmets faces paying about $3-million in damages to the motorcyclist and his mother. The estimate is based on a calculation by the trial judge, Mr. Justice Norman Dyson of the Ontario Court's General Division, that total damages suffered by Mr. Thomas amounted to $12 million.

Now 29, the plaintiff was a 19-year-old high-school student when his 1982 Honda motorcycle struck a car that made a sudden left turn as Mr. Thomas tried to overtake it. The impact sent Mr. Thomas hurtling about 55 metres through the air before he landed head first in a ditch. The permanent brain injuries he suffered have left him in a wheelchair.

Filed in 1988, the lawsuit also named the car driver and the Guelph dealership that sold the helmet as defendants. A settlement was reached in 1990 with the driver's insurer and the action was dismissed against the dealer. The Thomas family's lawyer, Jerome Morse, said the three main issues in the 42-day trial were whether the helmet was properly fastened, whether it would have prevented the head injuries, and whether Bell had adequately warned wearers of the risk of a loose helmet flying off.

Witnesses at the trial agreed that the full-face helmet flew off Mr. Thomas's head immediately upon impact. Mr. Morse said any doubt that the helmet had been done up was removed when a friend of Mr. Thomas testified that he clearly recalled seeing Mr. Thomas fasten it before getting on his motorcycle.

In holding Bell liable, the jurors found the firm "failed in their duty to the consumer." Exhibits at the trial included a memorandum Bell sent to dealers in 1984 explaining how helmets should work. It read in part: "In order for a helmet to do its job, it must stay on the head during an accident. Helmets that are too large have a greater chance of coming off than those which fit snugly." The jury also heard that in 1985, British regulators required warning labels on all helmets saying: "For adequate protection, this helmet must fit closely. Purchasers are advised to secure the helmet and to ensure that it cannot be pulled or rolled off the head."

Another Bell brochure shown to jurors described a "retention test" that it warned was "very important." It advised wearers to "reach over the top of the helmet, grabbing the bottom edge with your fingers. Then try to roll the helmet off your head. If it comes off, it is undoubtedly too large." Mr. Morse said a Bell helmet he purchased during the trial lacked a warning label, although advice on testing the fit was included in an owner's manual. The lawyer said the last witness he called was his client who, despite his physical handicaps, was able to demonstrate to the jurors that, when properly fastened, the helmet he had worn could be rotated off his head.

In its verdict, the jury set the car driver's portion of overall liability at 55 percent and Mr. Thomas's at 20 percent, finding that he had "proceeded at an excessive speed" (estimated by police at between 80 and 120 kilometres an hour) and "failed to exercise prudent judgment in passing."

The lawyer said one strange aspect of the trial was that he wound up arguing that the helmet was one of the safest made, and that had it stayed on it would have prevented any serious injury. He said Bell called witnesses in a bid to persuade the jurors that most of the damage to Mr. Thomas's brain was from a "diffuse axonal injury" that was inflicted before the helmet came off.

Mrs. Thomas is a registered nurse, who now spends most of her time caring for her son. She said her son spent 4½ years in hospital after the accident and will need round-the-clock supervision for the rest of his life. Bell Helmets is expected to file an appeal.[33]

It is useful to note that the helmet design itself was safe, as stated by the plaintiff's lawyer, but the lack of a warning label to ensure a proper fit prevented the helmet from performing properly in this case. Quality is not just the responsibility of the engineer; it requires a corporate commitment, as the ISO 9000 philosophy suggests.

TOPICS FOR STUDY AND DISCUSSION

1. A quality manual prepared under the ISO 9000 standard for a design/manufacturing company usually contains most of the advice for design engineers and for manufacturers provided earlier in this chapter. Obtain a copy of the ISO 9000 standard from your library, and examine the recommended table of contents for the quality manual. Under which manual headings would each of the twelve "advice" subheadings in this chapter (see pages 290–95) appear in the quality manual? Can you suggest what might appear under any of the other manual headings?

2. Cars are the most dangerous machines that most people use on a daily basis. Cars probably kill more people each year than trains, airplanes, and nuclear power plants have killed since their invention. Moreover, cars pollute: carbon dioxide emissions contribute to global warming, brake friction releases small asbestos fibres at every intersection, oil tends to leak, and carbon monoxide emissions, when not properly vented, can injure or kill. However, no serious campaign has ever been mounted to prevent cars from being manufactured. Prepare a cost–benefit analysis to decide whether the benefits of the vehicle outweigh its dangers and costs.

3. In his 1973 book *Unsafe at Any Speed*, Ralph Nader said engineers "subordinate whatever initiatives might flow from professional dictates in favor of preserving their passive roles as engineer-employees."[34] Nader's book was a major influence in changing practices in the automotive industry and putting more attention on safety and economy. Write a short essay evaluating his statement. Do you think it was justified in 1973? Is it justified now? Is the automotive industry any less ethical than other industries? Why did Nader attack the automotive rather than the aircraft industry? In preparing your essay, it might be wise to obtain a copy of Nader's book from your library so that you can see the full scope of his arguments.

4. In the discussion of negligence in this chapter, it was pointed out that conforming to a standard of care is important in engineering. In court, evidence regarding customary practices is often introduced to define the standard. List at least five instances in which customary practices that people follow regularly might not be acceptable to a court. Here are two examples:

   • On almost every major highway in North America, it is customary for a large fraction — maybe even a majority — of the motorists to exceed the speed limit. If you had an accident and were speeding at the time, do you believe a court would accept your argument that speeding is customary?

- Drug use has increased greatly in the past 30 years. Would a court accept that recreational use of marijuana is now customary?

5. A case once occurred in which the plaintiff's shoes slipped on a wet laundromat floor, causing injury. The plaintiff brought suit against the seller of the shoes, alleging that the shoes were easily inclined to slip and therefore dangerously unsafe, and claiming a large sum of money for damages. The plaintiff did not win this case because it was shown that the shoes were no more slippery than any others and thus were not defective. However, the executives of this shoe company want to avoid similar lawsuits in future.

   Imagine that you are an engineer being paid a large consulting fee to advise these executives. You know that most shoes are supposed to give firm, high-friction traction, but some shoes, such as dancing shoes, are supposed to be slippery so that they slide smoothly. Using your own knowledge of the shoe industry, what would you advise the executives to do? Would it be better to discontinue selling slippery shoes, test the shoes and scrap the slippery models, label the slippery shoes, or what? Does the manufacturer have an obligation to alert the purchaser about slippery shoes? If so, how would you do this? For example, would attention, caution, warning, or danger labels be appropriate? If so, which one would you use? Is there a manufacturing standard for shoes? If so, find the standard by contacting the Standards Council of Canada over the Internet. If not, discuss the benefits and disadvantages of creating a slipperiness standard to rate slippery shoes, thus helping the purchaser and protecting the manufacturer. How would you establish such a standard within the shoe industry? Prepare a brief report listing and explaining your recommendations.

6. A case once occurred in which an amphibious all-terrain vehicle being driven through a swamp by several hunters overturned; the occupants drowned. The relatives of the dead hunters sued the manufacturer of the ATV for the loss, claiming that the vehicle was clearly unsafe. The plaintiffs did not win this case because it was shown that the vehicle had been overloaded and that the hunters had been drinking. However, the manufacturer wanted to avoid similar lawsuits in future.

   Using your own knowledge of amphibious vehicles (water-tight land vehicles that float like boats in water), describe how you would advise the manufacturer. Does the manufacturer have an obligation to alert the purchaser or driver that the vehicle might tip if it is overloaded, or that it might tip if driven carelessly? If so, how would you do this? For example, would attention, caution, warning, or danger labels be appropriate? If so, which one would you use? Is there a manufacturing standard for amphibious vehicles, or would the standards for boats and vehicles both apply? Find at least one relevant standard by contacting the Standards Council of Canada over the Internet. If no standard exists for amphibious vehicles, discuss the benefits and disadvantages of creating such a standard by combining other marine and vehicle standards. How would a standard protect the manufacturer? What is the procedure for getting such a standard adopted by the Standards Council of Canada? Prepare a brief report listing and explaining your results and recommendations.

NOTES

1. Professional Engineers Act, Regulation 941/90, RSO 1990, c. P.28, s. 72.
2. D.L. Marston, *Law for Professional Engineers: Canadian and International Perspectives*, 3rd ed. (Whitby, ON: McGraw-Hill Ryerson, 1996), 37.
3. D.W. Noel and J.J. Philips, *Products Liability*, 1st ed. (St. Paul, MN: West Publishing, 1974).
4. Ibid., 19.
5. Marston, *Law for Professional Engineers*, 46.
6. Ibid., 51.
7. D.W. Noel and J.J. Philips, *Products Liability*, 2nd ed. (St. Paul, MN: West Publishing, 1981).
8. Noel and Philips, *Products Liability*, 1st ed., 116.
9. Ibid.
10. Ibid., 121–22.
11. *Matthews v. Lawnlite Co.*, 88 So. 2d 299 (Fla. S.C., 1956).
12. "Faults and Failures," *IEEE Spectrum* (December 1987): 16.
13. Noel and Philips, *Products Liability*, 1st ed., 142.
14. Noel and Philips, *Products Liability*, 2nd ed., 138.
15. Noel and Philips, *Products Liability*, 1st ed., 149, 151–52.
16. Ibid., 155–57.
17. S. Gibson-Harris, "Looking for Trouble," *Mechanical Engineering* (June 1987): 36–38.
18. Noel and Philips, *Products Liability*, 1st ed., 169–71.
19. Gibson-Harris, "Looking for Trouble."
20. A.S. Weinstein et al., "Product Liability: An Interaction of Law and Technology," *Duquesne Law Review* 12, no. 3 (Spring 1974): 434–38.
21. Noel and Philips, *Products Liability*, 1st ed., 150–51.
22. C.O. Smith, "Products Liability: Severe Design Constraint," *Structural Failure, Product Liability and Technical Insurance*, Proceedings, 2nd International Conference 1–3 July 1986 (Geneva-Interscience Enterprises, 1987), 59–75.
23. G.C. Andrews and H.C. Ratz, *Introduction to Professional Engineering*, 5th ed. (Waterloo, ON: University of Waterloo, 1996), 240.
24. Ibid., 258.
25. Ibid., 259.

26. Smith, "Products Liability: Severe Design Constraint," 71.

27. Society of Automotive Engineers (SAE), "Safety Signs," *SAE Handbook*, SAE Recommended Practice J115 SEP79 (Warrendale, PA: 1979).

28. Standards Council of Canada, "Serving Canadian Industry and Consumers," December 1997, http://www.scc.ca/about/index.html.

29. D.L. Goetsch and S.B. Davis, *Understanding and Implementing ISO 9000 and ISO Standards* (Toronto: Prentice-Hall, 1998), 182.

30. Ibid., 150.

31. Ibid., 119.

32. S.L. Jackson, *The ISO 14000 Implementation Guide* (Toronto: John Wiley & Sons, 1997), 1.

33. T. Claridge, "Jury Finds Helmet Maker Partly Responsible for Injuries," *The Globe and Mail* (15 June 1996): A11. Reprinted with permission from *The Globe and Mail*.

34. R.A. Nader, *Unsafe at Any Speed*, rev. ed. (New York: Bantam, 1973), 61.

CHAPTER THIRTEEN

Fairness and Equity in Engineering

A profession such as engineering attracts intelligent, creative, fair people who generally have (or should have) higher personal standards than mere adherence to the letter of the law as expressed in the provincial Professional Engineering Acts and the codes of ethics. This point is well expressed in the foreword to the code of ethics of the Association of Professional Engineers of New Brunswick, which says: "Honesty, justice and courtesy form a moral philosophy which, associated with mutual interest among people, constitute the foundation of ethics. Engineers should recognize such a standard, not in passive observance, but as a set of dynamic principles guiding their conduct and way of life.[1]

Unfair behaviour, such as discrimination, harassment, and racism, have no place in a civilized society nor in the engineering profession, yet these are not explicitly mentioned in the various codes of ethics. In Canada, these actions are illegal and are covered by the criminal code and by human rights legislation, so it may seem obvious that if anyone, especially a professional engineer, commits a criminal act, this behaviour is unethical. Nevertheless, some aspects of discrimination, harassment, and racism may exist unseen by the profession because they are subtle, systemic, or ingrained. These issues and some of their underlying causes are appropriate topics for this textbook on professional practice so that they can be recognized and avoided or corrected.

This chapter was written by Dr. Monique Frize, NSERC/Nortel Joint Chair in Women in Science and Engineering in Ontario, Faculty of Engineering, University of Ottawa and Carleton University.

DEFINITION OF DISCRIMINATION

In dictionaries, the word *discrimination* has a somewhat innocuous or neutral definition: it means the action of discerning, distinguishing things or people from others, and making a difference. However, in recent years, the term has come to be associated with the much more harmful meaning of segregating: the act of distinguishing one group from others, to its detriment. The legal definition of discrimination is more specific:

> In a 1985 decision, the Supreme Court of Canada described discrimination in this way: It is a result or the effect of the action complained of which is significant. If it does, in fact, cause discrimination; if its effect is to impose on one person or group of persons obligations, penalties, or restrictive conditions not imposed on other members of the community, it is discriminatory. The Court also describes two related concepts: Treating individuals in an identical manner when it is not appropriate is a denial of equality just as serious as to treat individuals differently when that is not appropriate.[2]

The more specific term *systemic discrimination* is defined as

> any result or effect of an apparently neutral and objectively stated or implied rule or practice implemented by the majority in place, which is to the detriment of an individual or a group who have, respectively, personal or collective characteristics which are different than those of the majority.[3]

DISCRIMINATION AND THE CANADIAN CHARTER OF RIGHTS AND FREEDOMS

Discrimination is unfair whether it is random or systemic, and, depending on the case, it may be subject to legal action:

> All contracts, collective agreements, work protocols, and handbooks are assumed to be consistent with the relevant provincial human rights legislation, the latter also being consistent with the Canadian Charter of Rights and Freedoms. Discrimination is not permitted under the law. Contracts, including collective agreements, can be rescinded and statutes and regulations can be nullified if found to be discriminatory. The main difference between the Canadian Charter of Rights and Freedoms and other federal and provincial human rights legislation is that the former applies to all levels of government, including agencies directly controlled by governments. On the other hand, provincial human rights legislation applies (al-

though not directly) to private parties which are not under federal jurisdiction. The Canadian Charter of Rights and Freedoms is the supreme law of Canada and has priority over any other types of legislation.[4]

DISCRIMINATION IN ENGINEERING?

Engineering is a profession that should appeal to creative people from all parts of our society, regardless of their sex or race. However, discrimination has not yet been eliminated from Canadian society, and since engineers are part of society, instances of discrimination are still evident from time to time in engineering education and practice.

In 1989, the murder of women engineering students at l'École Polytechnique in Montreal, an instance of incredibly senseless violence, focussed the attention of the nation on the issue of discrimination in engineering. Late in the afternoon of 6 December 1989, a man entered the school, separated the men from the women and, using a semi-automatic weapon, killed thirteen female students and one female employee and wounded thirteen others before taking his own life. To women engineers, this was a very personal event, and many remember exactly where they were and what they were doing when the terrible news was announced.

This massacre represents the ultimate expression of violence against women. There are various interpretations of the mental stability of the individual and what this event really meant, but there can be no doubt that the segregation and systematic murder of female students was his main goal. The murders could have resulted in discouraging women from entering the engineering profession, but the spontaneous outpouring of deep emotion and strong support of all fair-minded people in Canada ensured that this would not occur.

The massacre at l'École Polytechnique gives a special urgency to examining the problem of discrimination against women in engineering. It is time to look closely at the issue of discrimination against women and other underrepresented groups in the engineering profession and to guarantee, once and for all, that we have fairness and equity in the profession. Only then can the senseless tragedy of these murders have a positive, permanent outcome.

The participation of women in engineering is far less than it should be. While women are actually a majority in the Canadian population, they are very much underrepresented in the engineering profession, in spite of the fact that other professions, such as medicine, law, dentistry, and veterinary medicine, have reached and maintained gender-balanced enrollments for some years. The low representation of women in engineering can be explained by several factors: the perpetuation of socially defined stereotypical gender roles in the media and by some teachers, parents, and guidance counsellors; the perception that "brains and femininity" are incompatible; the low number of role models available to inspire young women to enter this field; and some aspects of the culture of engineering. The rest of this chapter is devoted mainly to discussing these factors, and describing methods to reduce or avoid their negative results.

SOCIALIZATION IN EARLY CHILDHOOD

A recent British Columbia provincial report explains how young girls and boys begin to form gender-role stereotypes and examines the effects of socialization and of self-esteem on the education and training of girls and women.

> From the moment we are born we begin the process of learning how to be human beings. We learn about attitudes, values, and behaviours which are acceptable in our society. We learn what is expected of us, what roles we can play, how to exercise self-control, how to live in a community. Social scientists call this learning process *socialization*. When this process is applied to how women and men are expected to behave, this is called gender socialization.[5]

The report mentions that

> significant gender inequalities persist in Canadian society and are reflected in and reinforced through the formal and informal processes of socialization. Gender socialization begins at birth and intensifies throughout childhood and adolescence. In some ways, gender socialization continues as part of lifelong learning.[6]

Gender socialization may have a harmful effect if, when children grow up, they feel excluded from professions, trades, or lifestyles because of their sex, and it appears that this may be the case for men choosing nursing or secretarial work and for women choosing engineering or science.

Gender socialization has a great impact on the education system from the elementary grades through the end of high school. For example, in their book *Failing at Fairness*,[7] the Sadkers report how teachers tend to give more attention to boys, not only in reprimands, but in getting them to answer questions, challenging them more than girls, and praising them more highly for their answers. Girls, on the other hand, tend to be praised for neatness and good behaviour. It is also common to hear young women describe how they were discouraged from studying mathematics and science.[8] In addition, social and health studies are often suggested as "more appropriate for young women," and then these courses are scheduled so that they conflict with other courses, such as physics. Although some guidance counsellors make a special effort to encourage girls and boys to consider nontraditional career choices, others — women and men — discourage young people from doing so. Thus, by the end of junior high school, less than one-third of young women select physics as an option, which is a required course to enter engineering. The few women teaching science in high school are usually in biology or chemistry, rarely in physics, and few have a background in engineering.

The result of gender socialization in schools is similar to systemic discrimination. Sadly, some young people may, through poor counselling, end up studying subjects that seem "appropriate" but are not what they would have chosen, based on their true interests and skills, if they had been given full information and encouragement. Clearly, there is a need to look for this sort of misdirection in the early years and to seek to eliminate or correct it.

ENCOURAGING FULL PARTICIPATION OF WOMEN IN ENGINEERING

Introducing science and engineering at an early age through activities that are fun to do, such as gender-balanced summer science camps, is a good way to stimulate the interest of youngsters in these subjects. The success of the camps can be measured by the proportion of young people who enrol in science and mathematics courses when entering high school; a second measure is the influence on the future career choices of girls and boys. The objective is not to have 100 percent engineers at the end of camp, but to have a reasonably similar proportion of girls and boys selecting such an option. This can be accomplished if the camp is gender-balanced in its leadership and attendance, and if the leaders ensure full hands-on participation by both genders in all the activities. Finally, the content should be of interest to both girls and boys.[9]

A study by Vickers et al. discussed whether extra-curricular science programs make a difference. They showed that several of the out-of-school activities had a slightly larger impact on boys than on girls, while one of three "girls-only" science programs seemed to have the highest impact on girls of any of the programs studied. This points to the importance of establishing reasonable goals for outreach programs (old and new) and assessing their effectiveness on a regular basis. Of the twenty summer science and engineering camps in Canada, a few are single-sex, such as the camp at Ryerson Polytechnic University, but most are mixed and are attended by a larger number of boys than girls (grades 5 to 8). In 1996, four of the camps had achieved a gender-balanced enrollment. One of these (Worlds Unbound, at the University of New Brunswick) had a simple and effective admission policy: if an imbalance existed in the applications for admission, applicants in the oversubscribed group were placed on a waiting list, and those who found a matching applicant from the underrepresented group were admitted.[10]

Here are some other initiatives taken around the country:

- Women engineers and engineering students visit classes in elementary and secondary schools, explaining how they chose their field, their career plans, and how they balance a career and personal life.
- Workshops for parents, teachers, and guidance counsellors on the appropriateness of an engineering career for young women have been helpful in obtaining their help to eradicate sexism and stereotypes based on gender roles.
- Public lectures, open houses, science fairs, "science Olympics," and industrial tours are other activities that have been shown to have various levels of success with girls and boys.[11]

More steps can be taken to ease the obstacles that girls and young women face after making the decision to pursue an engineering career. For example, in the Canada Scholarship program (until its demise in 1994), half of the awards were made to women and the other half to men. This had a positive effect on the enrollment of women in science and in engineering.

The Canadian Coalition for Women in Engineering, Science and Technology (CCWEST), a coalition of various organizations interested in promoting science, engineering, mathematics, and technology to girls and young women, was created in 1993 to provide an umbrella organization for uniting the various voices of women and to advocate for change. The Coalition has a Web site at http://www.ccwest.org.

BENEFITS OF GENDER BALANCE

The fight against discrimination based on the circumstances of birth, which are beyond a person's control, is an honourable fight. To accept discrimination as a necessary evil is to condone policies and attitudes that are unfair and can result in shattered dreams and ambitions.

Diversity encourages creativity and innovation. As yet, there is not a lot of research on whether women bring new perspectives to the engineering profession. However, a 1996 Canadian study by Ann van Beers,[12] in which she interviewed twenty female and twenty male engineers in the Vancouver area, revealed that several participants thought the presence of women would bring changes to the structure of the work environment, the culture, and the practice of engineering. Some of the responses lent support to the idea that women will introduce values and perspectives that have been more rarely represented so far, such as having a more contextual approach and better communication and interaction skills, and preferring a more consensual working relationship over a hierarchical structure. These results are supported by Vickers et al., who found that, at a young age (less than 21 years old), a substantially larger proportion of females, compared to males, stated that they wanted to make a contribution to society. This was also true for an older group (more than 21 years old).[13]

A recent survey carried out for the Women in Engineering Advisory Committee of Professional Engineers Ontario (WEAC/PEO), which polled both female and male engineers across Canada in most provinces, found that

> Workplace challenges continue to exist for female engineers. Women feel they face at least some attitudinal barriers from their superiors, and a substantial proportion of men share that view. In particular, women are concerned about opportunities to network and to gain entry to executive levels in their organizations.

The study concludes that "the workplace is changing in positive ways for women, but old, lingering beliefs held by even a few can act as barriers to full participation."[14]

In another large study in the United States, the authors interviewed recipients of prestigious fellowships in science and engineering to assess the career success of men and women.[15] The study showed that men received faculty positions at one level higher than women (except in biology), men published slightly more (2.4 papers per year versus 1.8 for women), but the average number of citations per paper was far higher for women than for men (24 versus 14). Several respondents also believed that the choice of research topics and the manner in which the research was approached would differ by gender.

These studies show that excellence in engineering and science is still defined in masculine terms; thus, it is a delicate balancing act for women engineers to retain their femininity while being perceived as good engineers. It is also apparent from these studies that women can bring new values to the profession, provided there are sufficient numbers of them to ensure a voice and they are allowed to be themselves — that is, valued for the contributions they make. Women's affinity for a consultative style of working is very much in tune with today's management philosophy. Many women also excel in verbal and interpersonal skills. These qualities, combined with a solid technical education, become a real asset, especially for smaller firms whose engineers must interact with suppliers, clients, and regulating agencies.

Similarly, individuals from other underrepresented groups who have been raised with different cultural influences may bring different and innovative solutions to engineering problems. Everyone benefits when diverse, gender-balanced teams design tomorrow's products.

STRATEGIES TO ACHIEVE A BETTER BALANCE

Several initiatives developed in Canada are fairly unique both in their concept and in the role they play in bringing about solutions to gender imbalance in the engineering profession.

NSERC/NORTEL CHAIR IN WOMEN IN ENGINEERING (WIE CHAIR)

In 1989, the Natural Sciences and Engineering Research Council (NSERC) created an Industrial Chair in Women in Engineering at the University of New Brunswick (UNB) with the support of an industrial partner, Nortel (Northern Telecom). The mandate of the Chair was to develop strategies to increase the participation of women in the engineering profession. The chairholder studied patterns of discrimination that could act as obstacles to women in the education system, in the workplace, and in the engineering associations by collecting information from surveys and personal testimonies from women and men in the profession. Successful strategies were shared with other organizations through speeches, workshops, and articles. In 1997, five new NSERC Chairs were created with the support of various industrial partners and host universities, covering every region in Canada (Atlantic and Northwest Territories, Quebec, Ontario, the Prairies, British Columbia and the Yukon). For more details, see the following Web site: http://www.carleton.ca/wise.

CANADIAN COMMITTEE ON WOMEN IN ENGINEERING (CCWE)

The massacre of female engineering students at l'École Polytechnique on 6 December 1989 and the low level of participation of women in undergraduate engineering programs across the country were key factors that led to the creation of the Canadian Committee on Women in Engineering. The sponsors were the federal department of Industry, Science & Technology Canada (now Industry Canada) and the Canadian Council of Professional Engineers (CCPE), supported by a contract with Employment & Immigration Canada (now Human Resources Development Canada). Other partners were the Association of Consulting Engineers of Canada (ACEC), the Canadian Manufacturer's Association (CMA), and the Association of Universities and Colleges of Canada (AUCC). These six signatories to the agreement invited others to join: the Canadian Association of University Teachers (CAUT), the National Committee of Deans of Engineering and Applied Science (NCDEAS), the Canadian Education Association (CEA), and the Canadian Federation of Engineering Students (CFES). The committee was chaired by the WIE Chair, ensuring that duplication of effort would be avoided and the possibility of synergy would be enhanced.

The mandate of the CCWE was to examine obstacles to women's participation in engineering and to identify strategies to increase this participation. Composed of nineteen engineer and non-engineer members, with a regional, bilingual, and gender balance, the CCWE studied issues in elementary and secondary school, universities, workplaces, and professional associations. The CCWE held six public forums across the country, received more than 200 testimonies and briefs from individuals and groups, and studied "best practices" in seven universities and six workplaces. Prior to the release of its report, over 200 participants reviewed the draft recommendations at a national conference in May 1991. The final report, *Women in Engineering: More Than Just Numbers*, was released in April 1992.[16] This was followed by a meeting in September in which all organizations represented on the CCWE endorsed the report. This event concluded the committee's work, although the Government of Canada, under the auspices of the National Advisory Board on Science and Technology, developed a second report to extend the CCWE work to science and technology. It was released by the prime minister on 8 March 1993 (International Women's Day).[17] These two reports provide a detailed overview of the issues and of their solutions. A national conference held in 1995 in Fredericton assessed the progress made since 1992, and recommendations were made by the participants on the work still to be done.[18]

ENSURING FAIRNESS AND EQUITY IN ENGINEERING SCHOOLS

Canadian universities have become more aware of gender and racial issues, and many have developed policies that attempt to ensure fairness and equity. Several CCWE recommendations were put in place by engineering schools/faculties between 1992 and 1995, and, in combination with their own additional efforts, these recommendations seem to have had a very positive impact on the enrollment of women in engineering programs.[19] In 1990, the average proportion of women in engineering undergraduate programs in Canada was 14 percent, but only 10 percent were in master's and 6 percent in doctoral programs; 2 percent of engineering faculty members were women. By 1995, the average proportion of women in engineering undergraduate programs in Canada had grown to just under 20 percent, with similar numbers at the master's level. Just under 14 percent of the enrollment in doctoral programs and just over 5 percent of the engineering faculty members were women.[20]

This increase is encouraging, but full gender balance will require further measures. There are wide variations among universities in the proportion of female undergraduate engineering students, ranging from under 7 percent to a high of 43 percent. There are also large differences by field of study, with higher enrollments in chemical, environmental, and industrial engineering, and lower enrollments in electrical, computer, and mechanical engineering.

Discrimination, sexism, and racism are unacceptable wherever they occur in society, but they are particularly offensive when they occur in places devoted to higher education. Universities have made special efforts to examine their policies to ensure fairness and equity, but policies are not enough. They need to be explained and enforced for real change to occur. This issue can be greatly influenced by individuals, and some professors and administrators have been effective in bringing about improvements. For example, there seems to be more sensitivity about using gender-inclusive language, avoiding sexist and racist remarks in class and in the office, and treating female students as seriously as male students. However, the learning experience of women and members of minority groups still varies from class to class and from university to university. Some inappropriate acts and behaviour still occur. More needs to be done to promote a positive educational experience for all students.[21] For example, a professor may ignore a female or minority student in a class, or may not intervene when the behaviour of some students is demeaning to others. This attitude on the part of the professor, even if inadvertent or negligent, adds to the problem instead of solving it. In the classroom, professors and instructors must do more than just show sensitivity to the diversity of our society: as role models, they must be active participants in advancing the concept of fairness and equity. (For more information on this topic, see "Criteria for Women-Friendly Engineering Schools" at http://www.carleton.ca/wise.)

The first-year experience is the most critical, and special attention must be paid to retaining students who enrol in engineering programs. Students who drop out because they did not receive adequate encouragement usually suffer reduced career expectations, but there is also a loss to the university and to the country. Some universities have increased levels of retention and morale by instituting peer mentoring programs for those who need support and encouragement. Entrance scholarships based on entrance qualifications for both genders, with a number of these reserved for high-achieving women, are effective in attracting more of these women to engineering instead of to medicine or law.

Universities have a duty to create an atmosphere of respect and tolerance for diversity. Studies have shown that teaching style is important. According to these studies, a co-operative style of teaching and learning has been shown to be far more effective than the traditional method for both female and male students alike.[22] It appears that relating topics and material to societal realities and needs is particularly appreciated by women, but this approach may also increase the interest of men.[23] Traditional engineering programs, such as mechanical and electrical engineering, have the lowest enrollments of women, whereas multidisciplinary programs related to the quality of the environment and to life sciences such as environmental, bio-resource, chemical, and water resource engineering have reached gender-balanced enrollments. The same is true for biomedical engineering where this program exists.

To reach beyond the current enrollment of women in engineering, which is still low, attention must be given to designing curricular content and using teaching styles that are more appealing to women. Such changes could result in a better and more natural fit between women and engineering. Attracting more women into graduate programs and hiring more women as faculty members will also have a positive impact. It is also pertinent to review the criteria for judging achievement and success, which affects decisions in hiring, tenure, and promotion, and to the creation of policies that allow young faculty members — female and male — to balance family and career. The criteria used to assess faculty performance, based on decades of tradition, should be re-examined from time to time to see if they are still relevant in a changing world. This applies not only to how merit is defined and measured, but also to how awards, appointments, and promotions are awarded.[24] Outdated stereotypes and biases can affect the success rates in competitions for scholarships, fellowships, grants, jobs, promotions, research grants, and awards. These biases can be reduced through education and sensitization programs, and by ensuring that there is gender-balanced representation on the committees that make the final decisions. A balance will be truly possible only when women represent a critical mass (defined as at least one-third) of engineering faculty members.

ENSURING FAIRNESS AND EQUITY IN THE ENGINEERING WORKPLACE

Everyone has a right to work in a fair and equitable workplace. Increased employee performance is one of the most obvious results of a positive work environment, and efforts by managers and employers to introduce fairness and equity will pay off by creating a more successful enterprise. The first steps are to recognize the special problems that women and members of minority groups may face within the organization and to ensure that effective policies are in place for hiring and promoting employees. Communicating these policies, their meaning, and their purpose to all staff is a critical step for their successful implementation. If some of the issues arouse anxiety or anger in a particular group, it is often because of misconceptions, which can be diffused and eliminated by open and frank dialogue. Creating a committee to identify issues of particular significance within an organization can be very effective.[25]

FAIR HIRING PRACTICES

Realistic objectives for hiring people from underrepresented groups should be based on achieving better than, or at least the level of, availability of each underrepresented group in the pool of candidates. Jobs should be advertised widely and externally in addition to being posted internally. A particular effort needs to be made to encourage qualified members of underrepresented groups to apply. This means finding and contacting them, as they may be few in number and may not consult the mainstream advertising channels. Training is needed for the people involved in hiring to recognize inappropriate and illegal interview questions and the importance of treating all applicants with fairness and respect. Unbiased interview techniques should be used; one way to test the appropriateness of a question is to ask whether everyone — whether male, female, or a member of a minority group — should be asked the same question. If the answer is no, it should not be asked.

Discriminatory practices in hiring are often evidenced by a predominantly female staff, but a predominantly male senior management. For example, in the 1940s, Canadian banks were staffed by tellers who were almost exclusively female, while branch managers and senior executives were exclusively male. Since then, women with management potential have been identified and assisted through proactive training, to qualify for promotions. This process provides female role models for younger women and can integrate women's values into the corporate culture. In turn, the women who are promoted must be committed both to the organization and to the goal of employment equity if real change is to occur. Unexpected new benefits and insights occur by hiring women, particularly if they are encouraged to introduce diversity and innovation and if they feel their attributes and values are respected.

EMPLOYMENT EQUITY

A Saskatchewan government brochure defines employment equity clearly:

> Employment equity is a comprehensive pro-active strategy designed to ensure that all members of society have a fair and equal access to employment opportunities. *It is a process for removing barriers that have denied certain groups equal job opportunities....* Employment equity programs encourage employers to hire, train, and promote members of these groups.[26]

Some employers, particularly in private industry, still fail to see the benefit of hiring women. A recent survey of engineering graduates from 1989 to 1992[27] shows that the majority of female graduates were hired by institutions committed to the principle of employment equity. Yet these organizations also hired male engineers in proportions very close to the existing pool of graduates. This is evidence that "employment equity" policies remove obstacles for underrepresented groups without adding barriers for the majority group.

AVOIDING AND REMOVING DOUBLE STANDARDS

Stereotypes often shape our perceptions of other people and can have a major impact on the career progress and success of women and members of minority groups. For example, Foschi et al.'s research, published in 1994, showed that even though groups of female and male candidates were equal in their background and skills, males were viewed as better qualified by a majority of the male subjects making the selection.[28] This was even more evident where the field was a nontraditional one for women. Foschi et al. also showed that when women participants made the hiring decisions, they rejected this stereotype by selecting close to half of the women candidates. This is in contrast to the 1968 study by Goldberg[29] where female names were put on essays written by males and male names were put on essays written by females. At that time, both female and male subjects selected essays with male names as superior, saying that they sincerely believed these essays were more important, authoritative, and convincing than those with female names. In Sweden, a 1997 study of postdoctoral awards showed that women had to be twice as productive as men to be perceived as equally competent.[30] Whether deliberate or not, stereotypes create a double standard in subjective evaluations. Hopefully, false assumptions about gender roles and capabilities will disappear as employers hire capable women and minority employees who, through commitment and experience, will succeed in dispelling these myths.[31]

FAIRNESS IN EMPLOYEE PERFORMANCE ASSESSMENT AND PROMOTIONS

In a large study of American scientists and engineers in high-tech companies conducted by DiTomaso and Farris, Caucasian men were rated by their managers as average on the attributes of innovation, usefulness, and promotability, and were rated a little lower on co-operativeness. These managers rated women lower on all these attributes except for co-operativeness. When the employees made a self-assessment, the Caucasian men rated themselves slightly higher than their managers' rating, but women rated themselves lower than their managers' rating for all attributes except co-operativeness. A possible interpretation of the results is that male Caucasian scientists and engineers understand the corporate culture better and interpret the feedback more accurately, since they come from a similar group as the managers for the most part. For women, the question of self-esteem arises, and the uneven understanding of feedback may be a problem.[32]

In the same study, women rated their managers lower than the men did on getting people to work together, letting people know where they stand, being sensitive to differences among people, and minimizing hassles with support staff. However, women rated managers more highly on communicating goals clearly, defining the problem, getting resources, and motivating commitment. These results show that managers must put more effort into improving the assessment process and developing objective and measurable criteria for assessment. They must focus more attention on the type of feedback they provide, communicate the rules clearly, and test whether these rules have been understood. Managers must especially work on understanding the different approaches and perspectives that women and minority groups bring to the organization and avoid underestimating the performance of particular groups of employees. If managers build teams with people from diverse backgrounds and perspectives, the organization's output will be enriched and improved.

ELIMINATING SEXUAL HARASSMENT

Several studies have shown that sexual harassment still occurs in the workplace, and engineering is no exception. Statistics show that approximately one woman in two working in an engineering environment will suffer or observe some type of harassment.[33] What makes this situation more difficult in engineering is that there are few women at each site or in each firm, so they may feel isolated and perhaps find little support or sympathy for what may be considered "normal behaviour" or a "boys will be boys" attitude.

It is essential that all engineers become familiar with the definition of harassment and how to prevent it. Some people trivialize the problem, and companies vary greatly in their policies and procedures for dealing with it. Education is at the heart of the solution, and a fair investigation procedure that does not victimize the complainant and provides checks and balances for verifying accusations is needed to eliminate harassment in the workplace.[34] Providing moral support for colleagues who are complainants in such a situation can do much to reduce the stress level. Choosing an informal investigation approach can be more successful, in many instances, than the more confrontational formal route. Many organizations consider both approaches in developing their policies and procedures.

FAIRNESS IN AWARDS AND TRIBUTES

In the past, it was rare to see women being nominated for awards or chosen as keynote speakers and for other honours provided by technical societies and professional engineering Associations. In the late 1990s, there are still some engineering events where only men appear in visible roles. Monitoring the proportion of women on executive committees, the number of awards given to women, and the number of women invited to serve as keynote speakers, plenary session speakers, and panelists on specialty topics will enable an assessment as to whether women have been integrated fully and fairly into the profession and whether their contributions are properly recognized. When the proportion of women invited to participate in these activities equals or surpasses their actual proportion in the society or association, then progress will be visible and real. At this time, if the quality of the candidates is really the criterion for rewards, women should make the lists more frequently.

IMPLEMENTING CHANGES

Many of these workplace issues and their solutions are discussed in greater detail in the report *Women in Engineering: More Than Just Numbers*.[35] Each organization must set its own goals based on where it currently stands, and draw up an achievable plan to create an environment where work is challenging and comfortable for all of its employees. The age of the industry is a key factor in determining the rate of change possible:

> Many anecdotes suggest that older industries are more likely to have an entrenched old-boy network that may not be receptive to female outsiders. Conversely, in some new sectors (biotech, for example) the old-boy network was built only over the past decade — and women helped build it.[36]

Moreover, as the culture of engineering changes from a predominantly male one to a culture that integrates diversity, many of the obstacles and challenges identified in this chapter will disappear.

CONCLUSION

This chapter focussed mainly on the issue of women's participation in engineering, because women are a majority in the Canadian population and their scarceness in an important field like engineering raises obvious questions. The culture of engineering, so far predominantly masculine,[37] needs to integrate feminine values to enrich its perspectives.[38] However, other groups are also underrepresented. For example, Aboriginal people have generally been excluded from the educational avenues that lead to engineering. Some universities have created access programs for such students to permit them to catch up on required courses. These programs are new, and their success will be assessed over time.

Fortunately, the engineering profession is becoming friendlier for women.[39] A number of universities and firms are in the process of creating a better climate for inclusion and fair policies. The extent of progress, however, depends very much on the particular organization. Some have progressed, some have done little, and others have regressed. There is some backlash from people who will not budge from the status quo position. The progress reports delivered at the 1995 update conference[40] showed that universities had achieved substantial progress, and workplaces generally had done the least. As for the professional engineering Associations, progress varied greatly among the provinces and territories. Much more needs to be done in the next few years, and each of us can make a difference.

CASES FOR DISCUSSION

The following case studies are based on real situations that have occurred in recent years. The names of the individuals and companies have been changed to protect the privacy of the people concerned.

CASE STUDY 13.1
DISCRIMINATION[41]

Assume that you are an engineer and Chief Executive Officer of a profitable company called the Exeter Corporation. You are contacted by Susan Smith, a highly valued sales manager at Exeter, who has been passed over for promotion to Director of Product Development. The promotion was awarded to Sam Brown on the recommendation of the Vice-President of Marketing, Peter Young. Smith sees this as a classic case of discrimination and is threatening to sue Exeter for unfair practices. She asks you to respond to her concerns within 24 hours or you will likely lose her, a valuable employee, and her lawyer will explore the possibility of a settlement through the Human Rights Commission or the courts.

You arrange a meeting with Peter Young and the Director of Human Resources to ask for more information as to why Brown was selected over Smith, and you are told that the difference between the two candidates was marginal. Young's explanation for his recommendation includes both objective and subjective criteria. His first comment is that Brown's experience in terms of seniority and familiarity with the industrial sector weighed slightly in his favour. Young adds that, through Brown's greater participation in company social events and in the squash ladder, he was better known to all of the vice-presidents, who said that Brown "looked like a winner." They could not say the same about Smith because she was less well known.

When prodded by the Human Resources Director, Young suggests an additional list of problems and shortcomings that he attributes to Smith:

- "Mark Tannen, Vice-President of Manufacturing, is thought to be having an affair with Smith, and he is pushing her for promotion."
- "If Smith was promoted, Exeter might be liable to discrimination charges placed by Brown because of Mark Tannen's push for promotion for his honey."
- "The Director of Product Development is a man's position. Human Resources — soft, person-to-person stuff — is for women. Factories are for men."
- "Exeter clients prefer to deal with men. They know how to relate to their wives, mothers, and girlfriends, but not to women Product Development Managers."
- "Women are undependable. They get married, get pregnant, want time off, and are less committed to the job."

Young provides no evidence to support these assertions; however, it is clear that they have influenced his decision to appoint Brown. He believes his decision made good business sense.

After the meeting, you reassess the situation. According to the objective data presented, Brown and Smith were both qualified for the position. Smith has shown excellent achievement as a product line manager. The same could be said about Brown. Choosing between the two would be understandably difficult. Ignoring Young's subjective evaluation of Smith's "shortcomings," you must make a choice between promoting a woman (Smith) to a higher management level or promoting a man (Brown) who has marginally more experience.

You review your company's existing employment equity policies and current equity situation. Although a quarter of the employees at Exeter are women, there are no women at the executive level and none on the Board of Directors. Recently, you and the Human Resources Director issued a policy stating that the company would make great efforts to ensure equity and fairness in the manner in which employees were recruited, trained, and promoted. Therefore, although you have no hard evidence, you worry that gender inequity may permeate the organization. Also, if Smith pursues her lawsuit, you wonder whether it may encourage other women to come forward and state similar experiences. You realize that if Peter Young's comments on Smith's "shortcomings" were repeated in a court, Exeter would surely be found guilty of discrimination. Thus, the firm would experience both a financial loss and the loss of an excellent employee.

QUESTIONS

1. What criteria should have been used to select the new Director of Product Development? Are these the same as Young's criteria?
2. Based on your criteria, who should have been appointed to the job: Sam Brown or Susan Smith? State and explain the reasons for your answer.
3. Since you are the CEO, what should you do in the next 24 hours regarding the potential lawsuit threatened by Smith?
4. As CEO, what long-term issues do you face if you want to ensure employment equity at Exeter, and what steps should you take to put this equity process in place?
5. Does your provincial code of ethics address this type of issue? Does it make a difference if Peter Young, Sam Brown, and Susan Smith are also engineers? If your answer is yes, quote the appropriate sections. If not, should the code provide guidance for dealing with this case? Alternatively, could you suggest the proper wording for an appropriate clause?

CASE STUDY 13.2
SEXUAL HARASSMENT

Michelle Kirkland has been a mechanical engineer in a consulting firm for four years. Recently, she wrote to a senior female engineer to discuss a serious work-related problem and to ask for advice on how to solve it. Here are major extracts from her letter:

In my academic years I never had any problems being a woman in a male-dominated environment, and therefore very naïvely entered the workforce with a very positive and healthy attitude toward men in engineering. Today, unfortunately, that is no longer the case. After four years of verbal abuse and three incidents of sexual harassment from my immediate supervisor, I have become so cynical about men that I no longer enjoy my work. Most men quite naturally treat women without respect and as second-class citizens without even being aware of it.

The worst part of the situation is that I feel I cannot talk to anyone about this. In our corporation, female managers are practically unheard of and men seem to stick together like glue. Their attitude is that everything seems to be my fault: "Women are more sensitive" and "Women are less reliable" are the most recent comments that I bluntly received from my manager.

I was considering leaving the profession at one point, but meeting other women engineers motivated me to fight back harder and try again. Should I transfer to another department? Should I leave the company? (But are there any better ones out there?) Should I leave the profession and let my daughter solve the problems? I really do not know what to do. Sticking it out means additional stress in an already stressful job, headaches, and more anger. On the other hand, leaving means letting "them" win.

The senior engineer has sufficient personal knowledge of Kirkland and the atmosphere in Kirkland's workplace to believe that these allegations are true.

QUESTIONS

1. What would you recommend that Kirkland do?
2. How does this work atmosphere of verbal abuse and harassment within her company affect the company's effectiveness and profitability? Is this a "professional" environment? Explain and justify your answer.
3. Was the manager's behaviour in violation of your province's code of ethics? If so, quote the appropriate sections. If not, what new clauses would you add to the code to deal with specific issues of harassment?

TOPICS FOR STUDY AND DISCUSSION

1. Prepare a display or talk for a local high school to present engineering careers at your company or engineering studies at your university. Ensure that your message conveys the idea that the career

paths and opportunities are open to all, regardless of race, sex, handicap, or other criteria. Consider using recent videos about careers in engineering and science and other means of attracting young people to these fields. You especially wish to portray how engineers and scientists apply their knowledge to the benefit of humankind, to solve problems, and to design the world in which we live and work. What will you say, and how will you say it?

2. Assume that you are the senior engineer who is responsible for employing and orienting new engineers in a large consulting firm. What policies would you expect the firm to have in place for interviewing, hiring, and promoting employees and resolving internal disputes in order to ensure fairness and equity?

3. The term "employment equity," which has become the norm in Canada, is frequently confused with the terms "affirmative action" and "equal employment opportunity," which are used more in the United States, even though the definitions of these three terms are significantly different. Do a search of human resources literature to obtain clear definitions of these three terms. What are the similarities and differences between them? Does the accusation of reverse discrimination, which is occasionally made against affirmative action, apply to employment equity? Is employment equity the best policy for Canada? Explain the basis for your answer in detail.

4. You have heard a rumour that a young person in your company is being harassed by a supervisor. Assume that you are the senior engineer responsible for the department in which these two people work. What will you do about this situation? What measures should be in place, or should be devised, to eliminate or to deal with such a situation in the workplace? Would your actions be the same or different for the following four cases:

   • a female employee being sexually harassed by her male immediate supervisor?
   • a male employee being sexually harassed by his female immediate supervisor?
   • a visible minority employee being harassed by a white supervisor?
   • a white employee being harassed by a visible minority supervisor?

5. Assume that you are a career counsellor in a university and you must give advice to first-year engineering students and to applicants who want to be admitted into engineering. What advice would you give, or what action would you take, to deal with the following situations?

a. A young woman in first-year engineering who is academically excellent has been told by her parents that they will not finance her education because she will just get married and quit her job when she has children, so it would be a waste of money to pay for her education.

b. A woman who had excellent marks in high school married immediately after finishing high school, then went to work as a secretary. Her husband was unfortunately abusive, the marriage has now ended in divorce, and she is stuck in a menial job at minimum wage. How can she afford to enrol in university to improve her skills and resume her dream of a professional career?

c. A young woman in high school wants to enter university. Her family has encouraged her to study engineering or science, but her school guidance counsellor has advised her to study art or nursing. Her confidence has been shaken by this contrasting advice from people whom she respects.

6. Many ideas have been suggested over the years to encourage people to live up to their full potential and not be deterred by artificial barriers in education and employment. For example, mentor programs, where young people meet role models in nontraditional occupations, create long-term support that helps to overcome obstacles. Another strategy has been to organize a nontraditional "career day" at a junior high school. Such a project was carried out very effectively in the Yukon Territory, where a nurse and a ballet dancer (both of whom were male) and a carpenter, a geologist, a firefighter, and an engineer (all of whom were female) organized workshops with grade 8 and grade 9 students. Introducing methods of engineering problem solving into science and mathematics courses at the secondary level and discussing the various engineering fields brings students closer to these fields at the critical time when they are making career choices.

   Conduct a brainstorming session and generate a list of at least four other techniques for encouraging people, especially young people, to ignore stereotypes and achieve their personal goals. Consider how you might act on these ideas in your university or in your local public school system.

NOTES

1. "Code of Ethics," Engineering Profession Act, SNB 1986, c.88, s.2B(1).
2. M. Blanchette, "What Is Systemic Discrimination? An Overview of the Legal Aspects" (CAUT), article in preparation.

3. Ibid.

4. Ibid.

5. R. Coulter, *Gender Socialization: New Ways, New World* (Victoria: Ministry of Equality, 1993).

6. Ibid.

7. M. Sadker and D. Sadker, *Failing at Fairness: How America's Schools Cheat Girls* (New York: Charles Scribner's Sons, 1994).

8. Sadker and Sadker, *Failing at Fairness*; W.H. Peltz, "Can Girls + Science − Stereotypes = Success? Subtle Sexism in Science Studies," *The Science Teacher* (December 1990): 44–49; personal testimonies provided to the author of this chapter; and M. Frize, "Impact of a Gender-Balanced Summer Engineering and Science Camp on Students' Future Course and Career Choices," abstract submitted to the 1998 National WEPAN Conference, Seattle, WA, June 1998.

9. Frize, "Impact of a Gender-Balanced Summer Engineering and Science Camp."

10. M. Vickers, H.L. Ching, and C.B. Dean, "Do Science Promotion Programs Make a Difference?" in M. Frize and J. McGinn, eds., *Papers and Initiatives*, More than Just Numbers Conference, University of New Brunswick, Fredericton, May 1995.

11. Ibid.

12. A. van Beers, *Gender and Engineering: Alternative Styles of Engineering*, MA thesis (Vancouver: Department of Sociology and Anthropology, July 1996).

13. Vickers et al., "Do Science Promotion Programs Make a Difference?"

14. Women in Engineering Advisory Committee, *National Report of Workplace Conditions for Engineers* (Toronto: Professional Engineers Ontario).

15. Sonnert and G. Holton, "The Career Patterns of Men and Women Scientists," *American Scientist* (January/February 1996).

16. Canadian Committee on Women in Engineering (CCWE), *Women in Engineering: More Than Just Numbers* (Fredericton: University of New Brunswick Bookstore, April 1992).

17. National Advisory Board of Science and Technology (NABST), *Winning with Women in Trades, Technology, Science, and Engineering* (Ottawa: Industry Canada, March 1993).

18. M. Frize and J. McGinn, eds., *Papers and Initiatives*, More Than Just Numbers Conference, University of New Brunswick, Fredericton, May 1995. Report released in October 1996 available at http://www.carleton.ca/wise.

19. Ibid.

20. Canadian Council of Professional Engineers (CCPE), *Statistics on Enrolments in Canadian Engineering Undergraduate Programs* (Ottawa: CCPE, 1996).

21. P. Caplan, *Lifting a Ton of Feathers: A Woman's Guide to Surviving in the Academic World* (Toronto: University of Toronto Press, 1992); and Natural Sciences and Engineering Research Council (NSERC), *Towards a New Culture: Report of the Task Force on How to Increase the Participation of Women in Science and Engineering Research* (Ottawa: NSERC, February 1996).

22. S. Tobias, *They're Not Dumb, They're Different* (Tucson, AZ: Research Corporation); C. Brooks, *Instructor's Handbook: Working with Female*

Relational Learners in Technology and Trades Training (Toronto: Ontario Ministry of Skills Development and Fanshawe College); P. Rogers, "Gender Differences in Mathematical Ability: Perception vs. Performance," *Proceedings*, ICME-6, Budapest, July 1988.

23. CCWE, *Women in Engineering: More Than Just Numbers*; NABST, *Winning with Women*; and Tobias, *They're Not Dumb.*

24. Caplan, *Lifting a Ton of Feathers.*

25. CCWE, *Women in Engineering: More Than Just Numbers.*

26. Women's Secretariat, *Employment Equity (Women in the Workplace)* (Regina: Government of Saskatchewan), brochure.

27. M. Frize and A. McLean, "Engineering: New Skills for a New Job Market," *Proceedings*, American Society for Engineering Education (ASEE) Annual Conference, Edmonton, June 1994.

28. M. Foschi, L. Lai, and K. Sigerson, "Gender and Double Standards in the Assessment of Job Applicants," *Social Psychology Quarterly* 17 (April 1994): 326–39.

29. P. Goldberg, "Are Women Prejudiced against Women?" *Transaction* (April 1968): 28–50.

30. C. Wenneras and A. Wold, "Nepotism and Sexism," *Nature* (May 1997).

31. M. Frize, "Reflections on the Engineering Profession: Is It Becoming Friendlier for Women?" CSME *Bulletin* (June 1993): 12–14.

32. N. DiTomaso and G.F. Farris, "Diversity in the High-Tech Workplace," *Spectrum* (June 1992): 21–32.

33. C.M. Caruana and C.F. Mascone, "Women Chemical Engineers Face Substantial Sexual Harassment: A Special Report," *Chemical Engineers Progress* (January 1992): 12–22.

34. M. Frize, "Eradicating Sexual Harassment in Higher Education and Non-Traditional Workplaces: A Model," *Proceedings*, 43–47, Canadian Association against Sexual Harassment in Higher Education Conference, Saskatoon, November 1995.

35. CCWE, *Women in Engineering: More Than Just Numbers.*

36. E. Culotta, "Women Struggle to Crack the Code of Corporate Culture," *Science* 260 (April 1993): 398–404.

37. J.G. Robinson and J.S. McIlwee, "Men, Women and the Culture of Engineering," *Sociological Quarterly* 32 (March 1991): 403–21.

38. Culotta, "Women Struggle to Crack the Code"; Robinson and McIlwee, "Men, Women, and the Culture of Engineering"; and J.F. Coates, "Engineering in the Year 2000," *Mechanical Engineering* (October 1990): 77–80.

39. Frize, "Reflections on the Engineering Profession."

40. Frize and McGinn, *Papers and Initiatives.*

41. This case is based on S. Seymour, "Case of the Mismanaged Ms.," *Harvard Business Review* (November/December 1987): 77–87. See also Alexander Mikalachki, Dorothy R. Mikalachki, and Ronald J. Burke, *Teaching Notes to Accompany Gender Issues in Management: Contemporary Cases* (Toronto: McGraw-Hill Ryerson, 1992), 5–8.

CHAPTER FOURTEEN

Disciplinary Powers and Procedures

The principal purpose of each provincial and territorial Association of Professional Engineers is to protect the public welfare by regulating the practice of professional engineering. To do so, each Association has been delegated the powers to prosecute people who unlawfully practise professional engineering and to discipline licensed engineers who are guilty of professional misconduct or incompetence. The methods of dealing with these two types of infractions are completely different.

PROSECUTION FOR UNLAWFUL PRACTICE

A person who is not a member or licensee of a provincial or territorial Association but who nevertheless

- practises professional engineering, or
- uses the title Professional Engineer, or
- uses a term or title to give the belief that the person is licensed, or
- uses a seal that leads to the belief that the person is licensed

is guilty of an offence under the Act. The procedure for prosecution and the penalties vary slightly depending on the province or territory. Each Association must initiate the action to prosecute offenders in the appropriate court under the authority of the Act, and the trial judge assesses the penalty, typically a fine proportional to the seriousness of the infraction. These prosecutions would typically be delegated to Association staff, and professional engineers generally would not be involved in these proceedings except perhaps as witnesses.

DISCIPLINE FOR PROFESSIONAL MISCONDUCT

In the case of licensed members, disciplinary action for professional misconduct or incompetence is conducted within the Association by a Discipline Committee formed of members of the governing council and other professional engineers. Under the authority of the provincial Act, the committee has the power to discipline members for professional misconduct, as defined below.

DEFINITION OF PROFESSIONAL MISCONDUCT

The various Acts have slightly different definitions of what constitutes grounds for disciplinary action. These definitions are reproduced in Appendix B. Although the definitions are not identical, they are very similar. In British Columbia, for example, incompetence, negligence, and unprofessional conduct are grounds for disciplinary action.[1] In Ontario, the Act states that professional misconduct and incompetence are grounds for disciplinary action,[2] but regulations made under the Ontario Act define professional misconduct in more detail, and include negligence.[3]

The provincial Acts typically identify six causes for disciplinary action: professional misconduct (or unprofessional conduct), incompetence, negligence, breach of the code of ethics, physical or mental incapacity, and conviction of a serious offence. Each of these terms is discussed briefly in the following paragraphs.

PROFESSIONAL MISCONDUCT

Professional misconduct (or unprofessional conduct, as it is called in some Acts) is the main source of complaints to provincial Associations. In about half of the Acts, the term is not defined, thus placing an additional burden of proof on the Association's legal counsel in any formal hearing, since alleged misconduct must be proven both to have been committed and to constitute professional misconduct.

However, Alberta, Newfoundland, and Prince Edward Island have much more general definitions. For example, the Alberta Act defines "Any conduct ... detrimental to the best interests of the public" or that "harms or tends to harm the standing of the profession generally" as unprofessional conduct.[4] While such clauses will stand the test of time because of their generality, they are really not specific enough to act as guidance in individual cases (the Association's code of ethics may give more specific guidance).

At the other extreme, Ontario's definition of professional misconduct includes some very specific acts, such as "signing or sealing a final drawing ... not actually prepared or checked by the practitioner."[5] In fact, the Ontario regulation is a fairly comprehensive definition of professional misconduct and may be of interest to readers whether they reside in Ontario or not (see Appendix B). Such specific guidance is clear and unambiguous, but since the regulations obviously cannot define every possible form of professional misconduct, they contain a general clause stating that professional misconduct includes "an act ... that ... would reasonably be regarded as ... unprofessional."[6] This circular definition is very general, but a complaint based on this clause would first have to demonstrate convincingly that a given action is "unprofessional."

INCOMPETENCE

Incompetence is typically defined in several Acts as a lack of knowledge, skill, or judgement or disregard for the welfare of the public of a nature or to an extent that demonstrates the member is unfit to carry out the responsibilities of a professional engineer. Depending on the Act, undertaking work that the engineer is not competent to perform may be considered either incompetence or professional misconduct. This rather subtle distinction covers the all-too-common occurrence of an engineer practising outside the area of his or her expertise, even though the engineer may be fully competent in his or her major field of practice.

NEGLIGENCE

In most Acts, "negligence" means "carelessness," or carrying out work that is below the accepted standard of care or performance. In many instances, it means the *omission* of an activity needed to ensure the proper care or safeguarding of life, health, and property. In fact, the omission of care or insufficient thoroughness in performing duties would probably be the most common complaint under this heading.

BREACH OF THE CODE OF ETHICS

In four provinces (Alberta, New Brunswick, Newfoundland, and Nova Scotia), a breach of the code of ethics is specifically defined in the Act to be equivalent to professional misconduct. These codes therefore have the full force of the Act. In other provinces (British Columbia, Manitoba, Prince Edward Island, Quebec, and Saskatchewan) and the territories, where the term "professional misconduct" is undefined or defined in very general terms, it would likely be understood to include the code of ethics, thus giving the code some enforceability under the respective Act.

In Ontario, the code of ethics is not clearly enforceable under the Act. However, there is an extensive definition of professional misconduct in the regulations that contains most of the code of ethics, such as "failure to act to correct or report a situation that the practitioner believes may endanger the safety or the welfare of the public," failure to disclose a conflict of interest, and about sixteen additional clauses (see Appendix B).[7] In other words, the Ontario definition of professional misconduct states in a *negative* way what the code of ethics states in a *positive* way. Therefore most, but not all, of the Ontario code of ethics is enforceable under the Act.

PHYSICAL OR MENTAL INCAPACITY

Most Acts also include a "physical or mental condition" as a definition of incompetence, providing it is of a nature and extent that, in the interests of the public or the member, the member no longer practise professional engineering.

CONVICTION OF AN OFFENCE

The provincial Acts also permit disciplinary action against a member who is guilty of an offence that is relevant to the member's suitability to practise. In other words, if a member should be found guilty of an offence under any other act, and the nature or circumstances of the offence affects the person's suitability to practise as a professional engineer, then the person can be found guilty of professional misconduct if proof of the conviction is provided to the Discipline Committee. This clause is used relatively rarely, since convictions for minor offences (traffic violations, local ordinance violations, etc.) do not affect one's suitability to practise engineering. However, convictions of serious offences such as fraud or embezzlement, which involve a betrayal of trust and questionable ethics, could be grounds for declaring a member unsuitable. This condition clearly imposes a standard of conduct on professional engineers that is somewhat higher than that expected of the average member of the public.

THE DISCIPLINARY PROCESS

Any member of the public can make a complaint against a licensed engineer, although most complaints are brought by building officials, government inspectors, or other engineers. Disciplinary procedures are very unpleasant, and since they may have a dramatic effect on an engineer's career, they must be fair and they must be *seen* to be fair. The Association's disciplinary procedures are defined in the provincial Act in very formal legal terms. When a complaint of negligence, incompetence, or professional misconduct is made against a licensed professional engineer, it sets in motion a three-stage process of gathering information, evaluating the complaint, and conducting a formal hearing that renders a judgement.

To ensure complete impartiality, the three stages of the disciplinary process are usually carried out by three different groups of people. No one who participates at an earlier stage is permitted to participate in the final hearing and judgement. The first stage is generally conducted by Association staff. The second stage is conducted by a Complaints Committee or an Investigations Committee (depending on the province) composed of members of the Association's governing council and other licensed engineers. The third stage is conducted by a Discipline Committee, which is composed of members of the governing council and people who have not previously been involved with the case.

A summary of the complaints process as carried out in Ontario is reprinted below with permission. The procedure in other provinces and territories is very similar, although not identical. Here's what happens when a member of the public has a complaint against an engineer.

STAGE 1 — GATHERING INFORMATION

- The complainant notifies the Association. A member of the Registrar's staff discusses the complaint, advising the complainant on the kind of evidence which will be necessary. Staff also assist on the wording of the complaint in relation to the Professional Engineers Act.
- Association staff do a preliminary investigation, hiring an expert witness if necessary.
- The complaint is [signed by the complainant and filed with the Registrar, who then sends it] to the engineer in question, who has a period of time for response.
- The complainant is given the opportunity for rebuttal.

STAGE 2 — EVALUATION OF THE COMPLAINT

- All of this material is presented in confidence to the Complaints Committee, which may:

  1. refer the complaint in whole or in part to the Discipline Committee;
  2. not refer the complaint;
  3. send a letter of advice to the engineer without referring the case to the Discipline Committee;
  4. take such action as it considers appropriate under the circumstances; or
  5. direct staff to obtain more information.

[6. Since 1994, a process called a "Stipulated Order" has been in use for less serious complaints. If the Complaints Committee decides to use a Stipulated Order, the discipline case is handled by a single member of the Discipline Committee, as described below.]

- [The reasons for the Complaints Committee decision are stated only if the matter is not referred to the Discipline Committee or as a Stipulated Order. However, if the matter is referred, no reasons are given.]
- If the complainant is dissatisfied with the way a complaint has been handled, it can be reviewed by the Complaints Review Councillor, who reviews the procedures only.

STAGE 3 — FORMAL HEARING

- If the case is referred to the Discipline Committee, a written notice of hearing is prepared by the Association's lawyer and served on the accused engineer, who usually hires legal counsel. A hearing date is set. A disclosure meeting is held between the respective lawyers, in which the Association makes its case known to the engineer's legal counsel.
- The hearing follows court procedure, with a court reporter present. The defendant can be represented by legal counsel. The Discipline Committee consists of five members, including a Lieutenant Governor-in-Council appointee. A written decision is given, with a copy to the complainant.
- Appeals are made through the civil courts.[8]

STIPULATED ORDER

A Stipulated Order process is a simpler form of disciplinary hearing for less serious cases. The Stipulated Order may be used instead of a formal hearing when the Complaints Committee has reason to believe, after reviewing the complaint, supporting materials, response of the accused engineer, and so on, that the Act, regulations, or by-laws have been breached but a formal hearing is not warranted. The written consent of the complainant and the accused engineer are required before the process can begin. The Stipulated Order process is carried out as follows:

A single representative of the Discipline Committee will review the Complaints Committee findings and meet separately with the member and the complainant to discuss the evidence. The Discipline Committee representative will determine — with staff and legal assistance if necessary — whether the Act, Regulations or Bylaws have been breached or not, and decide on an appropriate penalty, if applicable.

If the member is found guilty of breaching the Act, Regulations or Bylaws, the member will be given a Stipulated Order, comprising:

- a detailed description of the alleged offence;
- the finding of guilt; and
- the penalty.

The member [is then requested to] sign the Order. The penalty can include one or more of the penalties available to a full discipline hearing, as defined in the Act. By signing the Stipulated Order, the member agrees to guilt and accepts the penalty. There is no appeal.

If the member is found not guilty, the member and the complainant will be notified in writing that the matter is resolved and no further action will be taken.

If the matter cannot be resolved by way of a Stipulated Order, it proceeds to a discipline hearing. The entire Stipulated Order process is carried out "without prejudice," which means that it is not referred to during the hearing, nor can the Discipline Committee representative who participated sit on the discipline panel.[9]

Obviously, while this process is appropriate only for less serious and perhaps more clear-cut complaints, it could be very effective in avoiding the cost of formal hearings and reducing the time needed to obtain a judgement.

DISCIPLINARY POWERS

The penalties that can be meted out by the Discipline Committee are fairly general, including requiring the payment of fines and costs but not, of course, imprisonment. The disciplinary powers awarded under each provincial and territorial Act are summarized in Appendix B for reference. Typically, if a member or licensee should be found guilty, the Discipline Committee can

- revoke the licence of the member
- suspend the licence (usually for a period of up to two years)
- impose restrictions on the licence, such as supervision or inspection of work
- require the member to be reprimanded, admonished, or counselled and publish the details of the result, with or without names
- require the member to pay the costs of the investigation and hearing
- require the member to undertake a course of study or write examinations set by the Association
- have any order that revokes or suspends the licence of a member to be published, with or without the reasons for the decision
- impose a fine (up to $10 000 in Alberta; up to $5000 in Ontario; only hearing costs in the Yukon)

The severity of the penalty would, of course, vary with the circumstances of the case.

In the above discussion, and throughout this text, the terms "licensee" and "member" have been used interchangeably. The usage varies from province to province across Canada, and includes Temporary Licences, Certificates of Authorization, Limited Licences, Permit Holders, and (in Ontario) designation as a Consulting Engineer. The disciplinary actions described in this chapter also apply to these other categories. That is, other forms of permit or certificate may be revoked or suspended in the same manner, or in addition to, the revocation of a member's licence.

CASE HISTORY 14.1
DISCIPLINARY HEARING OVER INADEQUATE
STRUCTURAL STEEL DESIGN

This case history is a report of a disciplinary hearing conducted in Ontario and illustrates the outcome of a complaint that proceeded through all three stages of the complaint process. The report was published in the *Gazette*, the official publication of Professional Engineers Ontario, in 1990, and is reproduced with permission. Since the hearing resulted in an admonishment — a fairly light penalty — the engineer is not named.

DECISION AND REASONS THEREFOR
The Discipline Committee of the Association met in the offices of the Association to hear allegations of professional misconduct against a member (the engineer). The Association was represented by legal counsel. The engineer represented himself.

The allegations of professional misconduct are set out in Appendix "A" of the *Notice of Hearing* filed as an exhibit and relate to the engineer's involvement as a structural engineer employed by a steel fabricator who was retained by a development company to provide the structural steel frame and roof joists for a 50 000-square-foot building located at a retail mall in Ontario. It was alleged that joist details and calculations bearing the engineer's seal were inaccurate, incomplete, and failed to reflect acceptable engineering practice in that they failed to indicate appropriate member sizes for open web joists.

The joist shop drawings bearing the engineer's seal, when analyzed, disclosed that there would be overstress in tension and compression in both top and bottom chords. His design assumption for obtaining the necessary composite action between the mezzanine floor joists and the concrete floor slab was incorrect. Additionally, the engineer certified that the structural steel for the shopping centre would support the load shown on the structural drawings notwithstanding the fact that no design brief had been prepared. Therefore, in making such a statement in the absence of a design brief, it was alleged that this constituted a failure to meet acceptable structural engineering design practice.

From the evidence given by two members of the local building department, the consulting structural engineer on the project, and the engineer himself, the following scenario emerged.

1. In November of 1987, shop drawings prepared by the steel fabricator for the steel frame and roof joists were forwarded to the structural engineer for the project who rejected them as they were not stamped by a professional engineer. Despite further requests from the structural engineer, satisfactory shop drawings were not forthcoming, even though it appears that construction continued. After a site inspection report was forwarded to the owners in April 1988 by an engineering inspection company, a site meeting was held between the structural engineer, local building officials, and the engineer. The structural engineer reiterated that he could not certify the project without stamped shop drawings and calculations being made available. This resulted in a letter addressed to the municipality from the steel fabricator, signed and sealed by the engineer, certifying in part that: "The structural steel (columns, beams, open web steel joists, and connections) as supplied is capable of supporting the loads specified by the structural engineer on his drawing S9 dated August 1987."

   The structural engineer was not satisfied with this and again demanded stamped shop drawings. After the May meeting, the building department placed an *Order to Comply* on the project as it had discovered, among other things, that the fabricating company was not certified as an erector nor was it certified by The Canadian Welding Bureau. The engineer was then directly engaged by the owners to prepare shop drawings as required and he stamped joist details and calculations on June 3 and 10, 1988, and submitted them to the structural engineer. These calculations were found to be in error. Remedial work was eventually designed by an independent professional engineer and carried out by a certified steel fabricator.

2. A consulting structural engineer gave expert testimony on behalf of the Association. He indicated that he had reviewed the engineer's calculations and drawings and in his view they contained errors, the most serious of which were, [as] an analysis of four joists selected as typical samples disclosed, overstressing as follows: two bottom chords — 21% & 33%, three top chords — 14% to 20%, three web members in tension — 12% to 16%, nine web members in compression — 21% to 87%; a discrepancy in chord angle between the calculations and the shop drawings which resulted in a 12% overstress, and a failure to indicate properly the appropriate member sizes for open web steel joists.

3. With respect to the engineer's attempt to design the mezzanine floor joists to act compositively with the concrete floor slab, the [PEO's] engineering expert's view was that the design assumptions were incorrect in that the floor and joist cannot work together, therefore it was not acceptable. Further, one of the joists on the low roof was located in an area of significant snow accumulation, but the joist details did not indicate that snow accumulation was taken into account in the design.

   The engineer gave evidence on his own behalf with respect to this project and two major items arose as a result of this.

   - It was clear that this was a very difficult project and problems were encountered from day one. It is also clear that the engineer was not the only professional engineer involved in the project; however, he was the one who, by applying his stamp, assumed responsibility for the work.
   - He did not produce any evidence which contradicted any of that given by the [PEO's] engineering expert.

After reviewing the evidence and exhibits and hearing arguments, the Committee found the engineer guilty of:

[Regulation 941, Section 72(2)(a)] — Negligence in that he failed to maintain the standards that a reasonable and prudent practitioner would maintain in the circumstances in that he was carrying out work which he was not experienced enough to do. Testimony clearly indicated that the engineer lacked the qualifications and experience for the work he undertook on this project.

[Regulation 941, Section 72(2)(b)] — Failure to make reasonable provision for safeguarding life, health, or property of a person who may be affected by the work, in that this was a public place, an intended retail grocery store, the roof truss was under-designed, and expert testimony indicated that this could have resulted in a failure under design load.

[Regulation 941, Section 72(2)(d)] — Failed to make responsible provision for complying with applicable statutes, regulations, codes, standards, by-laws, and rules in connection with the work being undertaken, in that he failed to meet Building Code requirements and misread the composite standard steel deck manual produced by The Canadian Sheet Steel Building Institute.

[Regulation 941, Section 72(2)(j)] — Conduct or an act relevant to the practice of professional engineering that, having regard to all the circumstances, would reasonably be regarded by the engineering profession as unprofessional. It is clear from the engineer's testimony that this matter was not handled professionally and he admitted that he was guilty of poor judgment in this matter. The Committee was not of the view that his conduct was disgraceful or dishonourable.

Turning to the matter of penalty, the Committee took into consideration the fact that the engineer admitted to errors in judgment and reasonably cooperated with the Association throughout this matter.

It noted that the engineer was only one professional engineer on a project which employed many other professional engineers and it is clear that he was not the only one involved in the problems which arose. On the other hand, it was clear to the Committee that the engineer was not competent to perform this work by virtue of his training and experience. Therefore the Committee felt that an effort should be made to educate the engineer as to his responsibilities from both the professional and the technical points of view.

By virtue of the power vested in it by Section [28] of the Professional Engineers Act, the Committee directs:

1. That the engineer be counselled as to his professional responsibilities by the chairman of the Discipline Committee, such counselling not to be recorded on the Register.
2. That the engineer sit and pass the Association's Professional Practice Examination and the Association's Examination CIV-82 Advanced Structural Design, to be written at his own expense.
3. That the engineer provide a written undertaking to the Registrar of [PEO] that he will not offer and provide engineering services to the public until such time as he successfully completes the above examinations.
4. The written Decision and Reasons of the Committee will be published in the official journal of the Association without names.
5. In the event that the engineer does not provide the above written undertaking within 30 days of the written decision of this Committee being handed down, then his Certificate of Authorization will be revoked, and the written decision and reasons of the Committee will be published in the official journal of the Association in full, with names.
6. There will be no order as to costs.[10]

The engineer in question was served with the written decision and decided not to appeal. He provided the undertaking to the Registrar and passed the examinations, and his licence has been cleared.

TOPICS FOR STUDY AND DISCUSSION

1. Using Appendix B, "Excerpts from the Provincial and Territorial Acts and Regulations," compare the definitions of professional misconduct, negligence, and incompetence that form the basis for disciplinary action under each Act. Which province or territory has the most specific definitions? Which has the most general definitions? Would you say that the Acts are generally in agreement on the definitions, or do you see any major inconsistencies between them? Define and discuss these similarities and inconsistencies.

2. Would an infraction of the code of ethics in your province or territory be clearly enforceable under your provincial Act? Should all codes always be fully enforceable, or should they be purely voluntary codes of personal behaviour? Explain and justify the reasons for your decision. Regardless of your answer, check your provincial Act; does it agree with your viewpoint?

3. Using Appendix B, compare the disciplinary powers awarded under each provincial and territorial Act. Which Act provides the most severe fines and penalties? Would you say that the disciplinary powers in the Acts are generally similar, or are there serious inconsistencies between them? Define and discuss these similarities and inconsistencies.

4. Suppose that, in your employment as an engineer, you discover that some of your fellow employees who supervise the delivery and storage of material on the job site and are also professional engineers have been involved in "kickback" schemes with suppliers. The suppliers invoice your employer for materials that have not been delivered, your colleagues validate the invoices, and the suppliers pay them a hidden commission. These schemes obviously violate criminal law. Which clauses in your provincial code of ethics have been broken by your colleagues? To what types of disciplinary action have they exposed themselves as a result? Suppose that you confront them and they promise that they will discontinue these schemes if you agree not to reveal them. Would your silence be consistent with your provincial code of ethics? If not, could any disciplinary action be brought against you? Describe the course of action you should follow. Would it be different if your fellow employees were not professional engineers?

5. Imagine that you receive a registered letter from the Registrar of your provincial Association stating that you are the subject of a formal complaint made by a former client or employer. The letter contains a description of the complaint, which alleges that you are guilty of incompetence because advice included in a report you wrote was faulty, and the client or employer suffered a financial loss as a result of following your advice. A preliminary investigation by the Association appears to show that the complaint has some validity. The Registrar asks you to respond to the complaint. Describe the actions you would take to protect yourself.

NOTES

1. Engineers and Geoscientists Act, RSBC 1996, c. 116, s. 33(1)(c).
2. Professional Engineers Act, RSO 1990, c. P.28, s. 28.
3. Regulation 941/90, s. 72(2), Professional Engineers Act, c. P.28, RSO 1990.

4. Engineering, Geological and Geophysical Professions Act, SA 1981, c. E-11.1 (as amended), s. 43(1).
5. Regulation 941/90, s. 72(2)(e), Professional Engineers Act, c. P.28, RSO 1990.
6. Ibid., s. 72(2)(j).
7. Ibid., s. 72.
8. Professional Engineers Ontario (PEO), "Making a Complaint to APEO," *Engineering Dimensions* 11, no. 5 (September/October 1990): 46. Reprinted with permission of PEO.
9. PEO, "Alternatives to Discipline Hearings: Stipulated Order," *Gazette* 14, no. 1 (January/February 1995): 2. Reprinted with permission of PEO.
10. PEO, "Decision and Reasons Therefor of APEO Discipline Committee," *Gazette* 10, no. 4 (July/August 1990): 1–2. Reprinted with permission of PEO.

Part Four

Maintaining
Professionalism

CHAPTER FIFTEEN

Maintaining Professional Competence

Professional engineering is continually changing and evolving. New engineering theories, methods, and equipment are developed every day, and keeping abreast of these new developments is a challenge for practising engineers. Most professional people enjoy this rapid change and accept the challenge of maintaining professional competence. But, while personal satisfaction is a good motivator, there are more pressing reasons for maintaining competence — to meet legal requirements and to avoid professional obsolescence. This chapter discusses the importance of professional competence, examines several methods of maintaining competence and avoiding obsolescence, and comments on the roles of the universities, the engineering societies, and the provincial Associations in achieving these goals.

PERSONAL PERCEPTIONS AND MOTIVATION

Engineers are generally interested in new ideas and personal improvement. In fact, an early study showed that, soon after graduation, many engineers wish that they had been permitted or encouraged to study a broader range of subjects during their university years. Interestingly, the perceived gaps in education depended on the age group; thus

- Those who have been out of university 5 years or less wish that they had taken more courses of a practical nature.
- Those who have been out 5 to 15 years wish they had taken more math and science.

- Those who have been out between 15 and 25 years wish they had taken more courses in business and management.
- Those who have been out more than 25 years wish they had taken more humanities and fine arts.[1]

It is obvious that the undergraduate courses taken during an engineering education completed at the age of 22 or 23 years cannot cover all the needs of a person's 40-year professional career. There is, therefore, a basic motivation to expand one's knowledge and to fill in the perceived gaps, even for recent university graduates.

Moreover, new knowledge is continually being generated, and even a person at the top of the graduating class will need updating eventually. In engineering, most of this new knowledge results from rapid changes in computer technology and other high-tech areas. The pace of technological change is summarized in the popular but inaccurate cliché: "The half-life of an engineering education is ten years." The implication is that half of what an engineer learns in university today will be obsolete in ten years. This is untrue if one's education emphasized engineering *fundamentals*, since the fundamentals do not decay rapidly, although other subjects may. However, the level of competence for the entire engineering profession is constantly moving upward, and this will cause many engineers to experience obsolescence if they do nothing to keep up. In this era, engineers must meet the challenge of perpetual technological change by making a personal commitment to continual professional development.

AVOIDING OBSOLESCENCE

The perception of technical obsolescence in older engineers is always worrisome and sometimes unfair. As long as the economy is expanding rapidly, corporations tend to find room for everyone, and there is little mention of obsolescence. But as business cycles slow down and corporations and governments reduce costs by downsizing, the question of obsolescence arises. Older engineers, at higher salaries, are seen as expensive liabilities and obstacles to promoting less experienced but more energetic young people. As a result, the job of keeping up with change is a never-ending one, and engineers are coming more and more to accept the idea that a portion of their time throughout their entire careers will be engaged in fighting obsolescence.

The good news is that engineers can work productively over a much longer period if they participate in some form of effective continuing education.[2] In fact, while conventional folklore asserts that productivity declines with age, there is very little hard evidence to prove it. In a study of R&D organizations, evidence emerged that scientific productivity may actually increase after age 50, although the productivity may involve pulling together the ideas of one's life work rather than generating major new ideas.[3] The same study implied that engineers might actually improve with age if the work depends on experience and judgement and not as much on creativity.

An especially significant finding of this study was that engineers with advanced degrees were considered productive for as much as ten years longer than those with bachelor's degrees. Hence, the investigators made the recommendation that mid-career graduate work that is intensive enough to result in a degree might significantly prolong an engineer's productive life. For most engineers, enrolling in a graduate degree program would require company co-operation by means of time to attend classes, in-house graduate courses (either live or electronically), or provision of sabbatical leaves for self-renewal. Even though such programs cost money, the engineer's employers would likely lose even more if their technical personnel are not kept productive.

In fact, employers should view their engineers as corporate assets or resources, just as they view patents, processes, trademarks, and trade secrets — the employer has an investment in these assets that should be protected. Employers can maintain these resources very easily by giving engineers the opportunity for continuing study, good technical experience, or formal education. This benefits the corporation and helps engineers keep up with new developments in theories, methods, and equipment and remain on the frontiers of their profession. Engineers must, of course, support such initiatives and ensure that their continuing education is compatible with both their personal career aspirations and their employers' needs. Educational opportunities are available from universities, colleges, and engineering societies, and occasionally from groups within the industry, although finding suitable educational programs in the desired specialty may take some digging around.

LEGAL REQUIREMENTS FOR MAINTAINING COMPETENCE

Maintaining professional competence is not entirely a voluntary activity. As members of a profession, engineers have the privilege of self-regulation, a privilege accompanied by a corresponding duty to maintain competence and avoid professional obsolescence.

Almost all of the provincial and territorial professional engineering Acts that regulate engineering in Canada contain clauses requiring continuing competence, and while only a few Associations have *quality assurance (QA) programs* in place, most Associations have QA programs in the planning stage. Quebec has had a professional inspection program since 1980, Alberta has implemented a professional development plan, British Columbia has implemented a practice review program, and all jurisdictions except Yukon and Ontario are at various stages of implementing or developing such plans.[4] (As this text goes to press, Ontario is considering a new licensure model that might include a QA requirement to demonstrate continuing competence.)

The procedure for evaluating continuing competence, modelled on methods used in other professions, varies across Canada, and some procedures are obviously still being developed. However, the process typically requires engineers — particularly engineering corporations — to maintain competence through "creditable" activities, which might include

- *engineering practice or employment* involving work that falls within the definition of professional engineering
- *formal education,* such as courses provided by a university, college, or engineering society, for which there is typically a test or evaluation of the candidate's attendance and performance and a permanent record of the syllabus and standards
- *informal education,* such as short courses, seminars, conferences, and so on, for which there is typically a record of the event but no formal evaluation process or syllabus
- *publications,* such as papers, monographs, books, and so on, written by the engineer on engineering topics
- *participation in engineering societies* in activities such as editing engineering journals, organizing professional seminars, and presenting papers at conferences
- *benchmarking* — that is, reviewing the individual's (or the corporations) practices against the best practices currently used in the particular industry.

Obviously, some form of monitoring is required to ensure compliance. Self-assessment, accompanied by a random check of a small sample, is used in some provinces or has been proposed. When a licence is renewed, the engineer submits a statement declaring that he or she has engaged in a sufficient amount of the above activities to maintain competence. A small sample of engineers is selected and their records are verified. The process requires personal record keeping, but the vast majority of engineers are already maintaining their competence, and the verification process merely confirms this fact.

MAINTAINING COMPETENCE THROUGH PROFESSIONAL PRACTICE

The best method of keeping current as a professional engineer is to just do it! That is, keep yourself involved in interesting and advanced engineering projects. For most engineers, this is the preferred method of maintaining competence since it also generates income. This productive approach also benefits the province and the nation through taxes, the creation of wealth, interprovincial and international trade, and the formation of competent Canadian engineering capability.

However, engineers who maintain competence simply by "doing it" will soon learn that to remain competitive they must also remain current. That is, as they meet the challenges of new projects, they will intuitively seek the necessary information and will, informally and incrementally, make use of new theories, methods, and equipment. Fortunately, there is a wealth of information available to them, mainly from the various engineering societies; as a matter of professional common sense, they will want to join the appropriate societies to obtain this information. As their achievements mount, such engineers may also want to share the knowledge, techniques, and information that they have discovered and will find that engineering societies provide the best means of doing this. Because of the key role that engineering societies play, they are discussed at more length in the next chapter.

MAINTAINING COMPETENCE THROUGH CONTINUING EDUCATION

For a variety of reasons, some engineers will not be fully engaged in engineering practice or employment and will need some other method of maintaining their competence. For example, many engineers progress into management and find that most of their time is taken up with administrative demands. How can managers keep abreast of new developments when they have so little time? They probably cannot keep up to date when most of their time is spent "pushing paper." The solution might be simply to attend engineering conferences on a regular basis or to enrol in short courses or seminars on advanced engineering topics. Many short courses are offered by universities, colleges, and engineering societies either on a regular basis or as the demand arises.

A few engineers may face more serious challenges. As we move into a world dominated by computers, some engineers may find their areas of expertise suddenly overwhelmed by alternative technologies. For example, digital communication is quickly replacing analogue techniques. Similarly, some engineers may find their jobs, or perhaps their entire industries, disappearing suddenly for financial, political, or other nontechnical reasons. For example, many years ago the federal government cancelled several megaprojects for financial reasons, and many competent engineers found themselves unemployed. How do these engineers move out of unwanted areas of expertise into the more desirable "hot" areas of engineering? Although short courses may suffice, some engineers may find that formal education in a graduate engineering program either part-time or full-time is the key to maintaining or renewing competence.

Continuing education for engineers may take many forms depending on the topic, the duration, the method of presentation, whether the education has a lecture or hands-on format, and whether the organization can give continuing education units (CEUs) or formal degree credit. Frequently, courses can be arranged in-house by employers or may be given in the popular short-course format by universities, engineering societies, and some industries. In a short course, an overview of a subject, presented by leading authorities in the field, is packed into a full-time course that runs for a week or two.

The big advantage of in-house courses, typically given by engineering societies and some universities, is that working engineers can walk a short distance to a classroom at their place of work, take a class, and be back at work immediately afterward. No time is lost in commuting to a campus, yet the working engineer participates in a live classroom experience. Some societies and universities now have the equipment to provide such courses electronically using TV links, satellite hook-ups, or the Internet. Engineers at the remote location can even ask questions during class by means of microwave links or leased telephone lines (although eventually this may be possible by Internet audio). Such systems have been very successful at many locations in Canada and the United States, and as Internet capacity expands, there will be many more opportunities for electronic education in future.

Universities, colleges, and engineering societies are responding to the growing need for continuing education. Of course, mounting in-house courses requires the active co-operation of the engineers' employers, and the provincial Associations also have a role to play.

THE ROLE OF THE UNIVERSITIES

The engineering faculties of many universities offer evening, part-time, and short courses to satisfy the need for continuing education, and enrollment in postgraduate courses is usually possible on a part-time basis. Formal courses are an excellent form of continuing education and will lead to a recognized degree, as discussed in more detail later in this chapter. While formal education may not be the preferred route for most practising engineers, it is an option that appeals to some.

THE ROLE OF ENGINEERING SOCIETIES

An alternative source of continuing education is provided by engineering societies, which, because of their specialized interests, are well-equipped to keep practising engineers informed on specialized topics. In fact, many technical societies exist solely for the purpose of providing their members with regular publications, conferences, seminars, and courses, some of which are now being distributed by videotape, CD-ROM, satellite television, and the Internet. This electronic route has a great deal of potential value, and we must encourage its development for the vast majority of engineers.

Engineering societies are good sources of engineering information; they are typically organized by discipline, are highly specialized, and are oriented toward obtaining useful results. The Engineering Institute of Canada (EIC), in particular, has taken on the task of encouraging and co-ordinating continuing education activities. While engineering societies cannot award degrees, many societies have a less formal process of awarding continuing education units (CEUs). An extensive discussion of engineering societies is contained in the following chapter.

THE ROLE OF THE PROVINCIAL ASSOCIATIONS

In assisting engineers to remain competent, provincial Associations must be proactive and must not view their role as merely one of setting standards for their colleagues to meet. In particular, the Associations must recognize the many ways in which continuing competence can be achieved and give credit for this variety. They must make the documentation and verification process simple and unintrusive, and must also recognize that some forms of achievement, experience, and education may be very difficult to document. Moreover, in the engineering profession, there are very few organizations dedicated to assisting engineers, and the provincial Associations, with their established communication links, are in a unique position to encourage, advertise, and otherwise facilitate continuing education programs. In the specific area of continuing competence, where engineers have already satisfied the requirements for membership, provincial Associations must assist, motivate, and encourage colleagues even though these tasks may be at the limit of their regulatory duties.

MAINTAINING COMPETENCE THROUGH FORMAL EDUCATION

Historically, the length of higher education for the practising professional engineer has been four years in an undergraduate program or five years in the co-operative and internship programs. This is in contrast with that of most of the other professions, such as medicine and law, where professional education usually begins *after* a bachelor's degree has been earned. In the past, engineers who continued their formal education by enrolling in graduate studies were usually preparing for a career in research or teaching. The basic degree needed for research or teaching is the doctoral (PhD) degree, and the master's degree (MASc, MSc, or MEng) used to be viewed as merely a step in that direction. However, this attitude has changed, and the master's degree is now recognized as a valuable professional qualification. For example, a 1985 report on graduate education stated:

> The master's degree in some branches of the engineering profession has assumed the role of a "capstone" degree — the highest educational level to be sought, with no intention of proceeding to a higher degree. This produces a situation quite different from that prevailing in most fields of science, where full professional recognition by other scientists is usually accorded only to those with doctor's degrees.[5]

In fact, there have been several initiatives in Canada and the United States to make the master's degree the basic educational requirement for entry into the profession, or alternatively, to extend the bachelor's degree program into a five-year academic program (after university admission level). The reason for proposing this increase is understandable: computer concepts that were virtually nonexistent twenty years ago must now be included in a crowded curriculum. University faculty are constantly faced with the dilemma of deleting courses and concepts that were considered essential and fundamental only a few years ago to make room in the curriculum for new theories and methods, such as computer-aided design, solid modelling, artificial intelligence, solid-state physics, and a burgeoning list of computer applications in every discipline. Adding an extra year to the program would solve this problem.

Nevertheless, in spite of this constant problem of cramming new ideas into an already overburdened curriculum, it is extremely unlikely that a five-year requirement will ever be imposed. The reasons are simple:

- A large fraction of engineering graduates (as many as 30 or 40 percent) do not go into research, design, or development, but proceed to manufacturing, plant operations, sales engineering, and contracting. For them, the bachelor-level engineering degree already provides an ideal background, and an extra year would reduce enrollment in these key areas.
- Many engineering universities have co-operative work/study programs that require engineering students to spend several (usually six) four-month work terms in industry. This work requirement spreads the four-year program over five years. Any move to further lengthen these programs would be resisted.
- Many employers fear that five-year bachelor's graduates would be more specialized than four-year graduates, and the four-year gradu-

ates already meet their needs. Moreover, employers are not likely to pay more for five-year bachelor's graduates than for four-year graduates. Hence, if given a choice, most students are likely to avoid the longer programs, and those students who select a five-year program would probably want it to be rewarded with a master's degree, not a bachelor's degree.

So, while it is unlikely that the undergraduate program will be lengthened or that the master's degree will become the basic educational requirement for licensing, engineers should consider graduate study if they want to specialize, carry out research, or study advanced techniques, particularly computer-related topics.

ADMISSION TO GRADUATE STUDY

If you are thinking about graduate study, you should first examine the annual calendars or catalogues for the universities that interest you. Catalogues may now be viewed conveniently over the Internet; copies are also available in most libraries or may be obtained (usually free) by writing directly to the university. Admission requirements are clearly specified in the catalogue, and if you have been out of university for three or more years, check the admission requirements for mature students. Admission requirements may be reduced slightly for mature students, since they are usually more determined, more goal-oriented, and therefore more effective in their graduate studies.

TYPICAL ADMISSION REQUIREMENTS

Admission requirements for graduate study vary slightly from university to university. Some general advice about admission standards is given below.

MASTER'S APPLICANTS

To qualify for admission to a master's degree program, an engineer from an accredited undergraduate program must usually have ranked in the upper half of the undergraduate class (B average or better). However, admission requirements may be slightly different for master's programs that require a thesis (research master's) and those that require mainly courses (course-work master's). A master's degree usually requires a minimum of one academic year, although two years may be required if the student must make up any deficiencies or if the research project is particularly challenging or time-consuming. The engineering master's degree may be awarded in applied science (MASc), science (MSc), or simply engineering (MEng).

DOCTORAL APPLICANTS

Candidates for doctoral degrees must usually be ranked in the upper quartile of their undergraduate classes. To enter a doctoral program, applicants must usually have completed research at the master's level in a related area of study. It is important to become familiar with the research in progress, to make personal contact with the professors in your area of interest, and to discuss research opportunities before making a commitment, since the topic of the doctoral research frequently sets the direction for the rest of your career. A minimum of three years beyond the bachelor's degree (or two years beyond the master's degree) is usually stated for the doctoral degree, although the actual time is typically about a year longer than the minimum. One of the reasons for the longer time is that doctoral candidates usually work part-time as teaching assistants or research assistants. The usual engineering doctoral degree is the Doctor of Philosophy (PhD).

THESIS REQUIREMENTS

Some universities require a thesis to be written for the master's degree, and some allow courses to be taken in place of the master's thesis, but a thesis is universally required for a doctoral degree. The master's thesis is generally written under the guidance of the supervisor, and is expected to contribute to the supervisor's research goals and demonstrate the level of achievement of the candidate. However, the doctoral candidate is less dependent on the supervisor, and the doctoral thesis is expected to represent an original and independent contribution to the literature of the discipline.

DIRECT PhD ADMISSION

Although it is possible to be admitted directly into the doctoral program from the bachelor's degree in Canada, this is not common. The master's degree is usually required, although occasionally students who begin a master's degree and show exceptional ability in the first term or two may be permitted to transfer to the doctoral program, thus achieving the same result.

BENEFITS AND SACRIFICES

New employees with graduate degrees start at higher salaries than do those with bachelor's degrees, of course. However, since graduate school delays the time when those higher earnings begin, it takes many years before the higher earnings make up for the lost income. In the long run, the greatest benefit is that one's professional life may be considerably extended, permitting high salaries to be received for many years longer than might otherwise be possible. Therefore, advanced engineering degrees almost inevitably pay off, although the break-even point will vary from discipline to discipline. The greatest sacrifice in postgraduate study occurs in the early years, when family time is limited and major financial purchases, such as homes and cars, must be delayed. Most people enter graduate study because they like the novelty and challenge of learning; very few weigh the future benefits against the present sacrifices, but it is appropriate to give some thought, however brief, to this trade-off.

TOPICS FOR STUDY AND DISCUSSION

1. Most engineers do not think about entering graduate school until it is too late; that is, they delay until they have occupational, financial, or family commitments that take precedence. Since this textbook is intended for recent graduates and senior undergraduates, the best time to consider graduate studies is probably right now!

   a. Using the Internet or library facilities at your disposal, examine the postgraduate engineering opportunities at any five universities of your choice. Choose universities in at least two countries. Make a list of the programs that most inspire your interest and curiosity. Check the admission requirements. Upon graduating from your undergraduate program, did you have (or will you likely have) the required minimum academic record for admission? If so, continue with part b. If not, read the university's rules for make-up courses and mature or probationary students. What further action could you take to qualify for admission?

   b. For your own personal situation, write down the pros and cons of getting a postgraduate degree. Consider such matters as full-time or part-time study and the effects that either of these might have on your earnings, both present and future, your family life, your career satisfaction, and any other factors that you feel are relevant. Prepare a summary on a single sheet, with advantages on the left side of the page and disadvantages on the right. Does the summary confirm that you are following the proper path concerning graduate studies?

2. Using the Internet or the university catalogues in a local library (the annual *Maclean's* survey of universities may also help), compare graduate programs in your discipline at five universities of your choice. Prepare a summary in chart form, and rate the programs and the universities on the following: type of research in progress, research grants per faculty member, graduate courses provided, size of laboratories, number of books in the library, number of students in graduate and postgraduate programs, tuition fees, and any additional factors that you feel are relevant. Using your results, rank the universities in terms of attractiveness to you as a potential postgraduate engineering student.

3. As discussed in Chapter 4, many engineering graduates are interested in the Master of Business Administration (MBA) degree. Assuming that you have read Chapter 4 and that you have completed questions 1 and/or 2 above, consider whether an MBA would be a more appropriate degree for you. Make a summary of your decision on a single sheet, listing the advantages and disadvantages for the various courses of action as you see them.

4. If you are a senior student soon to graduate, consider organizing an exit evaluation survey of your engineering program to be done by all students in your class shortly before or after graduation. The exit survey is typically an objective evaluation of the relevance and content of the courses taken during your undergraduate engineering education. Many formats are possible, but the simplest is a chart for each academic year listing each subject taken. Students rate each subject under headings such as practical engineering subjects, mathematics, engineering science, management methods, and humanities, and each subject is given a rating: poor, adequate, good, very good, or excellent. The purpose of the survey must be clearly defined: it is an evaluation of program content. Avoid confusing course content with the personality of the professor teaching it. When you are finished, ask yourself: "Did this program prepare me properly to work as an engineer? If not, what needs to be changed?" A class representative must organize the collection and evaluation of the data, and would send the results to the chairperson of your university department. Although it may seem irrelevant to evaluate the program after you have left the university, such information is very valuable to your professors and to students who follow you. Moreover, as the quality of your alma mater rises, you may benefit indirectly.

NOTES

1. E.T. Cranch and G.M. Nordby, "Engineering at the Crossroads without a Compass," paper presented at 53rd Annual Meeting of Accreditation Board for Engineering and Technology, Phoenix, AZ, 15–18 October 1985.
2. M.A. Steinburg, *Continuing Education of Engineers* (Washington, DC: National Academy Press, 1985).
3. C.M. Van Atta, W.D. Decker, and T. Wilson, *Professional Personnel Policies and Practices of R&D Organizations* (Livermore, CA: Lawrence Livermore Laboratories, University of California, 1971).
4. Professional Engineers Ontario (PEO), "Professional Excellence: Proposed New Approach to Licensure," *The Link* 2, no. 2 (August/September 1997): 2–3.
5. J.D. Kemper, *Engineering Graduate Education and Research* (Washington, DC: National Academy Press, 1985).

CHAPTER SIXTEEN

Engineering Societies

Engineering societies play a key role in helping engineers learn about new theories, advanced techniques, and modern equipment. Engineering is evolving more rapidly now than ever before. In particular, the continuous introduction of new computer methods in engineering (such as computer-aided design and analysis, expert systems, and artificial intelligence) puts additional stress on the professional engineer. Fortunately, engineering societies can be very effective in helping engineers to remain up to date by distributing information about these new developments. Societies bring knowledge to engineers in many ways, such as through publications, conferences, seminars, and courses, some of which are now being distributed electronically. Engineering societies also provide a direct link to colleagues of similar interests, permitting members to meet and exchange useful information. This chapter provides some basic information about engineering societies and explains why every engineer should be a member of at least one society.

THE PURPOSE OF ENGINEERING SOCIETIES

The major purpose of engineering societies, which has not changed in 150 years, is to encourage research into new theories or methods, to collect and classify this new information, and to disseminate it to members so that it can be put to good use. Engineering societies are the major publishers of new research results in the forms of conference proceedings and monographs and are one of the leading groups in developing new standards for design. Over the years, the libraries of the world have received many useful publications that resulted from the efforts of the engineering societies, and everyone has benefited from this free exchange of information. The impact on engineering practice is immense.

The purpose of engineering societies is totally different from that of provincial and territorial Associations, yet some engineers confuse them. This may be due to the fact that, in other countries, the duties sometimes overlap; for example, British societies do perform a sort of regulation through the awarding of Chartered Engineer status, and many Associations distribute technical information from time to time. Conversely, many American engineering societies publish a code of ethics, and infractions of the code are grounds for expulsion from the society. In Canada, regulating professional behaviour is separate from disseminating engineering information, so a professional engineer must be a member of a provincial Association and should also be a member of an engineering society.

THE EVOLUTION OF ENGINEERING SOCIETIES

The first technical society for engineers was the Institute of Civil Engineers established in Britain in 1818, followed 30 years later by the Institution of Mechanical Engineers. Shortly thereafter, additional societies were established for naval architects and gas, electrical, municipal, heating, and ventilating engineers.[1]

In the United States, the first engineering society was the American Society of Civil Engineers, founded in 1852. Many others were established in the 1800s, including the American Society of Mechanical Engineers (1880), the American Institute of Electrical Engineers (1884), and the American Society of Heating and Ventilating Engineers (1894), to mention only a few.

In Canada, the formation of societies began in 1885 with the Engineering Society of the University of Toronto. The "Society was, indeed, a 'learned society' and published and disseminated technical information ... in addition to looking after the University undergraduates in engineering."[2] In 1882, the Canadian Institute of Surveys was formed, followed by the Engineering Institute of Canada in 1887 (although the EIC name was not adopted until 1917), the Canadian Institute of Mining and Metallurgy (1898), the Canadian Forestry Association (1900), and others. However, some Canadian engineering societies have been established very recently. This resulted from the realization that one of the oldest and most prestigious societies, the Engineering Institute of Canada (EIC), could not maintain the diverse specialties of engineering within a single organization. Several "constituent" societies were thus established. The Engineering Institute of Canada is now an executive or "umbrella" organization with five federated societies (CGS, CSME, CSCE, CSEM, IEEE-Canada), and arrangements were made with others (CSChE). Agreements signed between EIC and the Canadian Council of Professional Engineers (CCPE), which acts on behalf of the provincial associations when requested, clearly states the roles and duties of the organizations: the provincial Associations are responsible for regulating engineering, and the engineering societies are responsible for the traditional society role of collecting and disseminating technical information.[3] An engineer should be registered with a provincial Association (depending on the province of residence) and enrolled in an engineering society (depending on individual interests and branch of engineering).

CHOOSING A SOCIETY

The choice of which societies to join is influenced by one's engineering discipline, of course, and most major societies sponsor student chapters to acquaint undergraduate engineering students with their activities and get them involved. However, for most engineers, it is necessary to seek out the appropriate society through discussions with colleagues or senior engineers. A useful publication that summarizes the activities of hundreds of engineering societies throughout the world is the *International Directory of Engineering Societies and Related Organizations*, published annually by the American Association of Engineering Societies (AAES), and available in most university libraries.[4] This directory lists the purpose, membership, address, dues, and many other statistics for each society. For convenience, a brief list of some of the better-known societies is given in Table 16.1, and more information can be obtained by consulting the Web site for the society.

In the rapidly evolving technological environment, the professional engineer has an obligation to remain well informed. Engineering societies are one of the best means of doing so, and they are serving the same useful role today that they served during the Industrial Revolution. Each professional engineer should be a member of at least one society. For tax purposes, engineering society dues are deductible from personal income (for practising engineers, under Canadian income tax laws).

Although the Canadian engineering societies are more effective in dealing with problems that are typically Canadian, they do not have the many years of publications and the continuing series of journals, transactions, and so on that the older American and British societies have. To get access to these publications, several associate or secondary memberships have been negotiated that provide dual membership at a reduced fee. Agreements exist between the Canadian societies and most of the major American and British societies.

Table 16.1 COMPARISON OF SOME ENGINEERING SOCIETIES IN CANADA AND THE UNITED STATES (1995)

Society	EIC	Geotech CGS	Civil CSCE	Civil ASCE	Mechanical CSME	Mechanical ASME	Electrical and Computer IEEE-Canada	Electrical and Computer IEEE	Chemical CSChE	Chemical AIChE	Mining CIM	Mining AIME
Founding date	1887	1972	1972	1852	1970	1880	1973	1884	1966	1908	1898	1871
No. of staff	2	1	4	185	2	405	5	633	1	110	27	4
Local sections	32	16	16	155	12	202	7	292	8	109	65	—
Student chapters	Yes	None	27	210	27	333	4	849	20	147	59	—
Annual budget	$2000 K	$250 K	$600 K	$30 000 K	$150 K	$58 000 K	$150 K	$123 000 K	$200 K	$20 500 K	$2900 K	$570 K
Annual member dues	$85 to $125	$134	$144	$135	$100	$87	$85	$107	$95	$92	$70	—
No. of individual members	—	1378	3700	92 068	1000	101 000	800	274 778	1553	42 600	10 342	—
No. of student members	—	92	1300	13 737	400	24 000	50	39 524	334	9300	1021	—
No. of corporate members	6	None	None	None	None	None	None	None	None	None	219	4
Total membership	—	1470	5000	114 017	1400	125 000	900	314 316	1887	56 780	11 582	—
Publications (periodicals)	4	3+	2+	28+	2	30+	1+	95+	5	10+	6+	—
See Notes	(1,2)	(1,2)	(1,2)	(2)	(1,2)	(2)	(1,2)	(2)	(1989 data)	(2)	(2)	(2,3)

Notes:

(1) The Engineering Institute of Canada (EIC) is an "umbrella" organization that consists of the six constituent societies: CGS, CSME, CSCE, CSEM, CSChE, and IEEE-Canada. The dues for the constituent societies include membership in EIC.

(2) Dollar amounts are in U.S. dollars for American organizations and Canadian dollars for Canadian organizations.

(3) The American Institute of Mining, Metallurgical and Petroleum Engineers (AIME) is a corporation owned by its four member societies: The Metallurgical Society, Society of Mining Engineers, Iron and Steel Society, and Society of Petroleum Engineers.

SOURCE: American Association of Engineering Societies (AAES), *International Directory of Engineering Societies and Related Organizations* (Washington, DC: AAES, 1995). Reprinted with permission of AAES.

DATA ON SOME ENGINEERING SOCIETIES

Most people are simply not aware of the immense number and variety of engineering societies. As the *International Directory of Engineering Societies and Related Organizations* shows, there are literally hundreds of societies to assist information exchange in almost any specialty. For example, societies have been established by groups of engineers whose common interests are based on

- *Discipline* (agricultural, chemical, civil, computer, electrical, environmental, geological, geotechnical, manufacturing, marine, mechanical, mining, nuclear, petroleum, systems design, etc.)
- *Product* (abrasives, automation, automotive, computers, concrete, explosives, gears, illumination and lighting, lasers, machinery, paper, plastics, powder metallurgy, rubber, robotics, steel, textiles, vehicles, welding, wire, etc.)
- *Facility* (electric power plants, highways and bridges, hospitals, radio and television, railroads, shipbuilding, utilities, etc.)
- *Innovation* (design, inventions, human-powered vehicles, etc.)
- *Evaluation* (cost engineering, nondestructive testing, instrumentation, measurement, etc.)
- *Function* (consulting, engineering education, engineering management, human resources, quality control, etc.)
- *Environment* (conservation, environmental impact, glaciology, mapping, sustainable development, occupational health, remote sensing, etc.)
- *Language* (English, French, German, Japanese, Spanish, Portuguese, etc.)
- *Geographical area*[5]

Table 16.1 lists a few discipline-based engineering societies for engineers in North America. The older, larger, and better-established American and British societies have a greater storehouse of technical information and are usually able to offer a few more services to their members. The smaller, newer, Canadian societies are in the process of building up their reputation and membership. The names and addresses of some of these societies are listed in Table 16.2, although information about all of these societies is available over the Internet, and most of them permit an application for membership to be submitted electronically. The societies can be found on the Internet by a simple search using the acronyms.

The five U.S. "Founder Societies" (ASCE, AIME, ASME, IEEE, AIChE) are so called because they founded the United Engineering Trustees, Inc. in 1904, which provides a central building and office space for over twenty U.S. engineering societies and related organizations.

Table 16.2 ADDRESSES OF SOME ENGINEERING SOCIETIES IN CANADA AND THE UNITED STATES (1997)

Abbreviation	Society	Address
EIC	Engineering Institute of Canada	1980 Ogilvie Road P.O. Box 27078 RPO Gloucester Centre Gloucester, ON K1J 9L9
CGS	Canadian Geotechnical Society	P.O. Box 937 Alliston, ON L9R 1W1
CSCE	Canadian Society for Civil Engineering	2155 Guy, Suite 840 Montreal, QC H3H 2R9
CSME	Canadian Society for Mechanical Engineering	Suite 1500 One Nicholas Street Ottawa, ON K1N 7B7
CSChE	Canadian Society for Chemical Engineering	Suite 550 130 Slater Street Ottawa, ON K1P 6E2
CSEM	Canadian Society for Engineering Management	1353 Mountainside Crescent Orleans, ON K1N 3G5
IEEE – Canada	Institute of Electrical and Electronic Engineers Canada, formerly Canadian Society for Electrical and Computer Engineering (CSECE)	IEEE Operations Center 445 Hoes Lane P.O. Box 459 Piscataway, NJ 08855-0459
CIM	Canadian Institute of Mining and Metallurgy	400–1130 Sherbrooke Street West Montreal, QC H3A 2M8
CSAE	Canadian Society of Agricultural Engineering	Box 381, RPO University Saskatoon, SK S7N 4J8
ASCE	American Society of Civil Engineers	Head Office: 345 East 47th Street New York, NY 10017
AIME	American Institute of Mining, Metallurgical and Petroleum Engineers	
ASME	American Society of Mechanical Engineers	
IEEE	Institute of Electrical and Electronic Engineers	
AIChE	American Institute of Chemical Engineers	

Therefore, regardless of your engineering discipline, industry, or personal interests, there is an engineering society that is able to provide you with important information and advice and, conversely, to which you may be able to contribute. In order to remain competitive in our rapidly evolving world, you should tap into this source of useful knowledge.

TOPICS FOR STUDY AND DISCUSSION

1. Examine the periodical holdings of a good engineering library (this can be done very conveniently over the Internet), and list the periodicals in your area of engineering expertise. Copy the list (include journals, magazines, etc.), and note how many of them are published by engineering societies. Using the Internet, locate the Web site for the engineering society you encounter most frequently, and get its membership information.

2. There is a debate as to whether Canadian engineering societies are essential, or whether Canadian engineers should, for economy, belong to foreign-based societies that are already in existence and have a long history of transactions and an established membership base. Write a brief summary debating the pros and cons of the two alternatives. Does Canada need distinct engineering societies? Discuss the implications for Canadian sovereignty if Canadian engineering societies should be absorbed into the larger U.S.-dominated societies. Are these societies truly nonpolitical, or do national interests influence their policies and the content of journals and transactions? Are there uniquely Canadian conditions that would justify uniquely Canadian societies? Give examples. Write your summary as if you were sending it to a federal politician who has very little engineering knowledge.

3. Examine a copy of the *International Directory of Engineering Societies and Related Organizations* in a nearby library and, considering your own engineering discipline, employment or proposed employment, and other interests, make a list of engineering societies that are of interest to you in as many categories as possible (discipline, product, etc.). Compare the data on each society as given in the *Directory*, and examine the Web site for each of these societies. If you were to join only one (or two, or three), which would it be?

NOTES

1. L.C. Sentance, "History and Development of Technical and Professional Societies," *Engineering Digest* 18, no. 7 (July 1972): 73–74.
2. Ibid.
3. "Canadian Engineers Close the Ring," *Engineering Journal* 60, no. 1 (January 1977): 15–19.
4. American Association of Engineering Societies (AAES), *International Directory of Engineering Societies and Related Organizations* (Washington, DC: AAES, 1995).
5. Ibid.

Part Five

Exam
Preparation

CHAPTER SEVENTEEN

Writing the Professional Practice and Ethics Exam

Most of the provincial Associations of Professional Engineers require applicants for membership to write a short examination on professional practice and ethics. The purpose of the exam is to ensure that the applicant is familiar with the provincial Professional Engineering Act and code of ethics. These topics are rarely covered in university courses but are essential basic knowledge for a professional engineer.

The format of the exam varies widely from province to province. Some provinces have a formal one-and-a-half-hour written exam, usually administered in a three-hour session with an engineering law exam. Other provinces, such as Manitoba, have a different approach: the exam is a rather lengthy homework assignment that requires the applicant to review the provincial Act and code of ethics in minute detail. Both types of exam achieve the same goal: to determine whether the applicant is familiar with Canadian professional engineering practice and ethics.

This chapter gives some advice for readers preparing to write the Professional Practice Exam and illustrates four types of exam format with questions taken from previous exams. Readers who follow the advice in this chapter conscientiously will have little difficulty in passing the exam.

PREPARING FOR EXAMINATIONS

If you do not look forward to formal examinations, then you have company! Exams cause some anxiety for most people, but they were originally devised centuries ago to prevent favouritism; they ensure that people are admitted on knowledge and ability and not because of apple-polishing, bribery, or luck. There are no limits or quotas on the number of applicants who can pass; the exams are merely an impartial gauge applied to see that everyone measures up.

Exams are also learning experiences; in fact, the effort put into summarizing and organizing the subject matter in preparing for an exam is usually very efficient learning. However, even if you are well prepared, it is human nature to feel tense before an exam. Don't let it bother you — everyone else feels the same, even though they may not show it. The following suggestions may help you:

- Take a brisk walk before the exam. The mild exercise helps to combat anxiety and clear your mind. (A 10-km hike is *not* a brisk walk.)
- Arrive a little early, make sure you have an extra pen, and select a comfortable chair.
- *Read the examination paper!* It is amazing how many people waste time giving excellent answers to questions that were not asked.
- If you are faced with a really tough question, read it thoroughly, try to disengage your thoughts, and brainstorm answers. If you still cannot respond, then go on to the next question. Your mind will work on the tough question subconsciously, and when you come back to it, you may have the answer.
- Write clearly and arrange your answers in a logical order. This shows a methodical approach to problem solving. It frequently helps to jot down an outline before writing your response.
- Remember that the exam is a communication with the examiner. You may include any comments, references, or explanations that you would make verbally if you had the opportunity.

THE EGAD! STRATEGY FOR ESSAY-TYPE EXAM QUESTIONS

When preparing to write an essay-type Professional Practice Exam, it is important for applicants to recognize its purpose. Examiners are trying to determine whether the applicant is familiar with the Provincial Engineering Act and code of ethics, can make thoughtful decisions when faced with an ethical dilemma, and can explain the decision in a logical and convincing manner. This chapter helps applicants to respond to these requirements.

The strategy for resolving complex ethical problems discussed in Chapter 6 was written as a guide for real people who are participants in a real ethical dilemma. The examination setting, by contrast, is an artificial situation, so the strategy needs to be modified slightly.

For example, real problems usually develop slowly and unobtrusively; they are not presented in a numbered and typed format. Second, a key step in solving a real ethical problem is gathering the necessary information; in the exam setting, the information provided is all you get. It is important to read the questions thoroughly so that you don't miss anything, but it is not possible to use other sources for new insight as you would in real life. Finally, the strategy described in Chapter 6 emphasizes the importance of generating alternative solutions or courses of action and striving for an "optimum" solution. This creative aspect is drastically curtailed in the exam setting.

On the positive side, implementing real solutions to ethical problems usually requires time, money, and, occasionally, personal confrontations. In an exam, solutions merely need to be written down to be "implemented."

Therefore, the strategy described in Chapter 6 has been modified to serve as a more useful aid to readers preparing to sit the Professional Practice Exam (particularly an essay-type exam). In this modified form, the strategy has been renamed the *EGAD method*, as explained below.

The six-step EGAD problem-solving strategy can be applied to both ethics and law problems, although this chapter concentrates on the ethics applications. The six-step strategy has been adapted from the *Study Guide for the PEO Professional Practice Exam* by Andrews,[1] but it is also similar to a solution method taught to law students.[2]

An exam question is generally posed as an ethical dilemma in which the engineer is typically faced with two or more courses of action that are equally undesirable. For example, an engineer may be instructed by an employer to supervise the improper disposal of toxic wastes in the local dump. The engineer may not want to comply, but may be faced with retaliation by the employer if the instruction is not followed. Both alternatives are undesirable. The strategy for examining such problems and reaching a suitable decision is remembered easily by the three words *READ — EGAD! — WRITE*.

The term EGAD! (an old English exclamation of surprise) is an acronym or mnemonic for the four key steps in the solution strategy: **E**thical issues, **G**eneration of alternatives, **A**nalysis, and **D**ecision.

The six-step strategy is explained as follows:

Step 1: READ — *Read the problem thoroughly*. You must gather the facts and separate the key facts from unimportant details. Highlight or underline key facts on the exam paper, but *do not copy* the question into the exam book, since this takes valuable time which should be spent on your answer.

Step 2: E — ETHICAL ISSUES — *Identify the basic ethical issue*. In the Professional Practice Exam, the questions usually give you a hint. For example: "Has Mr. Smith broken the Code of Ethics? or "What obligations arise from this incident?" These questions help you to identify the ethical (or legal) area. However, some exam questions may say simply: "Explain and discuss this case." You must then review the various ethical issues which may apply, and compare the similarities (and differences) of the case with previous cases. If the ethical issue is not immediately clear, then analyze the facts by asking yourself the following "Who?/ What?/How?/Which?" questions:

- *Who is involved?* (Identify who has caused harm to whom.)
- *What type of harm or damage has occurred?* (or may potentially occur?)
- *How has this harm occurred?* (or may potentially occur?)
- *Which general area* of ethics (or law) appears to apply to this behaviour?

These questions should give you a clear statement of the problem; if so, the proper course of action may be clearly obvious and one can move immediately to the last step. However, you may still be faced with making a choice between courses of action which are equally undesirable. This ethical dilemma occurs frequently, and requires the creative effort explained in the next step.

Step 3: G — GENERATION — *Generate (or identify) possible courses of action.* The goal, in this step, is to generate (identify or imagine) all the alternatives. Sometimes, an exam question suggests an ethical dilemma with only two alternatives, both of which are undesirable. However, you may be able to suggest a third possibility, which is better. This step applies to most ethics problems, particularly if the question asks "What should you do?"

The new course of action may be a compromise, or a modification of one alternative to eliminate its negative aspects. This step requires creative thought and therefore it may be difficult. The creative techniques which are commonly used in engineering can be applied here, as well, such as brainstorming techniques, listing all the existing alternatives, etc. You might also try to imagine yourself as one of the protagonists. Engineers are creative people, and they are easily able to generate or imagine alternative courses of action. The goal is to find a new course of action without the undesirable aspects of the stated ethical dilemma.

Step 4: A — ANALYSIS — *Analyze the possible courses of action*. When several courses of action apply to a given situation, you must analyze the results for each course of action to find the best one. That is, you want the simplest course of action that solves the problem without leading to unacceptable side-effects or hardships. It may help to ask oneself questions concerning each possible course of action, such as the following:

- Is the suggested course of action legal?
- Is it consistent with human rights, employment standards, and labour law?
- Is it consistent with the code of ethics and/or ethical theories?
- What benefits will result, and who gets the most benefit?
- What hardships are involved, and are the benefits and hardships fairly distributed?

Step 5: D — DECISION — *Make a logical decision*. The analysis in the previous step should show that only one course of action is best.

However, in some cases it may appear that an acceptable solution *does not exist* or that the arguments for conflicting alternatives may be so equally balanced that no choice of action is clearly superior. In this case, it may be necessary to review the above steps. If you still face a dilemma, with two equal choices, you must decide which is *better* (or the *least undesirable*). It is usually best, in these cases, to select the course of action which does not yield a benefit to the person making the decision. If the choices are equally balanced, and no possibility of personal benefit exists, then this choice will ensure that the decision is *seen* to be morally defensible.

Step 6: WRITE — *Write a professional summary of your answer*. Finally, you must explain your answer clearly, logically, and neatly. You can't afford to waste time, so you must practise writing good answers. Your answer should start with your decision, which answers the question asked by the examiner, then explain why you came to that conclusion. That is, you may want to follow the (EGAD) steps listed above, to explain what alternatives were considered; why they were accepted or rejected; and what principles, regulations, precedents, or laws affected your decision. It is very important to cite sub-section numbers (from the Regulations) for the ethics questions, just as you would cite previous cases (or "precedents") for the law questions.

Do not copy clauses from the code of ethics; identify them by number, if possible. It is also important to write in a neat and legible style. The examiners will appreciate this courtesy.

AN IMPORTANT HINT

An examiner who sets one of the Professional Practice Exams has pointed out that, when applying the READ — EGAD! — WRITE procedure, too many candidates spend almost all of their time on the EGA part (ethical issues, generation of alternatives, and analysis) and not enough time on explaining step D (the decision) and writing a coherent summary of their answers. This lopsided use of time results in short, incomplete, and inadequate answers to the questions. Remember that the EGAD! process is intended merely to get you thinking about the problem in an orderly way. It is not necessary to write out all of these steps; your exam grade will be based only on your written summary, so make sure that your answer is complete, logical, and understandable.[3]

QUESTIONS FROM PREVIOUS EXAMINATIONS

This section contains 60 examination questions selected from previous Professional Practice and Ethics Exams in several provinces. The sources of the questions are not relevant, since similar ethical problems arise in every province and the answers will be essentially the same.⁴ The basic principles of ethics are universal. The questions have been chosen to show the various formats that might be encountered: essay-type, short-answer, multiple choice, and true–false. Readers are encouraged to attempt all questions. Solutions are suggested for a few of the questions, and an asterisk (*) indicates where the specific clause number(s) from the appropriate code of ethics or Act should appear.

ESSAY-TYPE EXAMINATION QUESTIONS

In the examination, the applicant would probably be asked to answer only four or five questions and would be permitted about twenty minutes per question. A copy of the code of ethics is usually provided for reference during the exam.

1. Professional Engineer A takes a job with a manufacturing company and almost immediately thereafter is given responsibility for preparing the draft of a bid for replacement turbine runners for a power corporation. While working on preparing the bid for the manufacturing company, Engineer A, as president and shareholder of his own company, which he has reactivated, writes to the power corporation requesting permission to submit a tender on the same project. A few days later, and while continuing to work on the bid for the manufacturing company, he receives word from the power corporation that a bid from his company would be considered. The day after learning this, he resigns his position with the manufacturing company and proceeds to finalize and submit a bid on behalf of his own company.

    Discuss Engineer A's actions from an ethical point of view.

    Suggested Answer: Engineer A is clearly unethical in his actions. He is not being fair or loyal to his employer, as required by the code of ethics.* He has taken advantage of inside information, betrayed the trust of his employer, and yielded to a conflict of interest. If his reactivated company was unknown to his employer, then he has failed to disclose his conflict of interest as required by the code of ethics.* By his actions, he has failed to show the necessary devotion to professional integrity required by the code of ethics.

    In his defence, it could be said that since he resigned before actually signing the contract, he did not compete with his employer, but this would be a technical point; the serious conflicts of interest occurred during the bid preparation stage. The only positive statement in his defence is that he provided an additional option for the power corporation in its selection of bids. Engineer A has exposed himself to the serious possibility of disciplinary action under the provincial or territorial Act.

2. You are a Professional Engineer with XYZ Consulting Engineers. You have become aware that your firm subcontracts nearly all the work associated with the set-up, printing, and publishing of reports, including artwork and editing. Your wife has some training along this line and, now that your children are at school, is considering going back into business. You decide to form a company to enter this line of business together with your neighbours, another couple. Your wife will be the president, using her maiden name, and you and your neighbours will be directors.

   Since you see opportunities for subcontract work from your company, you reason that there must be similar opportunities with other consulting firms. You are aware of the existing competition and the rates they charge for services and see this as an attractive sideline business. Can you do this ethically, and if so, what steps must you take?

   Suggested Answer: You *can* do this ethically, but there is a significant potential for a serious conflict of interest unless you scrupulously follow the code of ethics for your province or territory. You can undertake the sideline business providing it does not interfere with your regular employment and providing that your employer is fully informed, as required by the code of ethics.* Your wife, of course, is free to use any legal name in her business affairs; however, if the sole reason for using her maiden name is to conceal your participation in the company's ownership and operation, then your co-operation could be considered unethical. If your employer is fully informed, your interest in the company should not create a problem for you, although it might worry other clients, since a company that is publishing reports would usually be in a position of trust not to reveal the contents of the reports for the advantage of others.* Therefore, you must be seen *not* to be involved in handling sensitive engineering information submitted by other clients for publishing.

3. Brenda MacDonald, a Professional Engineer, is manager of a chemical plant in a northern Canadian town. Early this summer she noticed that the plant was creating slightly more water pollution in the lake into which its waste line drains than is legally permitted. If she contacts the provincial ministry of the environment and reveals the problem, the result will be a considerable amount of unfavourable publicity for the plant. The publicity will also hurt the lakeside town's resort business and may scare the community. Apart from that, solving the problem will cost her company well over $100 000. If she tells no one, it is unlikely that outsiders will discover the problem, because the violation poses no danger whatever to people. At the most, it will endanger a small number of fish.

Should MacDonald reveal the problem despite the cost to her company, or should she consider the problem little more than a technicality and disregard it? Discuss the ethical considerations affecting her decision.

Suggested Answer: MacDonald must, legally and ethically, take action to remedy this situation. She is obligated under the code of ethics to consider the public welfare as paramount.* The legal limit for pollution has been exceeded, and failure to take action could be considered professional misconduct under the Act.* If she has known about the excess for some time, she may already be considered negligent and therefore subject to disciplinary action under the Act.*

MacDonald must abide by the ministry's regulations, which would probably require her to submit a complete, factual report to inform officials about the pollution. Before sending the report, she should discuss it fully with her employer. If the employer reacts adversely, MacDonald must, nevertheless, forward the report to the ministry as required by law and the code of ethics.* If the employer attempts to dismiss her, MacDonald may find it useful to ask the provincial Association to mediate and to inform her employer of the requirements under the Act. Should MacDonald be dismissed while acting properly and in good faith, she would have grounds for a suit against the employer for wrongful dismissal to recoup lost wages and costs. It would be advisable for her to consult a lawyer in that event.

The engineer's concern over adverse publicity and the cost to the company must not obscure the requirement to act within the law. If the situation is permitted to continue unabated, the long-term consequences will be much more serious. The pollution could ruin the neighbouring resort industry, and MacDonald could find herself subject to disciplinary action for negligence or professional misconduct.

4. You are a Professional Engineer employed by a consulting engineering firm. Your immediate superior is also a Professional Engineer. You have occasion to check into the details of a recent invoice for work done on a project for which your boss is the project manager, but on which both you and members of your staff have done work.

   You are surprised to see how much of your time and the time of one of the senior engineers who reports to you are charged to the job. You decide to check further into this by reviewing the pertinent time sheets. The time sheets show that time charged to other work has been deliberately transferred to this job. You try to raise the subject with your boss but are rebuffed. You are quite sure something is wrong but are not sure where to turn. You turn to the code of ethics for direction.

What articles are relevant to this situation? What action must you take, according to the code of ethics?

Suggested Answer: According to the code of ethics, you must be loyal to the employer.* However, the code also states that you must be fair and loyal to clients.* This creates an ethical dilemma. The dilemma can be resolved by observing that the deliberate transfer of charges from one job or client to another could be a form of fraud or theft, which is illegal. Therefore, it is important to obtain a clear explanation or justification for this transfer. If your superior is completely unwilling to reassure you of the reasons for this action, you must expose this unprofessional, dishonest, or unethical conduct, in accordance with the code of ethics.* The information should be conveyed to the client who is being overcharged.

Should your superior threaten to dismiss you, consult the provincial Association and ask it to mediate or to explain the requirements placed upon you and your superior by the code of ethics.* If you are dismissed while following the requirements of the code of ethics in good faith, you should consult a lawyer about suing for wrongful dismissal.

5. A consulting engineering firm is preparing to submit a proposal to clean up an area contaminated by a chemical spill during a train derailment. From past experience, the engineers in the firm know the amount of work involved in doing the job properly. The experts will include people with training in ecology, water quality, ground water, soils, air pollution, and other areas. The methodology that they feel must be followed will result in an expenditure of about $5 million. Before their proposal is submitted, however, the federal government, which is the potential client, issues a news release saying that it has budgeted only $1 million for this work.

What can the consulting firm do? To reduce the level of work to one-fifth of what it thinks is necessary would infringe on the firm's perceived ethical responsibilities to the environment.

6. Engineer A enters into a consulting contract with a client to provide design and construction supervision of road surfaces in a partially completed land development project. He has taken over from another consultant, who was discharged partway through the job. Before Engineer A can finish the project, his contract also is terminated. Shortly thereafter it becomes obvious that there are deficiencies in the work done under A's supervision. Investigation shows that hastily placed road surfaces, completed under adverse late-fall weather conditions, are not up to specifications. It seems that A is aware of this and intended to require remedial work by the contrac-

tor in the spring, but his termination occurred before that time. Engineer A did not advise his client that he was expecting to reinspect in the spring and to have deficiencies corrected, nor did he inform his client of the existing state of the roads after he was released from his contract.

Did Engineer A act in an ethical way in his dealings with his client even though he may feel that he was unfairly terminated? Discuss the articles of the code of ethics that have a bearing on this case.

7. An engineer is employed by a large consulting engineering firm. Her work includes designing and specifying electrical equipment. She owns shares in a large, well-known electrical manufacturing company. Her shareholdings amount to only a very small fraction of 1 percent of all shares issued.

Is it a violation of the code of ethics for this engineer to select and specify equipment made by this company in which she holds stock?

8. Engineer X, a civil engineer and an employee of ABC Consultants Ltd., signed the 1982 Ontario Application for Renewal of Certificate of Authorization for that company as the engineer taking responsibility to see that the Professional Engineers Act, its by-laws, and its regulations would be complied with.

In 1982, ABC Consultants Ltd. prepared the electrical and mechanical designs for a multi-storey building, and, although Engineer X had very little to do with this project, the drawings bore his seal. These designs were found to be deficient in a number of respects. Contrary to the Ontario Building Code, fire walls were omitted, fire dampers were not shown, and sprinklers were improperly connected, among other things. Upon investigation it was found that both the electrical work and the mechanical work were done by professional engineers.

What is Engineer X's ethical position in this matter?

9. A building contractor engages a Professional Engineer to design and prepare drawings for the formwork and scaffolding for a reinforced concrete building to meet the requirements of construction safety legislation. The engineer does this and affixes his seal and signature to the original tracings, which he turns over to the contractor. Is this acceptable professional practice? Later the engineer is asked to inspect the scaffolding as built and finds that in many significant parts his design has been ignored and the contractor's superintendent has built it the way he thought it should be built.

What should the engineer do? Discuss this situation, with particular emphasis on the engineer's professional responsibility and the safety of the workers.

10. After having been employed by Consulting Engineer B for several years, Engineer A terminates her work with B and starts her own consulting practice. Later B learns that some of his subprofessional employees are doing work for A on their own time. B is of the opinion that the outside work by his subprofessional employees is so extensive that it diminishes their productivity.

    Did A act unethically by employing the subprofessional employees of B under the conditions stated?

11. John Doe, an engineer employed by a testing laboratory, represents his firm on a standards committee for automobile products. All but two of the members of this ten-person committee are engineers. After much deliberation on one standard, the committee arrives at a consensus, but Doe is violently opposed to the result and registers his objection. After carefully considering this objection, the committee passes the standard for formal publication. Subsequently, the laboratory receives a contract to test automobile products to this standard, and Doe is assigned the job of supervising the tests, compiling the final report, which indicates that the samples meet the requirements of the standard, and signing the report on behalf of the firm. He objects because he considers that his signature on a report attesting to the conformance of a product with a standard indicates that he endorses the standard.

    Is he correct in his assumption? What action should he take?

12. An engineer enters into a contract with a public body whereby he agrees to conduct such field investigations and studies as may be necessary to determine the most economical and proper method of designing and constructing a water supply system. He also agrees to prepare an engineering report, including an estimate of the cost of the project, and to estimate the amount of bond issue required. The contract provides that, if the bond issue passes, the engineer will be paid to prepare plans and specifications and supervise the construction, and he will be paid a fee for his preliminary services. If the bond issue should fail, the public body would not be obligated to pay for the preliminary work. The public body is prohibited by law from committing funds for the preliminary work until the bond issue is approved.

    May an engineer ethically accept a contingent contract under these conditions?

13. An owner retains an architect to prepare plans and specifications for a building, using a standard contract form. The architect, in turn, retains a structural engineer for the structural portion of the plans and specifications. The building is erected. Both professionals complete their respective portions of the contract except for the execution of

the required certificate of compliance. During the progress of the work, the owner makes progress payments to the architect, and the architect pays the appropriate amount from her payments to the structural engineer. However, when the building is completed and ready for occupancy, the owner still owes and refuses to pay the architect a substantial sum due under the contract, and the architect accordingly owes the structural engineer a proportionate amount. The owner alleges that there have been several deficiencies in the work of the architect and refuses to pay her the balance due. The owner requests city officials to issue him an occupancy permit, and they request the architect, who in turn requests the structural engineer, to certify that the structural system has been completed in compliance with the applicable building code and regulations. Such certification is required before the city may issue an occupancy permit to the owner. The structural engineer refuses to provide the certification until he has been paid for his services.

Is it unethical for the engineer to refuse to provide the certification that would enable the owner to secure the occupancy permit on the grounds that he has not been paid for his services?

14. An engineer in private practice is retained by a client to design and supervise the construction of a warehouse. Some time later he is asked by another client to provide professional engineering services for a warehouse almost identical to that previously designed by him, except for those minor changes necessary to adapt the building to the site. This client suggests that the fee be lower than that charged for the original design services, because the engineer could use his same design with only minor changes.

For the reuse of his design, would it be ethical for the engineer to charge a fee substantially less than that recommended by the Association?

15. Your firm is asked by the City of Townsville to assess the effects of a tidal wave. Located at the end of a long, narrow inlet, Townsville is in an earthquake zone, although the last one occurred in 1950 when the city was really only a fishing port. To make sure they have an adequate picture of the disaster that could result, they ask your firm to examine the effects of the 200-year earthquake. Your findings are so horrendous that the city authorities are appalled, and they feel that if the public were to realize the extent of impending damage, mass hysteria would result. As well, because many of the authorities are elected officials and have been in their positions for many years, people could ask why such a study was not carried out years ago and why adequate planning by-laws were never formulated. So you are asked to keep the findings of the 200-year quake confidential and to undertake another study of the effects of the 100-

year quake. The results are still frightening, and the city now asks you to study the 50-year quake.

Discuss this situation from an ethical point of view. What action will you take as a Professional Engineer? What advice will you give to the city council?

16. A Canadian Professional Engineer is working in a foreign country for a client building a power station. She is acting as technical adviser to the client. The client is directly supervising all construction labour. The client does not have any apparent safety procedures for his workers: no hard hats, no safety shoes, in some cases no shoes. Holes in floors do not have safety barricades. The conditions would be unacceptable in Canada. Even assuming the poor safety conditions will not affect the technical aspects of the power station, clearly they affect the safety of the workers.

Would it be ethical for the Canadian engineer not to take any action? What kind of action could she take? Do you consider it likely that the poor safety practices could affect only the safety of the workers and not have any relation to the technical aspects of the power station?

17. Consulting Firm A is preparing preliminary engineering and environmental impact studies for a client proposing an urban development project. The municipality has a planner on staff but has also engaged Consulting Firm B to assist with the review of A's submissions.

Firm A has made several submissions to secure approval, but each time some aspects are not satisfactory and the requirements are redefined. Finally the engineer from B offers, in the presence of the municipal staff planner, to complete the assignment for A, since he knows what is required. In addition to paying the fees of his original consultant, A, the developer must also pay the costs of the municipality's review, including its consultant from B.

Is it ethical for the engineer from Firm B to offer to complete the assignment for Firm A? If you were A, what would be your reaction to this situation from an ethics point of view?

18. Engineer A is employed by an industrial corporation. Her immediate supervisor is Engineer B, who is chairwoman of a civic committee responsible for retaining an architect to design a civic facility. When Engineer B receives the completed plans and specifications from the architect, she directs Engineer A to review them in order to gain knowledge, suggest improvements, and assure their compliance with the specified requirements.

Are Engineer B's instructions to Engineer A consistent with the code of ethics? Is Engineer A ethically permitted to carry out the instructions given her by Engineer B? Explain.

19. Engineer E, a member of a city council, is chairman of its finance committee, which deals with and makes recommendations regarding appropriations for projects undertaken by the city. One such project is a pollution abatement project, for which funds have been allocated. Engineer E is one of the principals in a consulting engineering firm, EPG, which has established a good reputation in the pollution control field. EPG has submitted to the council a proposal to provide the engineering services required for the project under consideration.

    Under these circumstances, is it ethical for EPG to offer to undertake this engineering work? Explain.

20. Because of a tight competitive market for engineering employees, the engineering department of a large manufacturing company has adopted a policy of paying a bonus to any member of its engineering staff who is successful in having an engineer or engineers working for other organizations recruited by the company. The theory is that the present engineering employee might know, or know of, other engineers who would consider employment with the company if approached by one of its engineering employees. The bonus offered is $200 per recruit.

    Is it ethical for the chief engineer to adopt this policy? Would it be ethical for the engineering employees of the company to participate in this recruitment program? Explain.

SHORT-ANSWER EXAMINATION QUESTIONS

21. Provide a definition of "ethics."

22. The code of ethics contains many clauses that describe the ethical responsibilities of the engineer with respect to professional life, relations with the public, relations with clients and employers, and relations with engineers. List at least half of these.

23. In a few sentences, describe what a "profession" is.

24. Is your province's code of ethics for engineers enforceable under your Professional Engineering Act? Explain.

25. Discuss a possible situation where an engineer's duty to his or her employer may be in conflict with his or her responsibility to the public.

26. What considerations and measures should an engineer take in the situation described in the previous question?

27. Explain what "conflict of interest" means.

28. Does your province's Professional Engineering Act explicitly restrict an engineer to practise in his or her branch of registration only? How does the code of ethics deal with the problem of practising outside of one's branch of registration?

29. a. The Association of Professional Engineers is the self-regulating organization responsible for the practice of engineering in your province. What is the principal objective of this organization?
    b. To become licensed to practise professional engineering in your province, you must meet certain requirements. Discuss briefly the five most significant of these.

30. a. What is the difference between a limited licence and a temporary licence in the practice of professional engineering? [This question does not apply in every province.]
    b. You are a practising Professional Engineer in a manufacturing company. Your division of the company has been transferred to Ontario from Manitoba. What must you do, if anything, to continue your engineering work?

MULTIPLE-CHOICE EXAMINATION QUESTIONS

Some provinces administer the examination in a multiple-choice format, as illustrated in the next five questions. A typical exam is two hours long, consists of 100 multiple-choice questions, and is "closed-book" (no aids permitted). Usually, half the questions concern practice and ethics, and half concern engineering law.

31. According to the code of ethics, which of the following activities by a professional member would be considered *unethical*?
    a. not charging a fee for presenting a speech
    b. signing plans prepared by an unknown person
    c. reviewing the work of another member with that member's consent
    d. providing professional services as a consultant

    Answer: b. It is unethical for professionals to sign plans not prepared by themselves or under their direct supervision or that they have not thoroughly checked. Since the plans were prepared by an unknown person, the analysis must be completely redone before they can be signed and sealed.

32. Which of the following is the most common job activity of top-level managers?
    a. writing and reading corporate financial reports
    b. developing and testing new products

    c. designing and implementing production systems

    d. directing and interacting with people

Answer: d. Most top managers spend most of their time interacting with other people.

33. The professional's standard of care and skill establishes the point at which a professional
    a. may or may not charge a fee for services
    b. has the duty to apply "reasonable care"
    c. may be judged negligent in the performance of services
    d. has met the minimum requirements for registration

    Answer: c. The standard of care is used to judge whether a professional has been negligent in the performance of services.

34. Which of the following is a minimum requirement for registration as a professional engineer?
    a. Canadian citizenship
    b. experience in engineering work
    c. course work in engineering
    d. residence in the province

    Answer: b. Of the items listed, engineering experience is the only requirement for registration that is absolutely essential. Each Act has provisions for people lacking one or more of the other three.

35. To effectively reduce liability exposure, the professional geologist should
    a. pursue continuing educational opportunities
    b. work under the supervision of a senior geologist
    c. maintain professional standards of practice
    d. provide clients with frequent progress reports

    Answer: c. Maintaining professional standards of practice is the most effective way of reducing liability exposure.

TRUE–FALSE EXAMINATION QUESTIONS

The following 25 questions illustrate the format for a typical true–false examination required by one of the provincial Associations. The examination has 50 questions and is performed as a homework assignment. The applicant therefore has access to the code of ethics, provincial by-laws and regulations, and the provincial Professional Engineering Act. Each answer must be justified by citing the appropriate clause from the code, by-law, regulations, or Act. The questions apply to every province or territory, and readers are urged to attempt to answer them; they are an excellent review for every engineer, regardless of his or her province of residence.

36. A person may assume the title "Professional Engineer" before being registered with the Association if working under the direct supervision of a registered Professional Engineer.

    True: _____ False: _____ Reference: _____

37. If an employer knowingly engages a person for work that requires the services of a Professional Engineer and that person is not registered or licensed with the Association, both the employer and the employee are in violation of the Professional Engineering Act.

    True: _____ False: _____ Reference: _____

38. A person convicted of a criminal offence under an act other than the Professional Engineering Act may be suspended from membership in the Association.

    True: _____ False: _____ Reference: _____

39. Members of the Armed Forces stationed in your province are subject to the provisions of the Professional Engineering Act.

    True: _____ False: _____ Reference: _____

40. A Professional Engineer must be aware of all the related facts before publicly expressing an opinion on an engineering subject.

    True: _____ False: _____ Reference: _____

41. A Professional Engineer must ensure that clients understand the full extent of his or her responsibilities.

    True: _____ False: _____ Reference: _____

42. It is voluntary, and not mandatory, for a Professional Engineer to strive to keep informed about new techniques in his or her field of endeavour.

    True: _____ False: _____ Reference: _____

43. A Professional Engineer may criticize the work of a fellow engineer publicly if he or she first advises the fellow engineer of the intent to do so.

    True: _____ False: _____ Reference: _____

44. If a person is working under the direct supervision of a Professional Engineer who assumes all responsibility, the subordinate is still required to be registered.

    True: _____ False: _____ Reference: _____

45. Unless a person is registered or licensed by the Association of Professional Engineers of any province, that person may not imply that he or she is entitled to engage in professional engineering.

    True: _____ False: _____ Reference: _____

46. As far as work is concerned, a Professional Engineer's first responsibility is to the employer.

    True: _____ False: _____ Reference: _____

47. A Professional Engineer may seal plans that have been prepared neither by himself or herself nor under his or her personal direction.

    True: _____ False: _____ Reference: _____

48. A Professional Engineer may be compensated by more than one interested party for the same service without the consent of all interested parties.

    True: _____ False: _____ Reference: _____

49. All specifications and reports must be sealed by the Professional Engineer who has done the work involved.

    True: _____ False: _____ Reference: _____

50. Council may ask witnesses to attend an inquiry on a discipline matter and has the power to ensure attendance.

    True: _____ False: _____ Reference: _____

51. If a member is found guilty of unprofessional conduct, the most severe penalty Council may mete out is a reprimand.

    True: _____ False: _____ Reference: _____

52. Council may initiate an inquiry where professional misconduct is suspected, even though no written complaint has been received.

    True: _____ False: _____ Reference: _____

53. A Professional Engineer, having first advised his or her fellow engineer of the intent to do so, may accept a commission to review the work of the fellow engineer.

    True: _____ False: _____ Reference: _____

54. It is not mandatory for a Professional Engineer to report a colleague he or she feels is engaged in unethical practice.

    True: _____ False: _____ Reference: _____

55. A Professional Engineer has no responsibility for the professional development of engineers in his or her employ.

    True: _____ False: _____ Reference: _____

56. If a Professional Engineer in charge of an assignment is overruled by his or her superior or client, the engineer should present clearly the consequences to be expected from the proposed deviations and then complete the assignment, provided that the ruling of the supe-

rior or client does not jeopardize public property, life, or the environment.

True: ____ False: ____ Reference: ____

57. The objects of the Association are primarily to
    a. ensure that the rights and interests of all engineers in the province are protected.

    True: ____ False: ____ Reference: ____

    b. ensure that the public interest is served and protected through the competent and ethical practice of engineering within the province.

    True: ____ False: ____ Reference: ____

58. Registration as a full member of the Association may be granted if the following education and experience requirements are met:
    a. The applicant has an accredited engineering degree and has experience that is satisfactory to Council.

    True: ____ False: ____ Reference: ____

    b. The applicant has passed the examinations required by Council and has a minimum of six years of satisfactory engineering experience.

    True: ____ False: ____ Reference: ____

59. The practice of engineering, as defined by the Act, covers a broad range of activities. However, the following persons are considered exempt from the provisions of the Act.
    a. those who were doing engineering work before the Act became law

    True: ____ False: ____ Reference: ____

    b. technicians working under the direct and personal supervision of professional engineers

    True: ____ False: ____ Reference: ____

    c. land surveyors, architects, electricians, and enginemen, provided they do not engage in the practice of engineering.

    True: ____ False: ____ Reference: ____

    d. those who have ten or more years of good engineering experience and feel they are competent have a basic right to work as an engineer.

    True: ____ False: ____ Reference: ____

60. Mr. X, a registered member of another provincial Association, is transferred to Manitoba by his company and has taken up perma-

nent residence in Winnipeg. His new title is Chief Design Engineer for Western Canada. For the past two months he has been the sole designer of a commercial building in Manitoba. He uses his title and P.Eng. after his name on his business card. He has not become a member of the Manitoba Association. Assess the following statements:

a. Membership in the other provincial Association allows Mr. X to practise engineering in Manitoba; therefore, he does not have to become a member of the APEM.

   True: _____ False: _____ Reference: _____

b. Mr. X should have applied for registration with APEM on arrival in Manitoba.

   True: _____ False: _____ Reference: _____

c. Mr. X should immediately apply for a temporary licence to practise in Manitoba.

   True: _____ False: _____ Reference: _____

d. Mr. X is in contravention of the Act since he is practising engineering and hence is liable for the penalty under the Act.

   True: _____ False: _____ Reference: _____

e. Mr. X is in contravention of the Act for using P.Eng. after his name.

   True: _____ False: _____ Reference: _____

NOTES

1. G.C. Andrews, *Study Guide for the PEO Professional Practice Exam*, 4th ed. (video/manual set), University of Waterloo, Distance Education Department, 1997.
2. J. Delaney, *How to Do Your Best on Law School Exams* (Bogota, NJ: J. Delaney Publications, 1990).
3. Andrews, *Study Guide for the PEO Professional Practice Exam*, 19–20. Reproduced with permission.
4. The authors would like to express their appreciation to provincial Associations in Alberta, Manitoba, New Brunswick, Newfoundland, and Ontario for their assistance in obtaining sample questions and their permission to reprint these questions from previous Professional Practice Exams.

APPENDIX A

Addresses for the Provincial and Territorial Engineering Associations

Canadian Council of Professional Engineers (CCPE)
116 Albert Street, Suite 401
Ottawa, ON K1P 5G3
Tel: (613)232-2474
Fax: (613)230-5759
Web site: http://www.ccpe.ca

Association of Professional Engineers, Geologists and Geophysicists of Alberta
#1500 Scotia Place, Tower One
10060 Jasper Avenue
Edmonton, AB T5J 4A2
Tel: (403)426-3990
Fax: (403)426-1877
Web site: http://www.apegga.com

Association of Professional Engineers and Geoscientists of British Columbia
4010 Regent Street, Suite 200
Burnaby, BC V5C 6N2
Tel: (604)430-8035
Fax: (604)430-8085
Web site: http://www.apeg.bc.ca

Association of Professional Engineers of Manitoba
850A Pembina Highway
Winnipeg, MB R3M 2M7
Tel: (204)474-2736
Fax: (204)474-5960
Web site: http://www.apem.mb.ca

Association of Professional Engineers of New Brunswick
535 Beaverbrook Court, Suite 105
Fredericton, NB E3B 1X6
Tel: (506)458-8083
Fax: (506)451-9629
Web site: http://ctca.unb.ca/APENB/APENB.HTML

Association of Professional Engineers, Geoscientists of Newfoundland
P.O. Box 21207
St. John's, NF A1A 5B2
Tel: (709)753-7714
Fax: (709)753-6131
Web site:
http://www.netfx-inc.com/apegn

**Association of Professional
Engineers, Geologists and
Geophysicists of the Northwest
Territories**
#5 – 4807, 49th Street
Yellowknife, NT X1A 3T5
Tel: (403)920-4055
Fax: (403)873-4058
Web site: http://www.napegg.nt.ca

**Association of Professional
Engineers of Nova Scotia**
P.O. Box 129, 1355 Barrington Street
Halifax, NS B3J 2M4
Tel: (902)429-2250
Fax: (902)423-9769
Web site: http://www.apens.ns.ca

Professional Engineers Ontario
25 Sheppard Avenue West, Suite 1000
North York, ON M2N 6S9
Tel: (416)224-1100
Fax: (416)224-8168
Web site: http://www.peo.on.ca

**Association of Professional
Engineers of Prince Edward Island**
549 North River Road
Charlottetown, PE C1E 1J6
Tel: (902)566-1268
Fax: (902)566-5551
Web site: http://www.isn.net/virtual/
apepei

Ordre des ingénieurs du Québec
2020 rue University, 18e étage
Montréal, PQ H3A 2A5
Tel: (514)845-6141
Fax: (514)845-1833
Web site: http://www.oiq.qc.ca

**Association of Professional
Engineers of Saskatchewan**
2255 Thirteenth Avenue
Regina, SK S4P 0V6
Tel: (306)525-9547
Fax: (306)525-0851
Web site: http://www.apegs.sk.ca

**Association of Professional
Engineers of the Yukon Territory**
P.O. Box 4125
Whitehorse, YT Y1A 3S9
Tel: (403)667-6727
Fax: (403)667-6727
Web site: As this book goes to press,
APEYT has no Web site.

APPENDIX B

Excerpts from the Provincial and Territorial Acts and Regulations

The engineering profession is regulated in each province or territory of Canada by Acts of the provincial legislatures (for the provinces) or legislative councils (for the territories). The Acts are listed below (except for Nunavut, since the territory is being established as this text goes to press, and no Act has yet been announced):

- Alberta: Engineering, Geological and Geophysical Professions Act
- British Columbia: Engineers and Geoscientists Act
- Manitoba: Engineering and Geoscientific Professions Act
- New Brunswick: Engineering Profession Act
- Newfoundland: Engineers and Geoscientists Act
- Northwest Territories: Engineering, Geological and Geophysical Professions Act
- Nova Scotia: Engineering Profession Act
- Ontario: Professional Engineers Act
- Prince Edward Island: Engineering Profession Act
- Quebec: Engineers Act
- Saskatchewan: Engineering and Geoscience Professions Act
- Yukon Territory: Engineering Profession Act

This appendix consists of five key excerpts from each of the above provincial and territorial Acts (or, where appropriate, from the regulations made under the Act) of direct interest to most engineers. The five excerpts include the

1. *definition of engineering,* or the practice of engineering (and, in those jurisdictions, where appropriate, the practice of geology, geophysics, and geoscience)
2. *admission criteria* for obtaining a licence or membership
3. *definition of professional misconduct,* or unprofessional conduct

4. *disciplinary powers* of the Association

5. *code of ethics* for the Association

These excerpts are included for quick reference on the above topics; the Acts themselves should be consulted for more detailed information. This information is believed to be correct as this text goes to press. Although this information changes very slowly and is therefore fairly stable, the Acts and regulations are amended on an irregular basis. Should there be a discrepancy between the information provided by the Association and the information in this text, the information provided by the Association should, of course, be followed.

Although more recent Acts are written to be gender neutral, some of the older ones use the personal pronouns "he," "him," and "his" to refer to both men and women and have been reproduced here in this form.

ALBERTA

ENGINEERING, GEOLOGICAL AND GEOPHYSICAL PROFESSIONS ACT (*STATUTES OF ALBERTA* 1981, C. E-11.1, AS AMENDED)

1. DEFINITION OF ENGINEERING (SECTION 1 OF THE ACT)

"practice of engineering" means
(i) reporting on, advising on, evaluating, designing, preparing plans and specifications for or directing the construction, technical inspection, maintenance or operation of any structure, work or process
 (a) that is aimed at the discovery, development or utilization of matter, materials or energy or in any other way designed for the use and convenience of man, and
 (b) that requires in the reporting, advising, evaluating, designing, preparation or direction the professional application of the principles of mathematics, chemistry, physics or any related applied subject, or
(ii) teaching engineering at a university

"practice of geology" means
(i) reporting, advising, evaluating, interpreting, geological surveying, sampling or examining related to any activity
 (a) that is aimed at the discovery or development of oil, natural gas, coal, metallic or non-metallic minerals, precious stones, other natural resources or water or that is aimed at the investigation of geological conditions, and
 (b) that requires in that reporting, advising, evaluating, interpreting, geological surveying, sampling or examining, the professional application of the principles of the geological sciences, or
(ii) teaching geology at a university

"practice of geophysics" means
(i) reporting on, advising on, acquiring, processing, evaluating or interpreting geophysical data, or geophysical surveying that relates to any activity
 (a) that is aimed at the discovery or development of oil, natural gas, coal, metallic or non-metallic minerals or precious stones or other natural re-

sources or water or that is aimed at the investigation of sub-surface conditions in the earth, and

(b) that requires in that reporting, advising, evaluating, interpreting, or geophysical surveying, the professional application of the principles of the geophysical sciences, or

(ii) teaching geophysics at a university

2. MEMBERSHIP CRITERIA (SECTION 21 OF THE ACT AND SECTION 13 OF ALBERTA REGULATION 244/81, AS AMENDED BY ALBERTA REGULATION 204/90, 61/96 AND 16/97)

[Applications as *professional members* may be made by persons resident in Alberta who are Canadian Citizens or lawfully admitted to Canada as permanent residents, or as *licensees* if resident outside Alberta, or resident in Alberta but neither Canadian citizens nor lawfully admitted to Canada as permanent residents. An applicant must also meet 1 of the following qualifications:]

(i) the applicant is a graduate of a university program in engineering, geology or geophysics or has completed university qualifications in a related program acceptable to the Board of Examiners and has had, since graduation or completion, at least 3 years of experience in engineering, geological or geophysical work,

(ii) the applicant has achieved an education satisfactory to the Board of Examiners consisting of

(a) the completion of at least 2 years of post secondary education in areas that relate to the science and technology of engineering, geology or geophysics, and

(b) the receipt of credit, or the equivalent, in an adequate number of related fundamental subjects satisfactory to the Board of Examiners

and the applicant has been engaged in work of an engineering, geological or geophysical nature for at least 3 years, or for 1 year following graduation from an engineering, geological or geophysical technology program recognized by the Board of Examiners, or

(iii) the applicant is registered as an engineer, geologist or geophysicist in a jurisdiction recognized by the Board of Examiners.

AR 244/81 s13; 204/90

3. DEFINITION OF PROFESSIONAL MISCONDUCT (SECTION 43 OF THE ACT)

43(1) Any conduct of a professional member, licensee, permit holder, certificate holder or member-in-training that in the opinion of the Discipline Committee or the Appeal Board

(a) is detrimental to the best interests of the public,

(b) contravenes a code of ethics of the profession as established under the regulations,

(c) harms or tends to harm the standing of the profession generally,

(d) displays a lack of knowledge of or lack of skill or judgement in the practice of the profession, or

(e) displays a lack of knowledge of or lack of skill or judgement in the carrying out of any duty or obligation undertaken in the practice of the profession,

whether or not that conduct is disgraceful or dishonourable, constitutes either unskilled practice of the profession or unprofessional conduct, whichever the Discipline Committee or the Appeal Board finds.

(2) If an investigated person fails to comply with or contravenes this Act, the regulations or the by-laws, and the failure or contravention is, in the opinion of the Discipline Committee, of a serious nature, the failure or contravention may be found by the Discipline Committee to be unprofessional conduct whether or not it would be so found under subsection (1).

4. DISCIPLINARY POWERS (SECTIONS 60, 61 OF THE ACT)

60. If the Discipline Committee finds that the conduct of the investigated person is unprofessional conduct or unskilled practice of the profession or both, the Discipline Committee may make any one or more of the following orders:

(a) reprimand the investigated person;

(b) suspend the registration of the investigated person for a specified period;

(c) suspend the registration of the investigated person either generally or from any field of practice until

(i) he has completed a specified course of studies or obtained supervised practical experience, or

(ii) the Discipline Committee is satisfied as to the competence of the investigated person generally or in a specified field of practice;

(d) accept in place of a suspension the investigated person's undertaking to limit his practice;

(e) impose conditions on the investigated person's entitlement to engage in the practice of the profession generally or in any field of the practice, including the conditions that he

(i) practise under supervision,

(ii) not engage in sole practice,

(iii) permit periodic inspections by a person authorized by the Discipline Committee, or

(iv) report to the Discipline Committee on specific matters;

(f) direct the investigated person to pass a particular course of study or satisfy the Discipline Committee as to his practical competence generally or in a field of practice;

(g) direct the investigated person to satisfy the Discipline Committee that a disability or addiction can be or has been overcome, and suspend the person until the Discipline Committee is so satisfied;

(h) require the investigated person to take counselling or to obtain any assistance that in the opinion of the Discipline Committee is appropriate;

(i) direct the investigated person to waive, reduce or repay a fee for services rendered by the investigated person that, in the opinion of the Discipline Committee, were not rendered or were improperly rendered;

(j) cancel the registration of the investigated person;

(k) any other order that it considers appropriate in the circumstances.

61(1) The Discipline Committee may, in addition to or instead of dealing with the investigated person in accordance with section 60, order that the investigated person pay

(a) all or part of the costs of the hearing in accordance with the by-laws,

(b) a fine not exceeding $10 000 to the Association, or

(c) both the costs under clause (a) and a fine under clause (b), within the time fixed by the order.

(2) If the investigated person ordered to pay a fine, costs or both under subsection (1) fails to pay the fine, costs or both within the time ordered, the Discipline Committee may suspend the registration of that person until he has paid the fine, costs or both.

(3) A fine or costs ordered to be paid to the Association under this section is a debt due to the Association and may be recovered by the Association by civil action for debt.

1981 cE-11.1 s61; 1984 c17s19

5. CODE OF ETHICS (ALBERTA REGULATION 204/90, SCHEDULE A, ESTABLISHED PURSUANT TO SECTION 18(1)(H) OF THE ACT)

PREAMBLE

Professional engineers, geologists and geophysicists shall recognize that professional ethics is founded upon integrity, competence and devotion to service and to the advancement of human welfare. This concept shall guide their conduct at all times. In this way each professional's actions will enhance the dignity and status of the professions.

Professional engineers, geologists and geophysicists, through their practice, are charged with extending public understanding of the professions and should serve in public affairs when their professional knowledge may be of benefit to the public.

Professional engineers, geologists and geophysicists will build their reputations on the basis of merit of the services performed or offered and shall not compete unfairly with others or compete primarily on the basis of fees without due consideration for other factors.

Professional engineers, geologists and geophysicists will maintain a special obligation to demonstrate understanding, professionalism and technical expertise to members-in-training under their supervision.

RULES OF CONDUCT

1. Professional engineers, geologists and geophysicists shall have proper regard in all their work for the safety and welfare of all persons and for the physical environment affected by their work.

2. Professional engineers, geologists and geophysicists shall undertake only work that they are competent to perform by virtue of training and experience and shall express opinions on engineering, geological or geophysical matters only on the basis of adequate knowledge and honest conviction.

3. Professional engineers, geologists and geophysicists shall sign and seal only reports, plans or documents that they have prepared or that have been prepared under their direct supervision.

4. Professional engineers, geologists and geophysicists shall act for their clients or employers as faithful agents or trustees; always acting independently and with fairness and justice to all parties.

5. Professional engineers, geologists and geophysicists shall not engage in activities or accept remuneration for services rendered that may create a conflict of interest with their clients or employers, without the knowledge and consent of their clients or employers.

6. Professional engineers, geologists and geophysicists shall not disclose confidential information without the consent of their clients or employers, unless the withholding of the information is considered contrary to the safety of the public.

7. Professional engineers, geologists and geophysicists shall present clearly to their clients or employers the consequences to be expected if their professional judgement is overruled by other authorities in matters pertaining to work for which they are professionally responsible.

8. Professional engineers, geologists and geophysicists shall not offer or accept covert payment for the purpose of securing an engineering, geological or geophysical assignment.

9. Professional engineers, geologists and geophysicists shall represent their qualifications and competence, or advertise professional services offered, only through factual representation without exaggeration.

10. Professional engineers, geologists and geophysicists shall conduct themselves toward other professional engineers, geologists and geophysicist, and toward employees and others with fairness and good faith.

11. Professional engineers, geologists and geophysicists shall advise the Registrar of any practice by a member of the Association that they believe to be contrary to this Code of Ethics.

Under section 43 of the Act, a contravention of this Code of Ethics may constitute unprofessional conduct or unskilled practice which is subject to disciplinary action.

AR 244/81 Sched. A; 386/85; 204/90

BRITISH COLUMBIA

ENGINEERS AND GEOSCIENTISTS ACT (*REVISED STATUTES OF BRITISH COLUMBIA* 1979, C. 109, AS AMENDED JUNE 1994)

1. DEFINITION OF ENGINEERING (SECTION 1 OF THE ACT)

"practice of professional engineering" means the carrying on of chemical, civil, electrical, forest, geological, mechanical, metallurgical, mining or structural engineering, and other disciplines of engineering that may be designated by the Council and for which university engineering programs have been accredited by the Canadian Engineering Accreditation Board or by a body which, in the opinion of the Council, is its equivalent, and includes the reporting on, designing, or directing the construction of any works that require for their design, or the supervision of their construction, or the supervision of their maintenance, such experience and technical knowledge as are required by or under this Act for the admission by examination to membership in the association, and, without limitation, includes reporting on, designing or directing the construction of public utilities, industrial works, railways, bridges, highways, canals, harbour works, river improvements, lighthouses, wet docks, dry docks, floating docks, launch ways, marine ways, steam engines, turbines, pumps, internal combustion engines, airships and airplanes, electrical machinery and apparatus, chemical operations, machinery, and works for the development, transmission or application of power, light and heat, grain elevators, municipal works, irrigation works, sewage disposal works, drainage works, incinerators, hydraulic works, and all other engineering works, and all buildings necessary to the proper housing, installation and operation of the engineering works embraced in this definition;

"practice of professional geoscience" means reporting, advising, acquiring, processing, evaluating, interpreting, surveying, sampling or examining related to any activity that

(a) is directed toward the discovery or development of oil, natural gas, coal, metallic or non-metallic minerals, precious stones, other natural resources or water or the investigation of surface or sub-surface geological conditions, and

(b) requires the professional application of the principles of geology; geophysics or geochemistry;

2. MEMBERSHIP CRITERIA (SECTION 13 OF THE ACT AND SECTION 11(E) OF THE BY-LAWS)

13(1) The Council must admit an applicant to membership in the association who is a Canadian citizen or permanent resident of Canada, and who has submitted evidence satisfactory to the Council of the following:

(a) that the applicant has either

(i) graduated in applied science, engineering or geoscience from an institute of learning approved by the Council in a program approved by the Council, or

(ii) passed examinations established by the bylaws of the association or passed examinations, requiring special knowledge in branches of learning specified by the Council, of an association or institute approved by the Council;

(b) that the applicant has passed special examinations required by the Council;

(c) that the experience in engineering or geoscience work established by the bylaws has been obtained;

(d) that the applicant is of good character and good repute;

(e) that all examination and registration fees have been paid to the association.

(2) Despite subsection (1) or (5), the Council may refuse registration or a licence to a person if the Council has reasonable and probable grounds to believe that the person has been convicted in Canada or elsewhere of an offence that, if committed in British Columbia, would be an offence under an enactment of the Province or of Canada, and that the nature or circumstances of the offence render the person unsuitable for registration or licensing.

(3) A person desiring to become a member must comply with the bylaws relating to application for membership, and, if required to qualify by examination, must comply with section 16(4) and (5).

(4) A person who is not a citizen or a permanent resident of Canada, whose qualifications are those required by subsection (1)(a) to (d), and who desires to engage temporarily in the practice of professional engineering or professional geoscience in the Province, must first obtain a licence from the Council which will entitle him or her to engage in the practice of professional engineering or professional geoscience in respect of a particular work or for a temporary period, or both, as the Council decides.

(5) On producing evidence satisfactory to the Council of his or her qualifications under subsection (4) and on payment of the prescribed fees, a person must be granted the licence.

(6) Neither corporations nor partnerships as such may become members of the association.

(7) If professional engineers or professional geoscientists are employed by corporations or are members of partnerships, they individually must assume the functions of and must be held responsible as professional engineers or professional geoscientists.

REGISTERED MEMBERS (BYLAW)

11(e) Registration as a full member of the Association may be granted to a Canadian Citizen or permanent resident of Canada when Council is satisfied that the applicant is of good character and repute and:

(1) Has graduated in applied science, engineering or geoscience from an institute of learning approved by the Council in a course approved by the Council and in addition has had 4 years' experience in engineering or geoscience satisfactory to the Council, or

(2) Has passed the examinations required by the Council or the equivalent examinations of an association or institute approved by the Council requiring special knowledge in branches of learning as may be specified by the Council and in addition has had five years' experience in engineering or geoscience satisfactory to the Council and has submitted as partial evi-

dence of this experience an engineering or geoscience report or thesis satisfactory to the Council, or

(3) Has passed the examinations required by the Council or the equivalent examinations of an association or institute approved by the Council requiring special knowledge in branches of learning as may be specified by the Council and in addition has had eight years of experience in engineering or geoscience satisfactory to the Council.

3. DEFINITION OF PROFESSIONAL MISCONDUCT

[Although the terms "incompetence," "negligence," and "unprofessional conduct" are used in the Act, sections 32 and 33, these terms are not defined in the Act.]

4. DISCIPLINARY POWERS (SECTIONS 33 TO 35 OF THE ACT)

33(1) After an inquiry under section 32, the discipline committee may determine that the member, licensee or certificate holder

(a) has been convicted in Canada or elsewhere of an offence that, if committed in British Columbia, would be an offence under an enactment of the Province or of Canada, and that the nature or circumstances of the offence render the person unsuitable for registration or licensing.

(b) has contravened this Act or the by-laws or the code of ethics of the association, or

(c) has demonstrated incompetence, negligence or unprofessional conduct.

(2) If the discipline committee makes a determination under subsection (1), it may, by order, do one or more of the following:

(a) reprimand the member, licensee or certificate holder;

(b) impose conditions on the membership, licence or certificate of authorization of the member, licensee or certificate holder;

(c) suspend or revoke the membership, licence or certificate of authorization of the member, licensee or certificate holder;

(d) impose a fine, payable to the association, of not more than $25 000 on the member, licensee or certificate holder.

(3) The discipline committee must give written reasons for any action it takes under subsection (2).

(4) If a member, licensee or certificate holder is suspended from practice,

(a) the registration or licence is deemed to be cancelled during the term of the suspension, and

(b) the suspended member, licensee or certificate holder is not entitled to any of the rights or privileges of membership and must not be considered a member while the suspension continues.

5. CODE OF ETHICS (SECTION 14(A) OF THE BY-LAWS, ESTABLISHED UNDER SECTION 10 OF THE ACT)

The purpose of the *Code of Ethics* is to give general statements of the principles of ethical conduct in order that Professional Engineers and Professional Geoscientists may fulfill their duty to the public, to the profession and their fellow members.

Professional Engineers and Professional Geoscientists shall act at all times with fairness, courtesy and good faith to their associates, employers, employees and clients, and with fidelity to the public needs. They shall uphold the values of truth, honesty and trustworthiness and safeguard human life and welfare and the environment. In keeping with these basic tenets, Professional Engineers and Professional Geoscientists shall:

(1) hold paramount the safety, health and welfare of the public, the protection of the environment and promote health and safety within the workplace;

(2) undertake and accept responsibility for professional assignments only when qualified by training or experience;

(3) provide an opinion on a professional subject only when it is founded upon adequate knowledge and honest conviction;

(4) act as faithful agents of their clients or employers, maintain confidentiality and avoid a conflict of interest but, where such conflict arises, fully disclose the circumstances without delay to the employer or client;

(5) uphold the principle of appropriate and adequate compensation for the performance of engineering and geoscience work;

(6) keep themselves informed in order to maintain their competence, strive to advance the body of knowledge within which they practice and provide opportunities for the professional development of their associates;

(7) conduct themselves with fairness, courtesy and good faith towards clients, colleagues and others, give credit where it is due and accept, as well as give, honest and fair professional comment;

(8) present clearly to employers and clients the possible consequences if professional decisions or judgments are overruled or disregarded;

(9) report to their association or other appropriate agencies any hazardous, illegal or unethical professional decisions or practices by engineers, geoscientists, or others; and

(10) extend public knowledge and appreciation of engineering and geoscience and protect the profession from misrepresentation and misunderstanding.

SOURCE: Engineers and Geoscientists Act, *Revised Statutes of British Columbia* 1979, c. 109, as amended (June 1994). Reproduced with permission of the Queen's Printer for the Province of British Columbia. (Not an official copy — for informational purposes only.)

MANITOBA

ENGINEERING AND GEOSCIENTIFIC PROFESSIONS ACT (*STATUTES OF MANITOBA* PROPOSED 1998 — SEE THE ACT FOR OFFICIAL WORDING)

1. DEFINITION OF ENGINEERING AND GEOSCIENTIFIC (SECTION 1 OF THE ACT)

"practice of professional engineering" or "practice of engineering" means carrying on for hire, gain or hope of reward, either directly or indirectly, of one or more of the following branches of the science of engineering, namely:

(i) agricultural,
(ii) biochemical,
(iii) chemical,
(iv) civil,
(v) electrical,
(vi) forest,
(vii) geological,
(viii) industrial,
(ix) mechanical,
(x) metallurgical,
(xi) structural,

or such other branch as hereafter may be recognized and adopted by by-law of the association as a branch of engineering and, without restricting the generality of the foregoing, includes the reporting on, advising on, valuing of, measuring for, laying out of, designing of, engineering inspection of (including the direction or supervision of any of the foregoing) or the construction, alteration, improvement or enlargement of, works or processes or any of them by reason of their requiring in connection with any of the operations above set forth, the skilled or professional application of the principles of mathematics, physics, mechanics, aeronautics, hydraulics, electricity, forestry, chemistry, geology, or metallurgy, but does not include the operation, execution or supervision of works as superintendent, foreman, inspector, road master, track master, bridge master, building master or contractor, where the works have been designed by and are constructed under the supervision of a professional engineer;

"practice of professional geoscience" or "practice of geoscience" is the application of the principles of geology, geophysics and geochemistry to analyze and manage the earth's materials, resources, forms and processes where the safeguarding of life, health, property or public welfare is concerned; it is to examine samples, conduct geoscience surveys, report, advise, acquire data on, evaluate, and interpret any aspect of metallic and non-metallic minerals, rocks, oil, natural gas, precious stones, water or other natural resources; it is to investigate surface and subsurface geological properties, conditions and processes to identify geologic hazards and to protect, enhance, sustain or remediate our natural environment;

2. MEMBERSHIP CRITERIA (SECTION 15 OF THE ACT)

15(1) No person is entitled to be registered as a professional engineer or as a professional geoscientist, unless he or she submits to the registrar an application in the prescribed form and such person
(a) is a natural person at least 18 years of age;
(b) submits to the registration board evidence that he or she is academically qualified;
(c) submits to the registration board evidence that he or she has acquired relevant engineering work experience or relevant geoscientific work experience;
(d) submits to the registration board evidence that he or she has successfully completed an educational program for engineers-in-training or geoscientists-in-training or is otherwise qualified;

(e) submits to the registration board evidence that he or she has subscribed to and agreed to abide by the code of ethics of the association;

(f) pays the dues and fees prescribed by the by-laws; and

(g) complies with such other terms and conditions as may be imposed, in accordance with this Act or the by-laws.

3. DEFINITION OF PROFESSIONAL MISCONDUCT ("OFFENSIVE CONDUCT") (SECTION 51 OF THE ACT)

UNSKILLED PRACTICE OR OFFENSIVE CONDUCT

51(1) Conduct of an investigated person that in the opinion of the panel

(a) is unprofessional conduct or professional misconduct;

(b) is detrimental to the public interest;

(c) is conduct unbecoming a professional engineer or professional geoscientist;

(d) is misconduct in the practice of engineering or geoscience;

(e) results in the conviction for a crime;

(f) contravenes that Act or the by-laws or the code of ethics adopted under section 11;

(g) displays a lack of knowledge of or lack of skill or judgment in the practice of engineering or geoscience;

(h) demonstrates incapacity or unfitness to practise engineering or geoscience or demonstrates that the person is suffering from an ailment that might, if the person is allowed to continue to practise engineering or geoscience, constitute a danger to the public;

constitutes either unskilled practice of engineering or geoscience or offensive conduct, whichever the panel finds.

CONVICTION IN ANOTHER JURISDICTION IS OFFENSIVE CONDUCT

51(2) Where any member, holder of a certificate of authorization, temporary licensee or restricted license is the subject of an order of any other association of professional engineers or association of professional geoscientists, having the same effect as any order made under section 53 or 54 of this Act, such member, holder of a certificate of authorization, temporary licensee or restricted licensee shall be deemed to be guilty of either unskilled practice of engineering or geoscience or offensive conduct, whichever the panel finds.

4. DISCIPLINARY POWERS (SECTION 24 OF THE ACT)

ORDERS OF THE PANEL

53(1) If the panel finds that the conduct of an investigated person constitutes

(a) unskilled practice of engineering or geoscience, or

(b) offensive conduct

or both, the panel may make any one or more of the following orders:

(a) reprimand the investigated person;

(b) suspend the certificate of registration, certificate of authorization, temporary licence, restricted licence or enrolment as an engineer-in-training or geoscientist-in-training of the investigated person for a stated period;

(c) suspend the certificate of registration, certificate of authorization, temporary licence, restricted licence or enrolment as an engineer-in-training or geoscientist-in-training of an investigated person until

(i) the investigated person has completed a specified course of studies or obtained supervised practical experience, or

(ii) the discipline committee is satisfied as to the competence of the investigated person to practise engineering or geoscience;

(d) accept in place of a suspension the investigated person's undertaking to limit his, her or its practice;

(e) impose conditions on the investigated person's entitlement to engage in the practice of engineering or geoscience, including the conditions that he, she, or it

(i) practise under supervision,

(ii) not engage in sole practice,

(iii) not function as a holder of a certificate of authorization for a specified period,

(iv) permit periodic inspections by a person authorized by the discipline committee to carry out inspections,

(v) permit periodic audits of records, or

(vi) report to the discipline committee or the council on specific matters;

(f) direct the investigated person to pass a particular course of study or satisfy the discipline committee or the council as to the person's competence;

(g) direct the investigated person to satisfy the discipline committee that a disability or addiction can be or has been overcome, and suspend the certificate of registration, certificate of authorization, temporary licence, restricted licence or enrolment as an engineer-in-training or geoscientist-in-training of the investigated person until the discipline committee is so satisfied;

(h) require the investigated person to take counselling that in the opinion of the panel is appropriate;

(i) direct the investigated person to waive, reduce or repay money paid to the investigated person that, in the opinion of the panel was unjustified for any reason; or

(j) cancel the certificate of registration, certificate of authorization, temporary licence, restricted licence or enrolment as an engineer-in-training or geoscientist-in-training of the investigated person.

PANEL MAY CONSIDER CAUTIONS

53(2) To assist the panel in making an order under this section, the panel may be advised of any formal, written caution previously issued to the member, holder of a certificate of authorization, temporary licensee, restricted licensee, engineer-in-training or geoscientist-in-training under clause 37(1)(e) and the circumstances under which it was issued.

CANCELLATION OR SUSPENSION OF CERTIFICATE, ETC.

53(3) When the certificate of registration, certificate of authorization, temporary licence or restricted licence is suspended or cancelled, the holder of such certificate of registration, certificate of authorization, temporary licence or restricted licence shall not practise engineering or geoscience during the period of the suspension or cancellation.

ANCILLARY ORDERS

53(4) The panel or the discipline committee may make any ancillary order that is appropriate or required in connection with an order mentioned in subsection (1) or may make any other order that it considers appropriate in the circumstances, including that

    (a) a further or new investigation be held into any matter; or

    (b) the discipline committee be convened to hear a complaint without an investigation.

CONTRAVENTION OF ORDER

53(5) If the discipline committee is satisfied that an investigated person has contravened an order under subsection (1), it may, without a further hearing, cancel or suspend the certificate of registration, certificate of authorization, temporary licence, restricted licence or enrolment as an engineer-in-training or geoscientist-in-training of the investigated person.

COSTS AND FINES

54(1) The panel or the discipline committee may, in addition to or instead of dealing with the conduct of an investigated person in accordance with section 53, order that the investigated person pay to the association

    (a) all or part of the costs of the investigation, hearing and appeal;

    (b) a fine not exceeding $25 000.00;

    (c) both the costs under clause (a) and the fine under clause (b);

within the time set by the order.

5. CODE OF ETHICS (ESTABLISHED UNDER SECTION 11 OF THE ACT)

INTRODUCTION

The Code of Ethics is a general guide to professional conduct. As such, it is intended to supplement, and is not intended to deny the existence of, other professional responsibilities equally important, though not specifically mentioned.

Each professional engineer is required by "The Engineering Profession Act" to subscribe to and follow this Code of Ethics. Non-compliance with this Code of Ethics, or failure to fulfill other professional responsibilities, by any member or licensee of this Association shall be considered inconsistent with honourable and dignified professional practice, and any such member or licensee may be deemed guilty of unprofessional conduct, and subject to discipline by Council under Section 24 of "The Engineering Profession Act."

Each professional engineer is encouraged to consult the Association for clarification as to appropriate conduct in specific cases.

FUNDAMENTAL PRINCIPLES

Honesty, justice and courtesy form a moral philosophy which, associated with mutual interest among all persons, constitutes the foundation of ethics. The professional engineer should recognize such a standard, not in passive observance, but as a set of dynamic principles guiding personal conduct and way of life. It is each professional engineer's duty to practise engineering according to this Code of Ethics.

As the keystone of professional conduct is integrity, the professional engineer should discharge all duties with fidelity to the public, employers, and clients, and with fairness and impartiality to all. As the obligation of the profession is to act in the public interest, the professional engineer should be ready to apply engineering knowledge for the benefit of humanity. The professional engineer should uphold the honour and dignity of the profession and avoid association with any enterprise of questionable character. All dealings with fellow professional engineers should be conducted in a spirit of fairness and tolerance.

FUNDAMENTAL CANONS
1. The professional engineer shall obey the laws of the land.
2. The professional engineer shall regard the physical, economic and environmental well-being of the public as the prime responsibility in all aspects of engineering work.
3. The professional engineer shall employ all reasonably attainable skill and knowledge to perform and satisfy the engineering needs of each client or employer in a professional manner.
4. The professional engineer shall uphold and enhance the honour, integrity and dignity of the engineering profession.
5. The professional engineer shall be fair to colleagues and shall support their professional development.

CANONS OF CONDUCT
1. The professional engineer shall obey the laws of the land.
Specifically, and without limiting the generality of this statement, the professional engineer shall:
1.a obey the laws of the land;
1.b be open and honest when engaged as an expert witness and give opinions conscientiously, only after an adequate study of the matter under review; and
1.c make responsible provision to comply with statutes, regulations, standards, codes, by-laws and rules applicable to all work.

2. The professional engineer shall regard the physical, economic and environmental well-being of the public as the prime responsibility in all aspects of engineering work.
Specifically, and without limiting the generality of this statement, the professional engineer shall:
2.a possess the training, ability and experience necessary to fulfill the requirements of any engineering work undertaken;
2.b guard against conditions that are dangerous or threatening to health, life, limb or property in engineering work for which he or she is professionally responsible, and, if aware of any adverse condition arising outside such bounds of professional responsibility, shall promptly call the condition to the attention of the authority having jurisdiction over the matter and, where possible, to the attention of the person having professional responsibility;
2.c ensure that designs and work for which he or she is professionally responsible are based on sound environmental principles;

2.d not knowingly associate with, or personally endorse, an enterprise of questionable character;

2.e not issue statements on engineering matters, or provide criticism or argument, or allow any publication of reports or any part of them, in a manner which might mislead;

2.f refrain from expressing an opinion publicly on an engineering matter without being qualified to do so, and without being aware of the pertinent facts;

2.g seal all plans and other engineering documents which "The Engineering Profession Act" stipulates shall be sealed, whether acting in the professionally responsible capacity of a consultant or an employee; and

2.h sign and seal only those plans and other engineering documents for which he or she has assumed professional responsibility and which he or she has prepared, or supervised the preparation of, or reviewed in detail and approved.

3. The professional engineer shall employ all reasonably attainable skill and knowledge to perform and satisfy the engineering needs of each client or employer in a professional manner.

Specifically, and without limiting the generality of this statement, the professional engineer shall:

3.a act as a faithful agent and trustee in professional matters for each client or employer;

3.b strive to maintain a high standard of competence by seeking opportunities to acquire knowledge of, and experience with, new techniques and developments;

3.c ensure that the extent of his or her professional engineering responsibility is understood by each client or employer before accepting an assignment;

3.d not disclose information concerning the lawful business affairs or technical processes of clients or employer without their consent;

3.e engage, or advise each client or employer to engage, and cooperate with other experts and specialists whenever the client's or employer's interests are best served by such service;

3.f inform each client or employer of the potential consequences which may result from deviations from an engineering judgement, should that judgement be disregarded or overruled by another authority;

3.g inform each client or employer of any interests, circumstances of business connections which the client or employer could deem as influencing his or her engineering judgement, or the quality of professional services, before accepting an assignment;

3.h not allow any interests, circumstances or business connections to inappropriately affect engineering decisions for which he or she is professionally responsible;

3.i not receive, directly or indirectly, any compensation, financial or otherwise, from other than a client or employer, for specifying the use of any materials, proprietary products, processes or systems for work for which he or she is professionally responsible, without the prior written authorization of the client or employer for the receipt of such compensation;

3.j not accept, directly or indirectly, any royalty or commission from any patented or protected article or process on which he or she holds any part of the rights, and specified for work undertaken on behalf of a client or employer, without the prior written authorization of the client or employer for the acceptance of such a royalty or commission;

3.k not accept compensation, financial or otherwise, from more than one interested party for services pertaining to the same work or works, without the prior written consent of all interested parties;

3.l not receive any gratuity from, or have any financial interest in, the bids of any business in work for which he or she is professionally responsible, without the prior written consent of the client or employer; and

3.m not accept an assignment outside of his or her regular employment which might interfere with regular duties, or make use of the employer's resources or facilities, without first notifying his or her employer, preferably with written confirmation.

4. The professional engineer shall uphold and enhance the honour, integrity and dignity of the engineering profession.

Specifically, and without limiting the generality of this statement, the professional engineer shall:

4.a cooperate in extending the effectiveness of the engineering profession by willingly participating in the exchange of information and experience with others in the profession;

4.b advertise only in a manner that serves the public interest by reporting accurate and factual information which does not exaggerate, mislead, or detract from the public image of the profession;

4.c endeavour to extend public knowledge of engineering, discourage the spreading of unfair or exaggerated statements regarding engineering, and strive to protect the engineering profession collectively and individually from misrepresentation and misunderstanding;

4.d present appropriate information to the Registrar of the Association if a professional colleague, or any other person or entity, is believed to be in violation of the Act, the By-laws or this Code of Ethics; and

4.e co-operate with the Association in the investigation of any complaint or other matter that is before the Association.

5. The professional engineer shall be fair to colleagues and shall support their professional development.

Specifically, and without limiting the generality of this statement, the professional engineer shall:

5.a take care that credit for engineering work is given to those to whom credit is properly due;

5.b support the standards of the engineering profession by upholding the principle that appropriate and adequate compensation for all those engaged in engineering work is in the public interest;

5.c endeavour to provide opportunities for the professional development and advancement of employees or subordinates in engineering and/or technical fields;

5.d encourage qualified engineering employees to become registered as professional engineers;

5.e not attempt to injure falsely or maliciously, directly or indirectly, the professional reputation, prospects, practise or employment of another engineer;

5.f notify an engineer, as soon as practicable, when giving an opinion on that engineer's work;

5.g not attempt to supplant another engineer in a particular employment if a definite commitment has been made toward the employment of that engineer; and

5.h not attempt to gain advantage over another engineer by offering a gratuity in order to secure professional engineering work.

NOTE

This Code of Ethics was adopted on April 13, 1992 following repeal on the same date of the Code adopted on February 28, 1968.

SOURCE: Engineering and Geoscientific Professions Act, *Statutes of Manitoba* Proposed 1998. Reproduced with permission of the Province of Manitoba. Users are reminded that this publication has no legislative sanction. It has been consolidated for convenience of reference only. The original Acts should be consulted for all purposes of interpreting and applying the law.

NEW BRUNSWICK

ENGINEERING PROFESSION ACT, 1986 (*STATUTES OF NEW BRUNSWICK* 1986, C. 88)

1. DEFINITION OF ENGINEERING (SECTION 2(1) OF THE ACT)

"engineering" means the application of scientific principles and knowledge to practical ends such as the investigation, design, construction, or operation of works and systems for the benefit of man;

"practice of engineering" means the provision of services for another as an employee or by contract; and such services shall include consultation, investigation, evaluation, planning, design, inspection, management, research and development of engineering works and systems.

2. MEMBERSHIP CRITERIA (SECTION 10(1) OF THE ACT)

10(1) Any applicant for registration who
    (a) is resident in New Brunswick;
    (b) is the age of legal majority;
    (c) is a graduate in engineering of an accredited university or other academic or technical institution recognized by the Council;
    (d) has fulfilled the requirements of approved engineering experience prescribed by the by-laws and satisfactory to the Council;
    (e) provides satisfactory evidence of good character; and
    (f) pays the fees prescribed by the by-laws;
upon approval of the Council, shall be entitled to become registered as a member of the Association.

3. DEFINITION OF PROFESSIONAL MISCONDUCT (SECTION 18 OF THE ACT)

18(8) A member, licensee, or the holder of a certificate of authorization may be found guilty of professional misconduct by the Discipline Committee if

(a) the member, licensee or holder of a certificate of authorization has been found guilty of an offence which, in the opinion of the Committee, is relevant to suitability to practice engineering; or

(b) the member, licensee, or holder of a certificate of authorization has been guilty, in the opinion of the Committee, of conduct relative to the practice of engineering which constitutes professional misconduct including, but not limited to, that defined in the by-laws.

18(9) The Discipline Committee may find a member or licensee incompetent if in its opinion,

(a) the member or licensee has displayed in his professional responsibility a lack of knowledge, skill, judgement, or disregard for the welfare of the public of a nature or to an extent that demonstrates the member or licensee is unfit to carry out the responsibilities of a professional engineer; or

(b) the member or licensee is suffering from a physical or mental condition or disorder of a nature and extent making it desirable in the interests of the public, or the member or licensee, that he no longer be permitted to engage in the practice of professional engineering, or that his practice of professional engineering be restricted.

4. DISCIPLINARY POWERS (SECTION 18 OF THE ACT)

18(10) When the Discipline Committee finds a member, licensee, or the holder of a certificate of authorization guilty of professional misconduct or incompetence it may, by order, do any one or more of the following

(a) revoke the right to practise engineering;

(b) suspend the right to practise engineering for a stated period, not exceeding twenty-four months;

(c) accept the undertaking of the member, licensee, or holder of a certificate of authorization to limit the professional work in the practice of engineering to the extent specified in the undertaking;

(d) impose terms, conditions or limitations on the membership, licence, or certificate of authorization, including but not limited to, the successful completion of a particular course or courses of study, as specified by the Committee;

(e) impose specific restrictions on the membership, licence, or certificate of authorization, including but not limited to,

(i) requiring the member, licensee or holder of the certificate of authorization to engage in the practice of engineering only under the personal supervision and direction of a member,

(ii) requiring the member or licensee to not alone engage in the practice of engineering,

(iii) requiring the member, licensee, or the holder of the certificate of authorization to submit to periodic inspections by the Committee, or its delegate, of documents, records and work of the member or the holder in connection with his practice of engineering,

(iv) requiring the member, licensee, or the holder of the certificate of authorization to report to the Registrar or to such committee of the Council as the Committee may name on such matters with respect to the member's or holder's practice of engineering for such period and times, and in such form, as the Committee may specify;

(f) reprimand, admonish or counsel the member, licensee, or the holder of the certificate of authorization, and if considered warranted, direct that the fact of the reprimand, admonishment or counselling be recorded on the register for a stated or unlimited period of time;

(g) revoke or suspend for a stated period of time the designation of the member or licensee as a specialist, consulting engineer or otherwise;

(h) impose such fine as the Committee considers appropriate, to a maximum of five thousand dollars, to be paid by the member, licensee, or the holder of the certificate of authorization;

(i) subject to subsection (11) in respect of orders of revocation or suspension, direct that the finding and the order of the Committee be published in detail or in summary and either with or without including the name of the member, licensee or the holder of the certificate of authorization in the official publication of the Association and in such other manner or medium as the Committee considers appropriate in the particular case;

(j) fix and impose costs of any investigation or procedures by the Professional Conduct Committee or the Committee to be paid by the member, licensee or the holder of the certificate of authorization to the Association;

(k) direct that the imposition of a penalty or order be suspended or postponed for such period, and upon such terms, or for such purpose, including but not limited to,

(i) the successful completion by the member or licensee of a particular course or courses of study,

(ii) the production to the Committee of evidence satisfactory to it that any physical or mental handicap in respect of which the penalty was imposed has been overcome.

18(11) The Discipline Committee shall cause an order of the Committee revoking or suspending a membership, licence, or certificate of authorization to be published, with or without the reasons therefore, in the official publication of the Association together with the name of the member or holder of the revoked or suspended licence or certificate of authorization.

5. CODE OF ETHICS (SECTION 2 OF PART B OF THE ACT)

1. FOREWORD

1 Honesty, justice and courtesy form a moral philosophy which, associated with mutual interest among people, constitute the foundation of ethics. Engineers should recognize such a standard, not in passive observance, but as a set of dynamic principles guiding their conduct and way of life. It is their duty to practise the profession according to the Act and the By-Laws including this Code of Ethics.

As the keystone of professional conduct is integrity, engineers shall discharge their duties with fidelity to the public, their employers and clients and with fairness and impartiality to all. It is their duty to interest themselves in public welfare and to be ready to apply their special knowledge for the benefit of humanity. They should uphold the honour and dignity of the profession and also avoid association with any enterprise of questionable character. In dealings with other engineers they should be fair and tolerant.

2. PROFESSIONAL LIFE

2.1 Engineers shall co-operate in extending the effectiveness of the engineering profession by interchanging information and experience with other engineers and students and by contributing to the work of engineering societies, schools and the scientific and engineering press.

2.2 Engineers shall encourage engineering employees to improve their knowledge and education.

2.3 Engineers shall strive to broaden their knowledge and experience by keeping abreast of new techniques and developments in their areas of endeavour.

2.4 Engineers shall report to the Association observed violations of the Engineering Profession Act or breaches of this Code of Ethics.

2.5 Engineers shall observe the rules of professional conduct which apply in the country in which they practise. If there are no such rules, they shall observe those established by this Code of Ethics.

2.6 Engineers shall not advertise their work or merit in a self-laudatory manner and will avoid all conduct or practice likely to discredit or do injury to the dignity and honour of the profession.

2.7 Engineers shall not advertise or represent themselves in an unprofessional manner by making misleading statements regarding their qualifications or experience.

3. RELATIONS WITH THE PUBLIC

3.1 Engineers shall endeavour to extend public knowledge of engineering and will discourage the spreading of untrue, unfair and exaggerated statements regarding engineering.

3.2 Engineers shall have due regard for the safety of life, health and welfare of the public and employees who may be affected by the work for which they are responsible.

3.3 Engineers, when giving testimony before a court, commission or other tribunal, shall express opinions only when they are founded on adequate knowledge and honest conviction.

3.4 Engineers shall not issue *ex parte* statements, criticisms or arguments on matters connected with public policy which are inspired or paid for by private interests, unless it is indicated on whose behalf the arguments are made.

3.5 Engineers shall refrain from expressing publicly opinions on engineering subjects unless they are informed of the facts relating thereto.

4. RELATIONS WITH CLIENTS AND EMPLOYERS

4.1 Engineers shall act in professional matters for clients or employers as faithful agents or trustees.

4.2 Engineers shall act with fairness and justice between the client or employer and the contractor when dealing with contracts.

4.3 Engineers shall make their status clear to clients or employers before undertaking engagements if they are called upon to decide on the use of inventions, apparatus, etc. in which they may have a financial interest.

4.4 Engineers shall ensure that the extent of their responsibility is fully understood by each client before accepting a commission.

4.5 Engineers shall undertake only such work as they are competent to perform by virtue of their training and experience.

4.6 Engineers shall not sign or seal drawings, specifications, plans, reports or other documents pertaining to engineering works or systems unless actually prepared or verified by them or under their direct supervision.

4.7 Engineers shall guard against conditions that are dangerous or threatening to life, limb or property. On work for which they are not responsible, they shall promptly call such conditions to the attention of those who are responsible.

4.8 Engineers shall present clearly the consequences to be expected if their engineering judgement is overruled.

4.9 Engineers shall engage, or advise clients or employers to engage and shall co-operate with other experts and specialists whenever the clients' or employers' interests are best served by such service.

4.10 Engineers shall not disclose information concerning the business affairs or technical processes of clients or employers without their consent.

4.11 Engineers shall not accept compensation, financial or otherwise, from more than one interested party for the same service, or for services pertaining to the same work, without the consent of all interested parties.

4.12 Engineers shall not accept commissions or allowances, directly or indirectly, from contractors or other parties dealing with clients or employers in connection with work for which they are responsible.

4.13 Engineers shall not be financially interested in bids as contractors on work for which they are engaged as engineers unless they have the consent of the client or employer.

4.14 Engineers shall promptly disclose to clients or employers any interest in a business which may compete with or affect the business of the client or employer. They shall not allow an interest in any business to affect their decisions regarding engineering work for which they are employed, or which they may be called upon to perform.

4.15 Engineers serving as members of any public body shall not act as vendors of goods or services to that body without disclosure of their interest.

4.16 Engineers shall respect the right of employees to voice their professional concerns in an appropriate manner about engineering works which they believe to be dangerous or threatening to life, limb, or property.

5. RELATIONS WITH ENGINEERS

5.1 Engineers shall endeavour to protect the engineering profession collectively and individually from misrepresentation and misunderstanding.

5.2 Engineers shall take care that credit for engineering work is given to those to whom credit is properly due.

5.3 When an engineer uses a design supplied by a client or by a consultant, the design remains the property of the client or consultant and should not be duplicated by the engineer without the express permission of the client or consultant.

5.4 Engineers shall uphold the principles of appropriate and adequate compensation for those engaged in engineering work, including those in subordinate capacity, as being in the public interest and maintaining the standards of the profession.

5.5 Engineers shall not accept financial or other considerations, including free engineering designs, from material or equipment suppliers in return for specifying their product.

5.6 Engineers shall not solicit or accept an engineering engagement that requires the engineer to give a preconceived conclusion or opinion.

5.7 Engineers shall endeavour to provide opportunity for the professional development and advancement of engineers in their employ.

5.8 Engineers shall not directly or indirectly injure the professional reputation, prospects or practice of other engineers. However, if they consider that an engineer is guilty of unethical, illegal or unfair practice, they shall present the information to the proper authority for action.

5.9 The engineer shall exercise due restraint in criticizing another engineer's work in public, recognizing that the engineering societies and the engineering press provide the proper forum for technical discussions and criticism.

5.10 An engineer shall not try to supplant another engineer in a particular employment after becoming aware that definite steps have been taken toward the other's employment.

5.11 An engineer shall not compete with another engineer by reducing normal fees after having been informed of the charges named by the other.

5.12 An engineer shall not use the advantages of a salaried position to compete unfairly with another engineer.

5.13 Engineers shall not provide a commission, a gift or other consideration in order to secure work.

5.14 An engineer shall not associate with any engineering or non-engineering enterprise that does not conform to ethical practices.

SOURCE: Engineering Profession Act, 1986, *Statutes of New Brunswick* 1986, c. 88. This publication contains reproductions of portions of the *Engineering Profession Act 1986*. The portions have been reproduced under licence from the Province of New Brunswick. Her Majesty the Queen in Right of the Province of New Brunswick, as represented by the Queen's Printer for the Province, retains title to all rights, including copyright, in the portions of the *Engineering Profession Act 1986* reproduced in the publication. The reproductions are provided for convenience only. For purposes of interpreting and applying the law, the *Engineering Profession Act 1986* as published by the Queen's Printer for the Province of New Brunswick should be consulted.

NEWFOUNDLAND

ENGINEERS AND GEOSCIENTISTS ACT (*STATUTES OF NEWFOUNDLAND* 1988, C. 48)

1. DEFINITION OF ENGINEERING (SECTION 2 OF THE ACT)

2. ...

(j) "practice of engineering" means reporting on, advising on, evaluating, designing, preparing plans and specifications for or directing the construction, technical inspection, maintenance or operation of a structure, work or process

(i) that is aimed at the discovery, except by the practice of geoscience, development or utilization of matter, materials or energy or is designed for the use and convenience of human beings, and

(ii) that requires in the reporting, advising, evaluating, designing, preparation or direction the professional application of the principles of mathematics, chemistry, physics or a related applied subject,

and includes providing educational instruction on the matters contained in this paragraph to a student at an educational institution but excludes practising as a natural scientist;

(k) "practice of geoscience" means reporting on, advising on, evaluating, interpreting, processing, geological and geophysical surveying, exploring, classifying reserves or examining activities related to the earth sciences or engineering-geology

(i) that is aimed at the discovery or development of oil, natural gas, coal, metallic or non-metallic minerals or precious stones, water or other natural resources or that is aimed at the investigation of geoscientific conditions, and

(ii) that requires in the reporting, advising, evaluating, interpreting, processing, geoscientific surveying, exploring, reserve classifying, or examining the professional application of mathematics, chemistry or physics through the application of the principles of geoscience,

and includes providing educational instruction on the matters contained in this paragraph to a student at an educational institution;

2. MEMBERSHIP CRITERIA (SECTIONS 20 TO 22 OF THE ACT AND SECTION 5 OF NEWFOUNDLAND REGULATION 209/89)

20. The Board of Examiners shall approve the registration as a professional engineer or geoscientist of a person who has applied to the Board and is eligible in accordance with this Act and the regulations to become a professional engineer or geoscientist.

21. The Board of Examiners shall approve the registration as a licensee of a person who has applied to the Board of Examiners and is eligible in accordance with this Act and the regulations to be registered to engage in the practice of engineering or of geoscience as a licensee.

22.

(1) The council shall approve the registration as a permit holder of a person, a partnership or other association of persons, or of a corporation which has applied to the Council and is eligible under this section and the regulations to be registered to engage in the practice of engineering or of geoscience as a permit holder.

(2) A person, a partnership or other association of persons or a corporation that applies to the Council is eligible to be registered as a permit holder entitled to engage in the practice of engineering or of geoscience, or both, if it satisfies the Council that it complies with this Act and the regulations.

(3) The Council may grant a permit under one or more of the following classifications to:

Class A: a person, a partnership or other association of persons or a corporation, which is primarily engaged in offering and providing professional services to the public; and

Class B: a person, a partnership or other association of persons or a corporation, one of whose customary functions is to engage in the practice of engineering or the practice of geoscience, but whose principal activity is not the offering and providing of professional services to the public.

REGULATION 5 (REGISTRATION AS PROFESSIONAL MEMBER)

5.

(1) In order to be eligible for registration as a professional member, an applicant shall, in addition to any other requirements of the Act, the regulations and the by-laws;

(a) have knowledge of the Act, the regulations and the by-laws, satisfactory to the Board of Examiners;

(b) have communication abilities in the English language satisfactory to the Board of Examiners or its Executive Committee as demonstrating the ability to competently practise engineering or geoscience in Newfoundland;

(c) have general knowledge of the practice of the professions demonstrated by successfully completing an examination in professional practice or by such other means as the Board of Examiners or its Executive Committee may require;

(d) be of good character and reputation; and

(e) satisfy all academic and practical experience requirements.

(2) In order to satisfy all academic and practical experience requirements an applicant shall have either:

(a) a Degree in Engineering or Geoscience from a university program approved by the Board of Examiners or its Executive Committee and at least two years of experience satisfactory to the Board of Examiners or its Executive Committee in the practice of engineering or geoscience subsequent to the conferral of the Degree; or

(b) academic standing equivalent to a Degree in Engineering or Geoscience demonstrated by successful completion of such confirmatory examinations as may be required by the Board of Examiners or its Executive Committee and at least two years of experience satisfactory to the Board of Examiners or its Executive Committee in the practice of engineering or geoscience subsequent to the attainment of such academic standing; or

(c) successfully completed such examinations as may be prescribed by the Board of Examiners or its Executive Committee and have a total of at least six years of experience satisfactory to the Board of Examiners or its Executive Committee in the practice of engineering or geoscience, one year of which must be obtained subsequent to successful completion of the prescribed examinations.

(3) In the event that an applicant for registration as a professional member seeks to transfer membership to this Association and is a registered member in good standing of an association of professional engineers or geoscientists acceptable to the Board of Examiners or its Executive Committee, the Board of Examiners or its Executive Committee may waive any or all of the requirements for registration provided in subsection (1) and (2).

3. DEFINITION OF PROFESSIONAL MISCONDUCT (SECTION 31 OF THE ACT)

31.

(1) Any conduct of a professional member, licensee, permit holder or member-in-training whether in Canada or elsewhere that in the opinion of the Discipline Committee or the Council

(a) is detrimental to the best interests of the public;

(b) contravenes a code of ethics of the professions as established under the regulations;

(c) harms or tends to harm the standing of the professionals generally; or

(d) displays a lack of knowledge or skill or judgement in the practice of the profession or in the carrying out of a duty or obligation undertaken in the practice of the profession,

constitutes either unskilled practice of the profession or unprofessional conduct, whichever the Discipline Committee finds.

(2) If an investigated person fails to comply with or contravenes this Act, the regulations or the by-laws, and the failure or contravention is, in the opinion of the Discipline Committee, of a serious nature, the failure or contravention may be found by the Discipline Committee to be unprofessional conduct whether or not it would be so found under subsection (1).

4. DISCIPLINARY POWERS (SECTION 44 OF THE ACT)

44. If the Discipline Committee finds that the conduct of the investigated person is unprofessional conduct or unskilled practice of the profession or both, the Discipline Committee may

(a) reprimand the investigated person;

(b) suspend the registration of the investigated person for a specified period;

(c) suspend the registration of the investigated person either generally or from any field of practice until

(i) the person has completed a specified course of studies or obtained supervised practical experience, or

(ii) the Discipline Committee is satisfied as to the competence of the investigated person generally or in a specified field of practice;

(d) accept in place of a suspension the investigated person's undertaking to limit his or her practice;

(e) impose conditions on the investigated person's practice of the profession generally or in a field of the practice, including the conditions that the person

(i) practise under supervision,

(ii) not engage in sole practice,

(iii) permit periodic inspections by a person authorized by the Discipline Committee, or

(iv) report to the Discipline Committee on specific matters;

(f) direct the investigated person to pass a course of study or satisfy the Discipline Committee of his or her practical competence;

(g) require the investigated person to take counselling or to obtain the assistance that in the opinion of the Discipline Committee is appropriate;

(h) direct the investigated person to waive, reduce or prepay a fee for services rendered by the investigated person that in the opinion of the Discipline Committee were improperly rendered;

(i) cancel the registration of the investigated person; or

(j) make any other order that it considers appropriate in the circumstances.

5. CODE OF ETHICS (NEWFOUNDLAND REGULATION 209/89, SCHEDULE A, ESTABLISHED IN ACCORDANCE WITH SECTION 17 OF THE ACT)

1. A professional engineer or geoscientist shall recognize that professional ethics are founded upon integrity, competence and devotion to service and to the advancement of human welfare. This concept shall guide the conduct of the professional engineer or geoscientist at all times.

DUTIES OF THE PROFESSIONAL ENGINEER OR GEOSCIENTIST TO THE PUBLIC

A professional engineer or geoscientist shall:

2. have proper regard in all his or her work for the safety, health and welfare of the public;

3. endeavour to extend public understanding of engineering and geoscience and their role in society.

4. where his or her professional knowledge may benefit the public, seek opportunities to serve in public affairs;

5. not be associated with enterprises contrary to the public interest;

6. undertake only such work as he or she is competent to perform by virtue of his or her education, training and experience;

7. sign and seal only such plans, documents or work as he or she has personally prepared or which have been prepared or carried out under his or her direct professional supervision;

8. express opinions on engineering or geoscientific matters only on the basis of adequate knowledge and honest conviction;

9. have proper regard in all his or her work for the well being and integrity of the Environment.

DUTIES OF THE PROFESSIONAL ENGINEER OR GEOSCIENTIST TO CLIENT OR EMPLOYER

A professional engineer or geoscientist shall:

10. act for his or her client or employer as a faithful agent or trustee;

11. not accept remuneration for services rendered other than from his or her client or employer;

12. not disclose confidential information without the consent of his or her client or employer;

13. not undertake any assignment which may create a conflict of interest with his or her client or employer without full knowledge of the client or employer;

14. present clearly to his or her clients or employers the consequences to be expected if his or her professional judgement is overruled by other authorities in matters pertaining to work for which he or she is professionally responsible.

DUTIES OF THE PROFESSIONAL ENGINEER OR GEOSCIENTIST TO THE PROFESSION

A professional engineer or geoscientist shall:

15. endeavour at all times to improve the competence, dignity and reputation of his or her profession;

16. conduct himself or herself towards other professional engineers and geoscientists with fairness and good faith;

17. not advertise his or her professional services in self-laudatory language or in any other manner derogatory to the dignity of the profession;

18. not attempt to supplant another engineer or geoscientist in an engagement after definite steps have been taken toward the other's employment.

19. when in salaried position, engage in a private practice and offer or provide professional services to the public only with the consent of his or her employer and in compliance with all requirements of such practice;

20. not exert undue influence or offer, solicit or accept compensation for the purpose of affecting negotiations for a engagement;

21. not invite or submit proposals under conditions that constitute only price competition for professional services;

22. advise the Council of any practice by another member of the profession which he or she believes to be contrary to the Code of Ethics.

Note: Item 9 was approved by a membership vote but has not been officially incorporated into the regulations.

SOURCE: Engineers and Geoscientists Act, *Statutes of Newfoundland* 1988, c. 48. Reproduced with permission of the Queen's Printer for the Province of Newfoundland.

NORTHWEST TERRITORIES

ENGINEERING, GEOLOGICAL AND GEOPHYSICAL PROFESSIONS ACT (*REVISED STATUTES OF NORTHWEST TERRITORIES* 1988, C. E-6, AS AMENDED BY STATUTES OF NORTHWEST TERRITORIES 1991–92, C. 21)

1. DEFINITION OF ENGINEERING (SECTION 1 OF THE ACT)

"professional engineer" means a member or licensee qualified to practise professional engineering;

"professional engineering" means

(a) the application of scientific principles and knowledge to practical ends as in the investigation, inspection, design, construction or operation of works, systems, structures or processes for the benefit of humans, and without restricting the generality of paragraph (a), includes

(b) the provision or performance of services including consultation, investigation, evaluation, reporting, planning, design, technical inspection, preparation of plans and specifications for, surveying for, supervision, management, research, co-ordination of design or construction, or both, and directing the construction, technical inspection, maintenance or operation of works, systems, structures or processes, but does not include

(c) the execution or supervision of the construction, technical inspection, maintenance or operation of works, systems, structures, or processes in the capacity of contractor, superintendent, work supervisor or supervisor or inspector or in a similar capacity, when the works, systems, structures or processes have been designed by and the execution or supervision is being carried out under the direct supervision of a professional engineer, or

(d) any of the services mentioned in paragraph (b) when they are carried out by a technician, technologist or engineer-in-training under the direct supervision of a professional engineer;

"professional geologist" means a member or licensee qualified to practise professional geology;

"professional geology" means reporting, advising, evaluating, interpreting, geological surveying, sampling or examining related to any activity

(a) that is aimed at the discovery or development of oil, natural gas, coal, metallic or non-metallic minerals, precious stones or other natural resources or water, or that is aimed at the investigation of geological conditions, and

(b) that requires the professional application of the principles of the geological sciences or any related subject including mineralogy, paleontology, economic geology, structural geology, stratigraphy, sedimentation, petrology, geomorphology, photogeology and similar fields,

but does not include any of the activities mentioned in paragraphs (a) and (b) that are normally associated with the business of prospecting when carried on by a prospector or any of the activities mentioned in paragraphs (a) and (b) carried on by a technician, technologist or geologist-in-training under the direct supervision of a professional geologist;

"professional geophysicist" means a member or licensee qualified to practise professional geophysics;

"professional geophysics" means reporting, advising, evaluating, interpreting or geophysical surveying related to any activity

(a) that is aimed at the discovery or development of oil, natural gas, coal, metallic or non-metallic minerals, precious stones or other natural resources or water, or that is aimed at the investigation and measurement of the physical properties of the earth, and

(b) that requires the professional application of the principles of one or more of the subjects of physics, mathematics or any related subject including elastic wave propagation, gravitational, magnetic and electrical fields, natural radioactivity, and similar fields, but does not include the routine maintenance or operation of geophysical instruments, or any of the activities mentioned in paragraphs (a) and (b) that are normally associated with the business of prospecting when carried on by a prospector or by a technician, technologist or geophysicist-in-training under the direct supervision of a professional geophysicist;

2. MEMBERSHIP CRITERIA (SECTIONS 11 AND 13 OF THE ACT)

11.

(1) This section does not apply to a person referred to in clause 13(1)(c)(iii)(B) or 13(2)(c)(ii)(B).

(2) Subject to this section, an applicant shall, before being approved for registration,

(a) be a graduate of a university program in engineering, geology or geophysics that is approved by the Board of Examiners; and

(b) have at least two years experience satisfactory to the Board, after university graduation, in the practice of engineering, geology or geophysics.

(3) Where an applicant for registration is a graduate of a university program in engineering, geology or geophysics that is not approved by the Board of Examiners, the applicant shall, before being approved for registration,

(a) pass confirmatory examinations set by the Board; and

(b) have at least two years experience satisfactory to the Board, after university graduation, in the practice of engineering, geology or geophysics.

(4) Where an applicant for registration has not graduated from a university program in engineering, geology or geophysics, the applicant shall, before being approved for registration,

(a) pass examinations set by the Board of Examiners; and

(b) have at least six years experience satisfactory to the Board in the practice of engineering, geology or geophysics.

. . .

13.

(1) The council shall register as a member every person, other than a licensee, who

(a) applies in accordance with the by-laws;

(b) pays the required fees; and

(c) satisfies the council that he or she

(i) is a resident of the Territories,

(ii) is a Canadian citizen, a permanent resident or is otherwise lawfully permitted to work in Canada, and

(iii) either

(A) satisfies the requirements set out in section 11, or

(B) is a member of an association or corporation in any other jurisdiction having requirements for registration considered by the council to be equivalent to those of the Association.

(2) The council shall register as a licensee every person, other than a member, who

(a) applies in accordance with the by-laws;

(b) pays the required fees; and

(c) satisfies the council that he or she

(i) is a Canadian citizen, a permanent resident or is otherwise lawfully permitted to work in Canada, and

(ii) either

(A) satisfies the requirements set out in section 11, or

(B) is a member or a licensee of an association or corporation in any other jurisdiction having requirements for registration considered by the council to be equivalent to those of the Association.

(3) Notwithstanding anything in this Act, the council may require an applicant for registration to write the professional practice examinations that the council considers necessary, and the council may refuse to register as a member or licensee any person who fails the examinations.

(4) Where an application for registration is rejected, the council shall provide the unsuccessful applicant with a written notice of that fact stating the reason for the rejection of the application.

(5) The Board of Examiners shall determine whether a successful applicant is to be designated as a professional engineer, a professional geologist or a professional geophysicist or any two or all of these, and on registration as a member or a licensee he or she shall be designated accordingly.

(6) Following registration, the council may, on application, and with the consent of the Board of Examiners, alter the designated profession in respect of which a member or licensee has been registered on the basis of extensive experience in a new field of registration.

3. DEFINITION OF PROFESSIONAL MISCONDUCT (SECTION 25 OF THE ACT)

25.

(1) The question of whether a person is guilty of conduct unbecoming a registrant or permit holder shall be determined by the council or, on appeal, by the Supreme Court.

(2) For the purposes of this Act, conduct that, in the judgement of the council, or the Supreme Court on appeal, constitutes professional misconduct, gross negligence, incompetence or misrepresentation or that is contrary to the best interests of the public or the profession of engineering, geology or geophysics shall be deemed to be conduct unbecoming a registrant or permit holder.

4. DISCIPLINARY POWERS (SECTIONS 35, 36 OF THE ACT)

35.

(1) Where on completion of a hearing, the council forms the opinion that the conduct under investigation is not conduct unbecoming a registrant of the relevant classification or a permit holder, the council shall

(a) dismiss the complaint or take no further action on the matter; and

(b) notify the complainant, if any, and the person whose conduct has been the subject-matter of the hearing.

(2) Where, on completion of a hearing, the council

(a) forms the opinion that the conduct under investigation is conduct unbecoming a registrant of the relevant classification or a permit holder, and

(b) considers that the conduct is not of such gravity or importance as to warrant suspension of the registrant or permit, the striking of the per-

son's name from the register or the revocation of the permit,
the council may reprimand the registrant or permit holder.

(3) Where, on completion of a hearing, the council

(a) forms the opinion that the conduct under investigation is conduct unbecoming a registrant of the relevant classification, or permit holder, and

(b) considers the conduct to be sufficiently grave to merit suspension of the registrant or permit, the striking of the person's name from the register, or the revocation of the permit,

the council may order that

(c) the name of a member or licensee be struck from the register of the Association;

(d) a member or licensee be suspended from practising professional engineering, professional geology or professional geophysics for such period as the council considers proper;

(e) the permit of a permit holder be suspended for such period as the council considers proper or revoked; or

(f) the complaint or matter, if it relates to a person-in-training or student, be disposed of in a manner set out in the by-laws.

(4) The council may order that a person whose name is to be struck from the register or whose permit is to be revoked under this section shall

(a) pass examinations set by the Board of Examiners, and

(b) pass a particular course of study or obtain experience generally or in a field of practice satisfactory to the Board of Examiners,

before the council shall reinstate that person as a member, licensee or permit holder.

(5) Notwithstanding anything in this Act, the council shall not register as a member or licensee, nor grant a permit to, any person whose name has been struck from the register or whose permit has been revoked under this section, unless the council is satisfied that the person has complied with any order made under subsection (4).

36.

The council, in addition to a reprimand, or in addition to or in the place of an order under subsection 35(3), may order

(a) the registrant or permit holder to pay a fine not exceeding $5000 to the Association within the time fixed by the order;

(b) the registrant or permit holder to pay to the Association the costs of the hearing in an amount and within a time fixed by the council; and

(c) that the registrant or the permit be suspended in default of payment of a fine or costs ordered to be paid until the fine or costs are paid.

5. CODE OF ETHICS (PART II OF THE BY-LAWS, ESTABLISHED UNDER SECTION 5 OF THE ACT)

PREAMBLE

Professional Engineers, Geologists, and Geophysicists shall recognize that professional ethics are founded upon integrity, competence, devotion to service, and to advancement of human welfare. These concepts shall guide their con-

duct at all times. In this way, each professional's actions will enhance the dignity and status of the professions.

Professional Engineers, Geologists, and Geophysicists, through their practice, are
charged with extending public understanding of the professions and should serve in
public affairs when their professional knowledge may be of benefit to the public.

Professional Engineers, Geologists, and Geophysicists will build their reputations on the basis of merit of their services, and shall not compete unfairly with
others or compete primarily on the basis of fees without due consideration of
other factors.

Professional Engineers, Geologists, and Geophysicists will maintain a special
obligation to demonstrate understanding, professionalism, and technical expertise to members-in-training under their supervision.

RULES OF CONDUCT

Professional Engineers, Geologists and Geophysicists:

1. shall have proper regard in all their work for the safety and welfare of all
persons and for the physical environment affected by such work.
2. shall undertake only such work as they are competent to perform by virtue
of training and expertise, and shall express opinions on engineering, geological and geophysical matters only on the basis of adequate knowledge and
honest conviction.
3. shall sign and seal only reports, plans or documents which they have prepared or which have been prepared under their direct supervision and control.
4. shall act for their clients or employers as a faithful agent or trustee; always
acting independently and with fairness and justice to all parties.
5. shall not engage in activities nor accept remuneration for services rendered, which may create a conflict of interest with their clients or employers,
without the knowledge and consent of their clients or employers.
6. shall not disclose confidential information without the consent of their
clients or employers, unless the withholding of such information is deemed
contrary to the safety of the public.
7. shall present clearly to their clients or employers the consequences to be
expected if their professional judgement is overruled by other authorities in
matters pertaining to work for which they are professionally responsible.
8. shall not offer or accept covert payment for the purpose of securing an engineering, geological, or geophysical assignment.
9. shall represent their qualifications and competence, or advertise professional services offered, only through factual representation without exaggeration.
10. shall conduct themselves toward other professional engineers, geologists,
geophysicists, employees and others with fairness and good faith.
11. shall advise the Executive Director of any practice by another member of
the Association, which they believe to be contrary to this code of ethics.

SOURCE: Engineering, Geological and Geophysical Professions Act, *Revised
Statutes of Northwest Territories* 1988, c. E-6, as amended by *Statutes of Northwest
Territories* 1991–92, c. 21 and *Statutes of Northwest Territories* 1994, c. 7. Reproduced with permission of the Territorial Printer, Yellowknife, Northwest Territories.

NOVA SCOTIA

ENGINEERING PROFESSION ACT (*REVISED STATUTES OF NOVA SCOTIA* 1989, C. 148)

1. DEFINITION OF ENGINEERING (SECTION 2 OF THE ACT)

2.

(g) "Engineering" means the science and art of designing, investigating, supervising the construction, maintenance or operation of, making specifications, inventories or appraisals of, and consultations or reports on machinery, structure, works, plants, mines, mineral deposits, processes, transportation systems, transmission systems and communication systems or any other part thereof.

2. MEMBERSHIP CRITERIA (SECTION 7 OF THE ACT)

7.

1. Any person shall be entitled to be registered as a member of the Association upon filing with the Registrar satisfactory proof that such person is a resident of the Province, has tendered the fees and dues prescribed by the By-laws, and

(a) has obtained a degree in engineering from a school, college or university, which degree is approved by the Council, and has had two years experience in engineering;

(b) has obtained a degree in science, other than engineering, from a school, college or university, which degree is approved by the Council, and has had four years experience in engineering;

(c) is a registered member of an association of engineers, which association in the opinion of the Council is similarly constituted and has similar membership requirements to this Association, and furnishes the Registrar with a certificate of membership in good standing in such other Association;

(d) has passed the examinations prescribed by the Council and has had sufficient number of years of experience in engineering to qualify such person in the opinion of the Council to practise professional engineering; or

(e) has had in the opinion of the Council outstanding experience in engineering.

2. Every person, who in the opinion of the Council, expressed by a resolution thereof, has complied with subsection (1), shall be registered as a member.

3. DEFINITION OF PROFESSIONAL MISCONDUCT (SECTION 24 OF THE BY-LAWS)

24. Members, persons licensed to practise and engineers-in-training shall conduct themselves in accordance with the Code of Ethics appended hereto, and

without restricting the meaning of unprofessional conduct, any breach of the Code of Ethics shall be deemed to be a form of unprofessional conduct.

4. DISCIPLINARY POWERS (SECTION 17 OF THE ACT)

17(1) The Council may, in the manner provided by the by-laws, reprimand and censure any member, person licensed to practise or engineer-in-training, or suspend or cancel the certificate of registration of any member or the license to practise of any person or the enrolment of any engineer-in-training, who is guilty of unprofessional conduct, negligence or misconduct in the execution of the duties of his office, or of any breach of this Act or of the by-laws, or who has been convicted of a criminal offence by any court of competent jurisdiction.

5. CODE OF ETHICS (APPENDIX TO THE BY-LAWS)

GENERAL
1. A Professional Engineer shall recognize that professional ethics are founded upon integrity, competence and devotion to service and to the advancement of public welfare. This concept shall guide his conduct at all times.

RELATIONS WITH THE PUBLIC
A Professional Engineer:
2. shall regard his duty to public welfare as paramount.
3. shall endeavour to enhance the public regard for his profession by extending the public knowledge thereof.
4. shall undertake only such work as he is competent to perform by virtue of his training and experience.
5. shall sign and seal only such plans, documents or work as he himself has prepared or carried out or as have been prepared or carried out under his direct professional supervision.
6. shall express opinions on engineering matters only on the basis of adequate knowledge, competence and honest conviction.
7. shall express opinions or make statements on engineering projects of public interest that are inspired or paid for by private interest only if he clearly discloses on whose behalf he is giving the opinion or making the statements.
8. shall not be associated with enterprises contrary to public interest or sponsored by persons of questionable integrity.

RELATIONS WITH CLIENTS AND EMPLOYERS
A Professional Engineer:
9. shall act for his client or employer as a faithful agent or trustee and shall act with fairness and justice between his client or employer and the contractor when contracts are involved.
10. shall not accept compensation, financial or otherwise, from more than one interested party for the same service, or for service pertaining to the same work, without the consent of all interested parties.
11. shall not disclose confidential information without the consent of his client or employer.

12. shall not be financially interested in bids on competitive work for which he is employed as an engineer unless he has the consent of his client or employer.
13. shall not undertake any assignment which may create a conflict of interest with his client or employer without the full knowledge of the client or employer.
14. shall present clearly to his clients or employers the consequences to be expected if his professional judgement is overruled by other authorities in matters pertaining to work for which he is professionally responsible.
15. shall refrain from unprofessional conduct or from actions which he considers to be contrary to the public good, even if expected or directed by his employer or client, to act in such a manner.
16. shall not expect or direct an employee or subordinate to act in a manner that he or the employee or subordinate considers to be unprofessional or contrary to the public good.
17. shall guard against conditions that are dangerous or threatening to life, limb or property on work for which he is responsible, or if he is not responsible, will promptly call such conditions to the attention of those who are responsible.

RELATIONS WITH THE PROFESSION
A Professional Engineer:
18. shall co-operate in extending the effectiveness of the engineering profession by interchanging information and experience with other engineers and students and by contributing to the work of engineering societies, schools and the scientific and engineering press.
19. shall endeavour at all times to improve the competence, and thus the dignity and prestige of his profession.
20. shall not advertise his work or merit in a self-laudatory manner and shall avoid all conduct or practice likely to discredit or do injury to the dignity and honor of his profession.
21. shall not attempt to supplant another engineer in an engagement after a definite commitment has been made toward the other's employment.
22. shall not exert undue influence or offer, solicit or accept compensation for the purpose of affecting negotiations for an engagement.
23. shall not compete with another engineer on the basis of charges for work by underbidding, through reducing his normal fees after having been informed of the charges named by the other.
24. shall not use the advantages of a salaried position to compete unfairly with another engineer.
25. shall advise the Discipline Committee of any practice by another member of his profession which he believes to be contrary to this Code of Ethics.
26. shall take care that credit for engineering work is given to those to whom credit is properly due.
27. shall uphold the principle of appropriate and adequate compensation for those engaged in engineering work including those in subordinate capacities as being in the public interest, and maintaining the standards of the profession.
28. shall endeavour to provide opportunity for the professional development and advancement of engineers in his employ.

SOURCE: Engineering Profession Act, *Revised Statutes of Nova Scotia* 1989, c. 148. Reproduced with permission of the Province of Nova Scotia.

ONTARIO

PROFESSIONAL ENGINEERS ACT, 1990 (*STATUTES OF ONTARIO*, C. P.28, AS AMENDED)

1. DEFINITION OF ENGINEERING (SECTION 1 OF THE ACT)

"practice of professional engineering" means any act of designing, composing, evaluating, advising, reporting, directing or supervising wherein the safeguarding of life, health, property or the public welfare is concerned and that requires the application of engineering principles, but does not include practising as a natural scientist; ("exercice de la profession d'ingénieur")

"professional engineer" means a person who holds a licence or a temporary licence; ("ingénieur")

2. MEMBERSHIP CRITERIA (SECTION 14 OF THE ACT)

14.

(1) The Registrar shall issue a licence to a natural person who applies therefor in accordance with the regulations and,

(a) is a citizen of Canada or has the status of a permanent resident of Canada;

(b) is not less than eighteen years of age;

(c) has complied with the academic requirements specified in the regulations for the issuance of the licence and has passed such examinations as the Council has set or approved in accordance with the regulations or is exempted therefrom by the Council;

(d) has complied with the experience requirements specified in the regulations for the issuance of the licence; and

(e) is of good character.

3. DEFINITION OF PROFESSIONAL MISCONDUCT (SECTION 72 OF ONTARIO REGULATION 941/90)

72.

(1) In this section, "negligence" means an act or an omission in the carrying out of the work of a practitioner that constitutes a failure to maintain the standards that a reasonable and prudent practitioner would maintain in the circumstances.

(2) For the purposes of the Act and this Regulation, "professional misconduct" means,

(a) negligence;

(b) failure to make reasonable provision for the safeguarding of life, health or property of a person who may be affected by the work for which the practitioner is responsible;

(c) failure to act to correct or report a situation that the practitioner believes may endanger the safety or the welfare of the public;

(d) failure to make responsible provision for complying with applicable statutes, regulations, standards, codes, by-laws and rules in connection with work being undertaken by or under the responsibility of the practitioner;

(e) signing or sealing a final drawing, specification, plan, report or other document not actually prepared or checked by the practitioner;

(f) failure of a practitioner to present clearly to his employer the consequences to be expected from a deviation proposed in work, if the professional engineering judgement of the practitioner is overruled by non-technical authority in cases where the practitioner is responsible for the technical adequacy of professional engineering work;

(g) breach of the Act or regulations, other than an action that is solely a breach of the code of ethics;

(h) undertaking work the practitioner is not competent to perform by virtue of his training and experience;

(i) failure to make prompt, voluntary and complete disclosure of an interest, direct or indirect, that might in any way be, or be construed as, prejudicial to the professional judgement of the practitioner in rendering service to the public, to an employer or to a client, and in particular without limiting the generality of the foregoing, carrying out any of the following acts without making such a prior disclosure:

    1. Accepting compensation in any form for a particular service from more than one party.

    2. Submitting a tender or acting as a contractor in respect of work upon which the practitioner may be performing as a professional engineer.

    3. Participating in the supply of material or equipment to be used by the employer or client of the practitioner.

    4. Contracting in the practitioner's own right to perform professional engineering services for other than the practioner's employer.

    5. Expressing opinions or making statements concerning matters within the practice of professional engineering of public interest where the opinions or statements are inspired or paid for by other interests;

(j) conduct or an act relevant to the practice of professional engineering that, having regard to all the circumstances, would reasonably be regarded by the engineering profession as disgraceful, dishonourable or unprofessional;

(k) failure by a practitioner to abide by the terms, conditions or limitations of the practitioner's licence, limited licence, temporary licence or certificate;

(l) failure to supply documents or information requested by an investigator acting under section 34 of the Act;

(m) permitting, counselling or assisting a person who is not a practitioner to engage in the practice of professional engineering except as provided for in the Act or the regulations

O. Reg. 538/84, s. 86.

4. DISCIPLINARY POWERS (SECTION 28 OF THE ACT)

28.

(1) The Discipline Committee shall,

(a) when so directed by the Council, the Executive Committee or the Complaints Committee, hear and determine allegations of professional misconduct or incompetence against a member of the Association or a holder of a certificate of authorization, a temporary licence or a limited licence;

(b) hear and determine matters referred to it under section 24, 27 or 37; and

(c) perform such other duties as are assigned to it by the Council.

(2) A member of the Association or a holder of a certificate of authorization, a temporary licence or a limited licence may be found guilty of professional misconduct by the Committee if,

(a) the member or holder has been found guilty of an offence relevant to suitability to practise, upon proof of such conviction;

(b) the member or holder has been guilty in the opinion of the Discipline Committee of professional misconduct as defined in the regulations.

(3) The Discipline Committee may find a member of the Association or a holder of a temporary licence or a limited licence to be incompetent if in its opinion,

(a) the member or holder has displayed in his professional responsibilities a lack of knowledge, skill or judgement or disregard for the welfare of the public of a nature or to an extent that demonstrates the member or holder is unfit to carry out the responsibilities of a professional engineer; or

(b) the member or holder is suffering from a physical or mental condition or disorder of a nature and extent making it desirable in the interests of the public or the member or holder that the member or holder no longer be permitted to engage in the practice of professional engineering or that his practice of professional engineering be restricted.

(4) Where the Discipline Committee finds a member of the Association or a holder of a certificate of authorization, a temporary licence or a limited licence guilty of professional misconduct or to be incompetent it may, by order,

(a) revoke the licence of the member or the certificate of authorization, temporary licence or limited licence of the holder;

(b) suspend the licence of the member or the certificate of authorization, temporary licence or limited licence of the holder for a stated period, not exceeding twenty-four months;

(c) accept the undertaking of the member or holder to limit the professional work of the member or holder in the practice of professional engineering to the extent specified in the undertaking;

(d) impose terms, conditions or limitations on the licence or certificate of authorization, temporary licence or limited licence, of the member or holder, including but not limited to the successful completion of a particular course or courses of study, as are specified by the Discipline Committee;

(e) impose specific restrictions on the licence or certificate of authorization, temporary licence or limited licence, including but not limited to,

(i) requiring the member or the holder of the certificate of authorization, temporary licence or limited licence to engage in the practice of professional engineering only under the personal supervision and direction of a member,

(ii) requiring the member to not alone engage in the practice of professional engineering,

(iii) requiring the member or the holder of the certificate of authorization, temporary licence or limited licence to accept periodic inspections by the Committee or its delegate of documents and records in the possession or under the control of the member or the holder in connection with the practice of professional engineering,

(iv) requiring the member or the holder of the certificate of authorization, temporary licence or limited licence to report to the Registrar or to such committee of the Council as the Discipline Committee may specify on such matters in respect of the member's or holder's practice for such period of time, at such times and in such form, as the Discipline Committee may specify;

(f) require that the member or the holder of the certificate of authorization, temporary licence or limited licence be reprimanded, admonished or counselled and, if considered warranted, direct that the fact of the reprimand, admonishment or counselling be recorded on the register for a stated or unlimited period of time;

(g) revoke or suspend for a stated period of time the designation of the member or holder by the Association as a specialist, consulting engineer or otherwise;

(h) impose such fine as the Discipline Committee considers appropriate, to a maximum of $5000, to be paid by the member of the Association or the holder of the certificate of authorization, temporary licence or limited licence to the Treasurer of Ontario for payment into the Consolidated Revenue Fund;

(i) subject to subsection (5) in respect of orders of revocation or suspension, direct that the finding and the order of the Discipline Committee be published in detail or in summary and either with or without including the name of the member or holder in the official publication of the Association and in such other manner or medium as the Discipline Committee considers appropriate in the particular case;

(j) fix and impose costs to be paid by the member or the holder to the Association;

(k) direct that the imposition of a penalty be suspended or postponed for such period and upon such terms or for such purpose, including but not limited to,

(i) the successful completion by the member or the holder of the temporary licence or the limited licence of a particular course or courses of study,

(ii) the production to the Discipline Committee of evidence satisfactory to it that any physical or mental handicap in respect of which the penalty was imposed has been overcome,
or any combination of them.

(5) The Discipline Committee shall cause an order of the Committee revoking or suspending a licence or certificate of authorization, temporary licence or limited licence to be published, with or without the reasons therefor, in the official publication of the Association together with the name of the member or holder of the revoked or suspended licence or certificate of authorization, temporary licence or limited licence.

5. CODE OF ETHICS (SECTION 77, ONTARIO REGULATION 941/90, ESTABLISHED UNDER SECTION 7 OF THE ACT)

77. The following is the Code of Ethics of the Association:

(1) It is the duty of a practitioner to the public, to the practitioner's employer, to the practitioner's clients, to other members of the practitioner's profession, and to the practitioner to act at all times with,

(i) fairness and loyalty to the practitioner's associates, employers, clients, subordinates and employees,

(ii) fidelity to public needs, and

(iii) devotion to high ideals of personal honour and professional integrity.

(2) A practitioner shall:

(i) regard the practitioner's duty to public welfare as paramount,

(ii) endeavour at all times to enhance the public regard for the practitioner's profession by extending the public knowledge thereof and discouraging untrue, unfair or exaggerated statements with respect to professional engineering,

(iii) not express publicly, or while the practitioner is serving as a witness before a court, commission or other tribunal, opinions on professional engineering matters that are not founded on adequate knowledge and honest conviction,

(iv) endeavour to keep the practitioner's licence, temporary licence, limited licence or certificate of authorization, as the case may be, permanently displayed in the practitioner's place of business.

(3) A practitioner shall act in professional engineering matters for each employer as a faithful agent or trustee and shall regard as confidential information obtained by the practitioner as to the business affairs, technical methods or processes of an employer and avoid or disclose a conflict of interest that might influence the practitioner's actions or judgement.

(4) A practitioner must disclose immediately to the practitioner's client any interest, direct or indirect, that might be construed as prejudicial in any way to the professional judgement of the practitioner in rendering service to the client.

(5) A practitioner who is an employee-engineer and is contracting in the practitioner's own name to perform professional engineering work for other than the practitioner's employer, must provide the practitioner's client with a written statement of the nature of the practitioner's status as

an employee and the attendant limitations on the practitioner's services to the client, must satisfy the practitioner that the work will not conflict with the practitioner's duty to the practitioner's employer, and must inform the practitioner's employer of the work.

(6) A practitioner must co-operate in working with other professionals engaged on a project.

(7) A practitioner shall,

(i) act towards other practitioners with courtesy and good faith,

(ii) not accept an engagement to review the work of another practitioner for the same employer except with the knowledge of the other practitioner or except where the connection of the other practitioner with the work has been terminated,

(iii) not maliciously injure the reputation or business of another practitioner,

(iv) not attempt to gain an advantage over other practitioners by paying or accepting a commission in securing professional engineering work, and

(v) give proper credit for engineering work, uphold the principle of adequate compensation for engineering work, provide opportunity for professional development and advancement of the practitioner's associates and subordinates, and extend the effectiveness of the profession through the interchange of engineering information and experience.

(8) A practitioner shall maintain the honour and integrity of the practitioner's profession and without fear or favour expose before the proper tribunals unprofessional, dishonest or unethical conduct by any other practitioner.

O. Reg. 538/84, s. 91.

SOURCE: Professional Engineers Act, 1990, *Revised Statutes of Ontario* 1990, c. P.28, as amended. © Queen's Printer for Ontario, 1998. Reproduced with permission.

PRINCE EDWARD ISLAND

ENGINEERING PROFESSION ACT (*STATUTES OF PRINCE EDWARD ISLAND* 1990, C. 12)

1. DEFINITION OF ENGINEERING (SECTION 1 OF THE ACT)

"engineer" means a person who is skilled, through specialized education, training and experience, in the principles and practice of professional engineering;

"professional engineer" means a member or an engineer having a licence to practise under the provisions of this Act;

"professional engineering" or the "practice of engineering" means the provision of services for another as an employee or by contract, and such services shall include consultation, investigation, instruction, evaluation, planning, design, inspection, management, research, development and implementation of engineering works and systems;

"engineering works and systems" includes

(i) transportation systems and components related to air, water, land or outer space, movement of goods or people,

(ii) works related to the location, mapping, improvement, control and utilization of natural resources,

(iii) works and components of an electrical, mechanical, hydraulic, aeronautical, electronic, thermic, nuclear, metallurgical, geological, mining or industrial character and other dependent on the utilization or the application of chemical or physical principles,

(iv) works related to the protection, control and improvement of the environment including those of pollution control, abatement and treatment,

(v) the structural, electrical, mechanical, communications, transportation and other utility aspects of building components and systems,

(vi) structures and enclosures accessory to engineering works and intended to support or house them, and

(vii) systems relating to surveying and mapping;

2. MEMBERSHIP CRITERIA (SECTION 5 OF THE ACT)

5.

(1) Every person who engages in the practice of engineering in the Province of Prince Edward Island must have a valid certificate of registration, licence to practise or certificate of engineer-in-training in accordance with this Act and by-laws.

(2) Any applicant for a certificate of registration who satisfies Council that he

(a) is a resident or is coming to reside in Prince Edward Island;

(b) is a graduate in engineering or applied science of an academic or technical institution recognized by the Council;

(c) has fulfilled the requirements of approved engineering experience as prescribed in the by-laws;

(d) has successfully completed any examinations that may be prescribed by Council;

(e) has provided evidence of good character; and

(f) has paid the fees as prescribed in the by-laws,

shall be entitled to become registered as a member of the Association.

. . .

(11) The Council may refuse to issue a licence to practise to an applicant where the Council is of the opinion, upon reasonable and probable grounds, that the past conduct of the applicant affords grounds for belief that the applicant will not engage in the practice of professional engineering in accordance with the law and in a manner consistent with the provision of good service to the public.

3. DEFINITION OF PROFESSIONAL MISCONDUCT (SECTION 13 AND SECTION 18 OF THE ACT)

13. Members, persons licensed to practise, and engineers-in-training shall conduct themselves in accordance with the Code of Ethics for Engineers, and without restricting the meaning of professional misconduct, any breach of the Code of Ethics shall be deemed to be a form of professional misconduct.

. . .

18.

(1) A member, licensee, engineer-in-training or holder of a certificate of authorization may be found guilty of professional misconduct by the Discipline Committee if

(a) the member, licensee, engineer-in-training or holder of a certificate of authorization has been found guilty of an offence which, in the opinion of the Committee, is relevant to suitability to engage in the practice of engineering; or

(b) the member, licensee, engineer-in-training or holder of a certificate of authorization has been guilty, in the opinion of the Committee, of conduct that is not in the best interest of the public or tends to harm the standing of the Association.

(2) The Discipline Committee may find a member, licensee, or engineer-in-training incompetent if, in its opinion

(a) the member, licensee, or engineer-in-training has displayed in his professional activities a lack of knowledge, skill or judgement, or disregard for the welfare of the public of a nature or to an extent that demonstrates the member or licensee is unfit to carry out the responsibilities of a professional engineer; or

(b) the member, licensee or engineer-in-training is suffering from a physical or mental condition or disorder of a nature and extent making it desirable in the interests of the public, the member, licensee, or engineer-in-training that he no longer be permitted to engage in the practice of professional engineering, or that his practice of professional engineering be restricted.

4. DISCIPLINARY POWERS (SECTION 20 OF THE ACT)

20.

(1) All findings of the Discipline Committee shall be based exclusively on evidence submitted to it.

(2) Upon completion of the hearing, the Discipline Committee may pass a resolution dismissing the complaint or, if the Discipline Committee finds a member, licensee, engineer-in-training or the holder of a certificate of authorization guilty of professional misconduct or incompetence, or in breach of any of the requirements of this Act or any bylaws made hereunder, the Committee may, by order, do any one or more of the following:

(a) revoke the right to practise professional engineering for a stated period of time after which time the person or holder of certificate of authorization may reapply for membership, license to practise, enrolment as an engineer-in-training or certificate of authorization;

(b) suspend the right to practise professional engineering for a stated period, not exceeding twenty-four months;

(c) accept the undertaking of the member, licensee, engineer-in-training or holder of a certificate of authorization to limit the professional work in the practice of engineering to the extent specified in the undertaking;

(d) impose terms, conditions or limitations on the member, licensee or engineer-in-training including, but not limited to the successful completion of a particular course of study, as specified by the Committee;

(e) impose specific restrictions on the member, licensee or engineer-in-training or holder of a certificate of authorization including

(i) requiring the member, licensee, or engineer-in-traiing to engage in the practice of engineering only under the personal supervision and direction of a member,

(ii) requiring the member, licensee, or engineer-in-training to not alone engage in the practice of engineering,

(iii) requiring the member, licensee, engineer-in-training or the holder of the certificate of authorization to submit to periodic inspections by the Committee, or its designate, of documents, records and work of the member, licensee, engineer-in-training or the holder of a certificate of authorization in connection with his practice of engineering,

(iv) requiring the member, licensee, engineer-in-training or the holder of the certificate of authorization to report to the Discipline Committee or its designate on such matters with respect to the member's, licensee's, engineer-in-training or holder's practice of engineering for such period and times, and in such form, as the Committee may specify;

(f) reprimand, admonish or counsel the member, licensee, engineer-in-training or the holder of certificate of authorization, and if considered warranted, direct that the fact of the reprimand, admonishment or counselling be recorded on the register for a stated or unlimited period of time;

(g) direct that the imposition of a penalty or order be suspended or postponed for such period, and upon such terms, or for such purpose, including

(i) the successful completion by the member, licensee, or engineer-in-training of a particular course of study,

(ii) the production to the Committee or its designate of evidence satisfactory to it that any physical or mental handicap in respect of which the penalty was imposed has been overcome.

5. CODE OF ETHICS (SECTION 14 OF THE BY-LAWS)

14.1 Foreword

Honesty, justice, and courtesy from a moral philosophy which, associated with mutual interest among people, constitute the foundation of ethics. Engineers should recognize such a standard, not in passive observance, but as a set of dynamic principles guiding their conduct and way of life. It is their duty to practise the profession according to this Code of Ethics.

As the keystone of professional conduct is integrity, engineers will discharge their duties with fidelity to the public, their employers, and clients, and with fairness and impartiality to all. It is their duty to interest themselves in public welfare, and to be ready to apply their special knowledge for the benefit of mankind. They should uphold the honour and dignity of the profession and also avoid association with any enterprise of questionable character. In dealings with other fellow engineers they should be fair and tolerant.

14.2 Professional Life

14.2.1 Engineers will co-operate in extending the effectiveness of the engineering profession by interchanging information and experience with other engineers and students and by contributing to the work of engineering societies, schools, and the scientific and engineering press.

14.2.2 Engineers will not advertise their work or merit in a self-laudatory manner, and will avoid all conduct or practice likely to discredit or do injury to the dignity and honour of the profession.

14.3 Relations with the Public

14.3.1 Engineers will endeavour to extend public knowledge of engineering.

14.3.2 Engineers will have due regard for the safety of life and health of the public and employees who may be affected by the work for which they are responsible.

14.3.3 Engineers will not issue *ex parte* statements, criticisms, or arguments on matters connected with public policy which are inspired or paid for by private interests, unless it is indicated on whose behalf the statements are made.

14.3.4 Engineers will refrain from expressing publicly opinions on engineering subjects unless they are informed as to the facts relating thereto.

14.4 Relations with Clients and Employers

14.4.1 Engineers will act in professional matters for clients or employers as faithful agents or trustees.

14.4.2 Engineers will act with fairness and justice between the client or employer and the contractor when dealing with contracts.

14.4.3 Engineers will make their status clear to clients or employers before undertaking engagements if they may be called upon to decide on the use of inventions, apparatus, or any other thing in which they may have a financial interest.

14.4.4 Engineers will guard against conditions that are dangerous to life, limb, or property on work for which they are responsible, or if they are not responsible, will promptly call such conditions to the attention of those who are responsible.

14.4.5 Engineers will present clearly the consequences to be expected from deviations proposed if their engineering judgement is overruled by non-technical authority in cases where they are responsible for the technical adequacy of engineering work.

14.4.6 Engineers will engage, or advise clients or employers to engage, and will co-operate with, other experts and specialists whenever the clients' or employers' interests are best served by such service.

14.4.7 Engineers will not accept compensation, financial or otherwise, from more than one interested party for the same service, or for services pertaining to the same work, without the consent of all interested parties.

14.4.8 Engineers will not accept commissions or allowances, directly or indirectly, from contractors or other parties dealing with clients or employers in connection with work for which they are responsible.

14.4.9 Engineers will not be financially interested in bids as or of contractors on competitive work for which they are employed as engineers unless they have the consent of the client or employer.

14.4.10 Engineers will promptly disclose to clients or employers any interest in a business which may compete with or affect the business of the client or employer. They will not allow an interest in any business to affect their decisions regarding engineering work for which they are employed, or which they may be called upon to perform.

14.5 Relations with Engineers

14.5.1 Engineers will endeavour to protect the engineering profession collectively and individually from misrepresentation and misunderstanding.

14.5.2 Engineers will take care that credit for engineering work is given to those to whom credit is properly due.

14.5.3 Engineers will uphold the principle of appropriate and adequate compensation for those engaged in engineering work, including those in subordinate capacities, as being in the public interest and maintaining the standards of the profession.

14.5.4 Engineers will endeavour to provide opportunity for the professional development and advancement of engineers in their employ.

14.5.5 Engineers will not directly or indirectly injure the professional reputation, prospects, or practice of other engineers. However, if they consider that an engineer is guilty of unethical, illegal, or improper practice, they will present the information to the proper authority for action.

14.5.6 An engineer will not compete with another engineer on the basis of charges for work by underbidding, through reducing normal fees after having been informed of the charges named by the other.

14.5.7 An engineer will not use the advantages of a position to compete unfairly with another engineer.

14.5.8 An engineer will not become associated in responsibility for work with engineers who do not conform to ethical practices.

SOURCE: Engineering Profession Act, *Statutes of Prince Edward Island* 1990, c. 12.

QUEBEC

ENGINEERS ACT (*REVISED STATUTES OF QUEBEC 1986, C. I-9*)

1. DEFINITION OF ENGINEERING (SECTIONS 1, 2, AND 3)

1.

(d) "Engineer": a member of the Order.

2. Works of the kinds hereinafter described constitute the field of practice of an engineer:

(a) railways, public roads, airports, bridges, viaducts, tunnels and the installations connected with a transport system the cost of which exceeds three thousand dollars;

(b) dams, canals, harbours, lighthouses and all works relating to the improvement, control or utilization of waters;

(c) works of an electrical, mechanical, hydraulic, aeronautical, electronic, thermic, nuclear, metallurgical, geological or mining character and those intended for the utilization of the processes of applied chemistry or physics;

(d) waterworks, sewer, filtration, purification works to dispose of refuse, and other works in the field of municipal engineering, the cost of which exceeds one thousand dollars;

(e) the foundations, framework and electrical and mechanical systems of buildings the cost of which exceeds one hundred thousand dollars and of public buildings within the meaning of the Public Buildings Safety Act;

(f) structures accessory to engineering works and intended to house them;

(g) temporary framework and other temporary works used during the carrying out of works of civil engineering;

(h) soil engineering necessary to elaborate engineering works;

(i) industrial work or equipment involving public or employee safety.

3. The practice of the engineering profession consists in performing for another any of the following acts, when they relate to the works mentioned in section 2:

(a) the giving of consultations and opinions;

(b) the making of measurements, of layouts, the preparation of reports, computations, designs, drawings, plans, specifications;

(c) the inspection or supervision of the works.

2. MEMBERSHIP CRITERIA (SECTIONS 15 TO 17)

15.

(2) The Bureau shall also, subject to its regulations, admit as a member of the Order any Canadian citizen and any candidate who fulfils the conditions prescribed by section 44 of the Professional Code who establishes:

(a) he is domiciled in Québec;

(b) he has passed an examination before the committee of examiners on the theory and practice of engineering and especially in one of the following branches at his option: civil, mechanical, electrical, agricultural, geological, industrial, mining, metallurgical or chemical engineering or, at the discretion of the committee of examiners, in any combination or subdivision thereof; and

(c) he has paid the required fees fixed by regulation of the Bureau.

16. The Bureau, in all cases and notwithstanding the method of admission provided, may refuse admission to any candidate who cannot provide evidence of good character to the Bureau's satisfaction.

17. The Bureau, upon a written report by the committee of examiners to the effect that the candidate possesses the required knowledge and qualifications, may admit as a member of the Order any Canadian citizen domiciled in Québec, or domiciled in an adjacent province and practising his profession continuously and exclusively in Québec, provided that such candidate:

(a) holds a diploma in engineering or the degree of bachelor of applied sciences, or an equivalent diploma from a school or university recognized by the Gouvernement, or is a member of an engineering society recognized by the bureau;

(b) pays the requisite fee for admission to practice.

3. DEFINITION OF PROFESSIONAL MISCONDUCT

[Although the Act refers to "professional competence," the term is not defined. Conviction in a Canadian court of an indictable offence (or in a foreign court, of an offence of equal seriousness) is considered equivalent to an offence against the Act.]

4. DISCIPLINARY POWERS (SECTIONS 156 AND 188 OF THE PROFESSIONAL CODE, RSQ C. C-26)

156. The committee on discipline shall impose on a professional convicted of an offence referred to in section 116, one or more of the following penalties in respect of each count contained in the complaint:
(a) reprimand;
(b) temporary or permanent striking off the roll, even if he has not been entered thereon from the date of the offence;
(c) a fine of not less than $600 nor more than $6000 for each offence;
(d) the obligation to remit to any person entitled to it a sum of money the professional is holding for him;
(d.1) the obligation to transmit a document or the information contained in any document, and the obligation to complete, delete, update or rectify any document or information;
(e) revocation of his permit;
(f) revocation of his specialist's certificate;
(g) restriction or suspension of his right to engage in professional activities.
[For the purpose of subparagraph (c) of the first paragraph, when an offence is continuous, its continuity shall constitute a separate offence, day by day. The committee on discipline may decide on the terms and conditions of the penalties it imposes.]

188. Every person who contravenes a provision of this Code or the act or letters patent constituting a corporation is guilty of an offence and is liable on summary proceeding to a fine of not less than $600 nor more than $6000.

5. CODE OF ETHICS (RSQ C. I-9, R. 3, AS REVISED IN 1983)

DIVISION I — GENERAL PROVISIONS
1.01. This Regulation is made pursuant to section 87 of the Professional Code (RSQ, c. C-26).
1.02. In this Regulation, unless the context indicates otherwise, the word "client" means a person to whom an engineer provides professional services, including an employer.
1.03. The Interpretation Act (RSQ, c. I-16), with present and future amendments, applies to this Regulation.

DIVISION II — DUTIES AND OBLIGATIONS TOWARD THE PUBLIC

2.01. In all aspects of his work, the engineer must respect his obligations toward man and take into account the consequences of the performance of his work on the environment and on the life, health and property of every person.

2.02. The engineer must support every measure likely to improve the quality and availability of his professional services.

2.03. Whenever an engineer considers that certain works are a danger to public safety, he must notify the Ordre des ingénieurs du Québec (Order) or the persons responsible for such work.

2.04. The engineer shall express his opinion on matters dealing with engineering only if such opinion is based on sufficient knowledge and honest convictions.

2.05. The engineer must promote educational and information measures in the field in which he practises.

DIVISION III — DUTIES AND OBLIGATIONS TOWARD CLIENTS

1. General Provisions

3.01.01. Before accepting a mandate, an engineer must bear in mind the extent of his proficiency and aptitudes and also the means at his disposal to carry out the mandate.

3.01.02. In cases where it is in his client's interest, the engineer shall retain the services of experts after having obtained his client's authorization, or he shall advise the latter to do so.

3.01.03. An engineer must refrain from practising under conditions or in circumstances which could impair the quality of his services.

3.01.04. An engineer must at all times acknowledge his client's right to consult another engineer and, in such cases, he must offer his cooperation to the latter.

2. Integrity

3.02.01. An engineer must fulfil his professional obligations with integrity.

3.02.02. An engineer must avoid any misrepresentation with respect to his level of competence or the efficiency of his own services and of those generally provided by the members of his profession.

3.02.03. An engineer must, as soon as possible, inform his client of the extent and the terms and conditions of the mandate entrusted to him by the latter and obtain his agreement in that respect.

3.02.04. An engineer must refrain from expressing or giving contradictory or incomplete opinions or advice, and from presenting or using plans, specifications and other documents which he knows to be ambiguous or which are not sufficiently explicit.

3.02.05. An engineer must inform his client as early as possible of any error that might cause the latter prejudice and which cannot be easily rectified, made by him in the carrying out of his mandate.

3.02.06. An engineer must take reasonable care of the property entrusted to his care by a client and he may not lend or use it for purposes other than those for which it has been entrusted to him.

3.02.07. Where an engineer is responsible for the technical quality of engineering work, and his opinion is ignored, the engineer must clearly indicate to his client, in writing, the consequences which may result therefrom.

3.02.08. The engineer shall not resort to dishonest or doubtful practices in the performance of his professional activities.

3.02.09. An engineer shall not pay or undertake to pay, directly or indirectly, any benefit, rebate or commission in order to obtain a contract or upon the carrying out of engineering work.

3.02.10. An engineer must be impartial in his relations between his client and the contractors, suppliers and other persons doing business with his client.

3. Availability and Diligence

3.03.01. An engineer must show reasonable availability and diligence in the practice of his profession.

3.03.02. In addition to opinion and counsel, the engineer must furnish his client with any explanations necessary to the understanding and appreciation of the services he is providing him.

3.03.03. An engineer must give an accounting to his client when so requested by the latter.

3.03.04. An engineer may not cease to act for the account of a client unless he has just and reasonable grounds for so doing. The following shall, in particular, constitute just and reasonable grounds:

(a) the fact that the engineer is placed in a situation of conflict of interest or in a circumstance whereby his professional independence could be called in question;

(b) inducement by the client to illegal, unfair or fraudulent acts;

(c) the fact that the client ignores the engineer's advice.

3.03.05. Before ceasing to exercise his functions for the account of a client, the engineer must give advance notice of withdrawal within a reasonable time.

4. Seal and Signature

3.04.01. An engineer must affix his seal and signature on the original and the copies of every engineering plan and specification prepared by himself or prepared under his immediate control and supervision by persons who are not members of the Order.

An engineer may also affix his seal and signature on the original and the copies of documents mentioned in this section which have been prepared, signed and sealed by another engineer.

An engineer must not affix his seal and signature except in the cases provided for in this section.

5. Independence and Impartiality

3.05.01. An engineer must, in the practice of his profession, subordinate his personal interest to that of his client.

3.05.02. An engineer must ignore any intervention by a third party which could influence the performance of his professional duties to the detriment of his client.

Without restricting the generality of the foregoing, an engineer shall not accept, directly or indirectly, any benefit or rebate in money or otherwise from a supplier of goods or services relative to engineering work which he performs for the account of a client.

3.05.03. An engineer must safeguard his professional independence at all times and avoid any situation which would put him in conflict of interest.

3.05.04. As soon as he ascertains that he is in a situation of conflict of interest, the engineer must notify his client thereof and ask his authorization to continue his mandate.

3.05.05. An engineer shall share his fees only with a colleague and to the extent where such sharing corresponds to a distribution of services and responsibilities.

3.05.06. In carrying out a mandate, the engineer shall generally act only for one of the parties concerned, namely, his client. However, where his professional duties require that he act otherwise, the engineer must notify his client thereof. He shall accept the payment of his fees only from his client or the latter's representative.

6. Professional Secrecy

3.06.01. An engineer must respect the secrecy of all confidential information obtained in the practice of his profession.

3.06.02. An engineer shall be released from professional secrecy only with the authorization of his client or whenever so ordered by law.

3.06.03. An engineer shall not make use of confidential information to the prejudice of a client or with a view to deriving, directly or indirectly, an advantage for himself or for another person.

3.06.04. An engineer shall not accept a mandate which entails or may entail the disclosure or use of confidential information or documents obtained from another client without the latter's consent.

7. Accessibility of Records

3.07.01. An engineer must respect the right of his client to take cognizance of and to obtain copies of the documents that concern the latter in any record which the engineer has made regarding that client.

8. Determination and Payment of Fees

3.08.01. An engineer must charge and accept fair and reasonable fees.

3.08.02. Fees are considered fair and reasonable when they are justified by the circumstances and correspond to the services rendered. In determining his fees, the engineer must, in particular, take the following factors into account:

(a) the time devoted to the carrying out of the mandate;
(b) the difficulty and magnitude of the mandate;
(c) the performance of unusual services or services requiring exceptional competence or speed;
(d) the responsibility assumed.

3.08.03. An engineer must inform his client of the approximate cost of his services and of the terms and conditions of payment. He must refrain from demanding advance payment of his fees; he may, however, request a deposit.

3.08.04. An engineer must give his client all the necessary explanations for the understanding of his statement of fees and the terms and conditions of its payment.

DIVISION IV — DUTIES AND OBLIGATIONS TOWARD THE PROFESSION

1. Derogatory Acts

4.01.01. In addition to those referred to in sections 57 and 58 of the Professional Code, the following acts are derogatory to the dignity of the profession:

(a) participating or contributing to the illegal practice of the profession;

(b) pressing or repeated inducement to make use of his professional services;

(c) communicating with the person who lodged a complaint, without the prior written permission of the syndic or his assistant, whenever he is informed of an inquiry into his professional conduct or competence or whenever a complaint has been laid against him;

(d) refusing to comply with the procedures for the conciliation and arbitration of accounts and with the arbitrators' award;

(e) taking legal action against a colleague on a matter relative to the practice of the profession before applying for conciliation to the president of the Order;

(f) refusing or failing to present himself at the office of the syndic, of one of his assistants or of a corresponding syndic, upon request to that effect by one of those persons;

(g) not notifying the syndic without delay if he believes that an engineer infringes this Regulation.

2. Relations with the Order and Colleagues

4.02.01. An engineer whose participation in a council for the arbitration of accounts, a committee on discipline or a professional inspection committee is requested by the Order, must accept this duty unless he has exceptional grounds for refusing.

4.02.02. An engineer must, within the shortest delay, answer all correspondence addressed to him by the syndic of the Order, the assistant syndic, or a corresponding syndic, investigators or members of the professional inspection committee or the secretary of the said committee.

4.02.03. An engineer shall not abuse a colleague's good faith, be guilty of breach of trust or be disloyal towards him or willfully damage his reputation. Without restricting the generality of the foregoing, the engineer shall not, in particular:

(a) take upon himself the credit for engineering work which belongs to a colleague;

(b) take advantage of his capacity of employer or executive to limit in any way the professional independence of an engineer employed by him or under his responsibility, in particular with respect to the use of the title of engineer or the obligation of every engineer to commit his professional responsibility;

(c) induce a colleague to commit an offence against the laws and regulations governing the practice of the profession.

4.02.04. Where a client requests an engineer to examine or review engineering work that he has not performed himself, the latter must notify the engineer concerned thereof and, where applicable, ensure that the mandate of his colleague has terminated.

4.02.05. Where an engineer replaces a colleague in engineering work, he must notify that colleague thereof and make sure that the latter's mandate has terminated.

4.02.06. An engineer who is called upon to collaborate with a colleague must retain his professional independence. If a task is entrusted to him and such task goes against his conscience or his principles, he may ask to be excused from doing it.

4.02.07. An engineer may not refuse to collaborate with a member of the Order, in professional dealings, on the basis of race, colour, sex, religion, national, ethnic or social origin and for any ground mentioned in section 10 of the Charter of human rights and freedoms (RSQ, c. C-12).

3. Contribution to the Advancement of the Profession
4.03.01. An engineer must, as far as he is able, contribute to the development of his profession by sharing his knowledge and experience with his colleagues and students, and by his participation as professor or tutor in continuing training periods and refresher training courses.

SOURCE: Engineers Act, *Revised Statutes of Quebec* 1986, c. I-9.

SASKATCHEWAN

ENGINEERING AND GEOSCIENCE PROFESSIONS ACT (*REVISED STATUTES OF SASKATCHEWAN 1996*)

1. DEFINITION OF ENGINEERING (SECTION 2 OF THE ACT)

2(1)(k) "practice of professional engineering" means reporting on, advising on, valuing, measuring for, laying out, designing, directing, constructing or inspecting any works or processes that are mentioned in subsection (2) or any works or processes similar to those mentioned in subsection (2) by reason of their requiring the skilled application of the principles of mathematics, physics, mechanics, aeronautics, hydraulics, electricity, chemistry or geology in their development and attainment;
(l) "practice of professional geoscience" means the application of principles of geoscience that include, but are not limited to, principles of geology, geophysics and geochemistry, to any act of acquiring or processing data, advising, evaluating, examining, interpreting, reporting, sampling or geoscientific surveying, that is directed toward:
> (i) the discovery or development of oil, natural gas, coal, metallic or non-metallic minerals, precious stones, water or other natural resources; or
> (ii) the investigation of surface or sub-surface geological conditions;
(m) "professional engineer" means an individual who is registered with the association as a professional engineer;
(n) "professional geoscientist" means an individual who is registered with the association as a professional geoscientist;
[Note: Section 2, mentioned above, is an extensive list of engineering works, structures, and activities. As the above definition indicates, Section 2 nevertheless does not include all of the works or activities that are included in the definition.]

2. MEMBERSHIP CRITERIA (SECTION 20 OF THE ACT)

20(1) The council may register a person as a member where the person produces evidence establishing to the satisfaction of the council that he or she:

(a) has paid the prescribed fees;

(b) has complied with the bylaws with respect to registration as a member;

(c) produces evidence establishing to the satisfaction of the council that the individual is eligible according to the bylaws to be a member of the association; and

(d) has successfully completed:

(i) in the case of a person who applies for registration as a professional engineer, a bachelor level university program of study in engineering recognized by the council; or

(ii) in the case of a person who applies for registration as a professional geoscientist, a four-year bachelor level university program of study in geoscience recognized by the council.

(2) Notwithstanding that a person does not comply with the requirements in subsection (1), the council may register the person as a member and issue a restricted licence to the person to practise professional engineering or professional geoscience, as the case may be, where the person:

(a) is eligible, according to the bylaws, to be a member;

(b) has paid the prescribed fees; and

(c) has complied with the bylaws with respect to registration.

3. DEFINITION OF PROFESSIONAL INCOMPETENCE AND MISCONDUCT (SECTIONS 29 AND 30 OF THE ACT)

PROFESSIONAL INCOMPETENCE

29. Professional incompetence is a question of fact, but the display by a member of:

(a) a lack of knowledge, skill or judgement; or

(b) a disregard for the welfare of members of the public served by the professions;

of a nature or to an extent that demonstrates that the member is unfit to continue in the practice of the profession is professional incompetence within the meaning of this Act.

PROFESSIONAL MISCONDUCT

30. Professional misconduct is a question of fact, but any matter, conduct or thing, whether or not disgraceful or dishonourable, is professional misconduct within the meaning of this Act if:

(a) it is harmful to the best interests of the public or the members;

(b) it tends to harm the standing of the professions;

(c) it is a breach of this Act or the bylaws; or

(d) it is a failure to comply with an order of the investigation committee, the discipline committee or the council.

4. DISCIPLINARY POWERS (SECTION 35 OF THE ACT)

35(1) Where a discipline committee finds that a member's conduct constitutes professional misconduct or professional incompetence, it may make one or more of the following orders:

(a) an order that the member be expelled from the association and that the member's name be struck from the register;

(b) an order that the member be suspended from the association for a specified period;

(c) an order that the member be suspended pending the satisfaction and completion of any conditions specified in the order, which may include, but are not restricted to, an order that the member:

(i) successfully complete specified classes or courses of instruction;

(ii) obtain medical treatment, counselling or both;

(iii) undergo treatment for substance addiction.

(d) an order that the member may continue to practise only under conditions specified in the order, which may include, but are not restricted to, an order that the member:

(i) not do specified types of work;

(ii) successfully complete specified classes or courses of instruction;

(iii) restrict his or her practice in the manner ordered by the discipline committee;

(iv) practise only as a partner with, or as an associate or employee of, one or more members of the association that the discipline committee may specify;

(v) obtain medical treatment, counselling or both;

(vi) undergo treatment for substance addiction;

(e) an order reprimanding the member;

(f) any other order that the discipline committee considers just.

(2) In addition to any order made pursuant to subsection (1), the discipline committee may order:

(a) that the member pay to the association, within a fixed period:

(i) a fine in a specified amount not exceeding $15 000; and

(ii) the costs of the investigation and hearing into the member's conduct and related costs, including the expenses of the investigation committee and the discipline committee and costs of legal services and witnesses; and

(b) where a member fails to make payment in accordance with an order pursuant to clause (a), that the member be suspended from the association.

(3) The registrar shall send a copy of an order made pursuant to this section to the member whose conduct is the subject of the order, to the member's employer, if any, and to the person, if any, who made the complaint.

(4) Where a member is expelled or suspended, the registrar shall strike the name of the member from the register or indicate the suspension on the register, as the case may be.

CRIMINAL CONVICTION

36. A discipline committee may, by order, expel a member from the association where:

(a) the member has been convicted of an indictable offence pursuant to the Criminal Code;

(b) a report of the investigation committee is made to the discipline committee respecting the conviction mentioned in clause (a);

(c) the discipline committee has given the member mentioned in clause (a) an opportunity to be heard; and

(d) the discipline committee finds that the conduct of the member giving rise to the conviction makes the member unfit to continue to be a member.

5. CODE OF ETHICS (SECTION 37 OF THE BY-LAWS ESTABLISHED UNDER SECTION 13(B) OF THE ACT)

37. For the governance of the conduct of the members and licences of the Association, the following Code of Ethics is hereby established;

PREAMBLE
Honesty, justice, and courtesy form a moral philosophy, which associated with mutual interest among men, constitute the foundation of ethics. The Professional Engineer should recognize such a standard, not in passive observance but as a set of dynamic principles guiding his conduct and way of life. It is his duty to practise his profession according to this Code of Ethics.

GENERAL
1. A professional engineer owes certain duties to the public, his employers, other members of his profession and to himself and shall act at all times with,
    (a) fidelity to public needs;
    (b) fairness and loyalty to his associates, employers, subordinates and employees; and
    (c) devotion to high ideals of personal honour and professional integrity.

DUTY OF PROFESSIONAL ENGINEER TO THE PUBLIC
2. A professional engineer shall,
    (a) regard his duty to the public as paramount;
    (b) endeavour to maintain public regard for his profession by discouraging untrue, unfair or exaggerated statements with respect to professional engineering;
    (c) not give opinions or make statements on professional engineering projects of public interest that are inspired or paid for by private interests unless he clearly discloses on whose behalf he is giving the opinions or making the statements;
    (d) not express publicly or while he is serving as a witness before a court, commission or other tribunal, opinions on professional engineering matters that are not founded on adequate knowledge and honest conviction;
    (e) make effective provisions for the safety of life and health of a person who may be affected by the work for which he is responsible; and shall act to correct or report any situation which he feels may endanger the public.
    (f) not knowingly associate with, or allow the use of his name by, an enterprise of doubtful character, nor shall he sanction the use of his reports, in part or in whole, in a manner calculated to mislead, and if it comes to his knowledge they have been so used, shall take immediate steps to correct any false impression given by them;
    (g) on all occasions sign or seal reports, plans and specifications which legally require sealing and for which he is professionally responsible;

(h) not offer his services for a fee without first notifying the Council of the Association of his intent to do so and of the area of specialty in which he proposes to practise, and receiving from the Council of the Association permission to do so.

DUTY OF PROFESSIONAL ENGINEER TO EMPLOYER

3. A professional engineer shall,

(a) act for his employer as a faithful agent or trustee and shall regard as confidential any information obtained by him as to the business affairs, technical methods or processes of his employer, and avoid or disclose any conflict of interest which might influence his actions or judgement;

(b) present clearly to his employers the consequences to be expected from any deviations proposed in the work if his professional engineering judgement is overruled in cases where he is responsible for the technical adequacy of professional engineering work;

(c) advise his employer to engage experts and specialists whenever the employer's interests are best served by so doing;

(d) have no interest, direct or indirect in any materials, supplies or equipment used by his employer or in any persons or firms receiving contracts from his employer unless he informs his employer in advance of the nature of the interest;

(e) not act as consulting engineer in respect of any work upon which he may be the contractor unless he first advises his employer; and

(f) not accept compensation, financial or otherwise, for a particular service, from more than one person except with the full knowledge of all interested parties.

DUTY OF PROFESSIONAL ENGINEER TO OTHER PROFESSIONAL ENGINEERS

4. A professional engineer shall,

(a) not attempt to supplant another engineer after definite steps have been taken toward the other's employment;

(b) not accept employment by a client, knowing that a claim for compensation or damages, or both, of a fellow engineer previously employed by the same client, and whose employment has been terminated, remains unsatisfied, or until such claim has been referred to arbitration or issue has been joined at law, or unless the engineer previously employed has neglected to press his claim legally, or the Council of the Association gives its consent;

(c) not accept any engagement to review the work of another professional engineer for the same employer except with the knowledge of that engineer, or except where the connection of that engineer with the work has been terminated;

(d) not maliciously injure the reputation or business of another professional engineer;

(e) not attempt to gain an advantage over other members of his profession by paying or accepting a commission in securing professional engineering work, or by reducing his fees below the approved minimums;

(f) not advertise in a misleading manner or in a manner injurious to the dignity of his profession;

(g) give proper credit for engineering work;

(h) uphold the principle of adequate compensation for engineering work;

(i) provide opportunity for professional development and advancement of his professional colleagues;

(j) extend the effectiveness of the profession through the interchange of engineering information and experience.

DUTY OF PROFESSIONAL ENGINEER TO HIMSELF

5. A professional engineer shall,

(a) maintain the honour and integrity of his profession and without fear or favour expose before the proper tribunals unprofessional or dishonest conduct by any other member of the profession;

(b) undertake only such work as he is competent to perform by virtue of his training and experience; and

(c) constantly strive to broaden his knowledge and experience by keeping abreast of new techniques and developments in his field of endeavour.

SOURCE: Engineering and Geoscience Professions Act, *Revised Statutes of Saskatchewan* 1996. Reproduced with permission of the Queen's Printer for Saskatchewan.

YUKON TERRITORY

ENGINEERING PROFESSION ACT (*REVISED STATUTES OF THE YUKON* 1995, C. 9)

1. DEFINITION OF ENGINEERING (PART 1, SECTION 1 OF THE ACT)

1. "practice of engineering" means

(a) reporting on, advising on, evaluating, designing, preparing plans and specifications for, or directing the construction, technical inspection, maintenance, or operation of, any structure, work, or process

(i) that is aimed at the discovery, development, utilisation, storage, or disposal of matter, materials, or energy, or is in any other way designed for, the use and convenience of persons; and

(ii) that, for the protection of persons, requires in the reporting, advising, evaluating, designing, preparation, or direction, the professional application of the principles of engineering or any related applied subject; and

(b) teaching engineering at a university or college

. . .

"professional engineer" means an individual who holds a certificate of registration to engage in the practice of engineering under this Act

2. MEMBERSHIP CRITERIA (PART 4, SECTIONS 21 AND 23 OF THE ACT)

21. The Board of Examiners shall approve for registration as a professional engineer, or a holder of a limited licence, as the case may be, an individual who has

applied to the Board and is eligible to be registered under this *Act* as a professional engineer or a holder of a limited licence.

. . .

23.(1) The Council shall approve the registration as a permit holder of a corporation, partnership or other entity that has applied to the Council and is eligible under this section and the regulations to become registered to engage in the practice of engineering as a permit holder.

A corporation, partnership, or other entity that applies to the Council is eligible to become registered as a permit holder entitled to engage in the practice of engineering if it satisfies the Council that it complies with the Act and the regulations.

3. DEFINITION OF PROFESSIONAL MISCONDUCT (PART 5, SECTIONS 25 AND 27 OF THE ACT)

25.
   (a) "conduct" includes an act or omission:

. . .

27.(1) Any conduct of a professional engineer, holder of a limited licence, permit holder or engineer-in-training that
   (a) is detrimental to the public interests,
   (b) contravenes a code of ethics of the profession as established under the regulations,
   (c) harms or tends to harm the standing of the profession generally,
   (d) displays a lack of the knowledge or of the skill or judgement reasonably to be expected in the practice of the profession, or
   (e) displays a lack of the knowledge or the skill or judgement reasonably to be expected for the carrying out of any duty or obligation undertaken in the practice of the profession,
whether or not that conduct is disgraceful or dishonourable, constitutes either unskilled practice of the profession or unprofessional conduct, whichever the Discipline Committee finds.

4. DISCIPLINARY POWERS (PART 5, SECTION 44 OF THE ACT)

44. If the Discipline Committee finds that the conduct of the investigated person is unprofessional conduct or unskilled practice of the profession or both, the Discipline Committee may make any one or more of the following orders:
   (a) reprimand the investigated person;
   (b) suspend the registration of the investigated person for a specified period;
   (c) suspend the registration of the investigated person either generally or from any field of practice until
      (i) he or she has completed a specified course of studies or obtained supervised practical experience, or

(ii) the Discipline Committee is satisfied as to the competence of the investigated person generally or in a specified field of practice;

(d) accept in place of a suspension the investigated person's undertaking to limit his or her practice;

(e) impose conditions on the investigated person's entitlement to engage in the practice of the profession generally or in any field of the practice, including the conditions that they

(i) practice under supervision

(ii) not engage in sole practice

(iii) permit periodic inspections by a person authorized by the Discipline Committee, or

(iv) report to the Discipline Committee on specific matters;

(f) direct the investigated person to pass a particular course of study or satisfy the Discipline Committee as to their practical competence generally or in a field of practice;

(g) direct the investigated person to satisfy the Discipline Committee that a disability or addiction can be overcome, and suspend the person until the Discipline Committee is so satisfied;

(h) require the investigated person to take counselling or to obtain any assistance that in the opinion of the Discipline Committee is appropriate; or

(i) direct the investigated person to waive, reduce or repay a fee for services rendered by the investigated person that, in the opinion of the Discipline Committee, were not rendered or were improperly rendered;

(j) cancel the registration of the investigated person.

5. CODE OF ETHICS

(a) General

A professional engineer owes certain duties to the public, to his employers, to his clients, to other members of his profession and to himself and shall act at all times with:

(1) fairness and loyalty to his associates, employers, clients, subordinates and employees;

(2) fidelity to public needs and

(3) devotion to high ideals of personal honour and professional integrity.

(b) Duty of Professional Engineer to the Public

A professional engineer shall:

(1) regard his duty to the public welfare as paramount;

(2) endeavour at all times to enhance the public regard for his profession by extending the public knowledge thereof and discouraging untrue, unfair or exaggerated statements with respect to professional engineering;

(3) not give opinions or make statements, on professional engineering projects of public interest, that are inspired or paid for by private interests unless he clearly discloses on whose behalf he is giving the opinions or making the statements;

(4) not express publicly, or while he is serving as a witness before a court, commission or other tribunal, opinions on professional engineering matters that are not founded on adequate knowledge and honest conviction;

(5) make effective provisions for the safety of life and health of a person who may be affected by the work for which he is responsible; and at all times shall act to correct or report any situation which he feels may endanger the safety or welfare of the public;

(6) make effective provisions for meeting lawful standards, rules, or regulations relating to environmental control and protection in connection with any work being undertaken by him or under his responsibility;

(7) sign or seal only those plans, specifications and reports made by him or under his personal supervision and direction or those which have been thoroughly reviewed by him as if they were his own work, and found to be satisfactory and

(8) refrain from associating himself with or allowing the use of his name by an enterprise of questionable character.

(c) Duty of Professional Engineer to Employer

A professional engineer shall:

(1) act in professional engineering matters for each employer as a faithful agent or trustee and shall regard as confidential any information obtained by him as to the business affairs, technical methods or processes of an employer and avoid or disclose any conflict of interest which might influence his actions or judgement;

(2) present clearly to his employers the consequences to be expected from any deviations proposed in the work if he is informed that his professional engineering judgement is overruled by nontechnical authority in cases where he is responsible for the technical adequacy of professional engineering work;

(3) have no interest, direct or indirect, in any materials, supplies or equipment used by his employer or in any persons or firms receiving contracts from his employer unless he informs his employer in advance of the nature of the interest;

(4) not tender on competitive work upon which he may be acting as a professional engineer unless he first advises his employer;

(5) not act as consulting engineer in respect of any work upon which he may be the contractor unless he first advises his employer and

(6) not accept compensation, financial or otherwise, for a particular service from more than one person except with the full knowledge of all interested parties.

(d) Duty of Professional Engineer in Independent Practice to Client

A professional engineer in private practice, in addition to all other sections, shall:

(1) disclose immediately any interest, direct or indirect, which may in any way be constituted as prejudicial to his professional judgement in rendering service to his client;

(2) if he is an employee-engineer and is contracting in his own name to perform professional engineering work for other than his employer, clearly advise his client as to the nature of his status as an employee and the attendant limitations on his services to the client. In addition he shall ensure that such work will not conflict with his duty to his employer;

(3) carry out his work in accordance with applicable statutes, regulations, standards, codes, and by-laws and

(4) co-operate as necessary in working with such other professionals as may be engaged on a project.

(e) Duty of Professional Engineer to Other Professional Engineers

A professional engineer shall:

(1) conduct himself toward other professional engineers with courtesy and good faith;

(2) not accept any engagement to review the work of another professional engineer for the same employer or client except with the knowledge of that engineer, or except where the connection of that engineer with the work has been terminated;

(3) not maliciously injure the reputation or business of another professional engineer;

(4) not attempt to gain an advantage over other members of his profession by paying or accepting a commission in securing professional engineering work;

(5) not advertise in a misleading manner or in a manner injurious to the dignity of his profession, but shall seek to advertise by establishing a well-merited reputation for personal capability; and

(6) give proper credit for engineering work, uphold the principle of adequate compensation for engineer work, provide opportunity for professional development and advancement of his associates and subordinates; and extend the effectiveness of the profession through the interchange of engineering information and experience.

(f) Duty of Professional Engineer to Himself

A professional engineer shall:

(1) maintain the honour and integrity of his profession and without fear or favour expose before the proper tribunals unprofessional or dishonest conduct by any other members of the profession; and

(2) undertake only such work as he is competent to perform by virtue of his training and experience, and shall, where advisable, retain and co-operate with other professional engineers or specialists.

SOURCE: Engineering Profession Act, *Revised Statutes of the Yukon* 1995, c. 9.

APPENDIX C

The Ritual of the Calling of an Engineer

THE IRON RING

Most engineers in Canada wear the Iron Ring and have solemnly obligated themselves to an ethical and diligent professional career through the Ritual of the Calling of an Engineer. This Ritual is the result of efforts by the Corporation of the Seven Wardens, started in 1922 when a group of prominent engineers met in Montreal to discuss a concern for the general guidance and solidarity of the profession. These seven prominent engineers formed the nucleus of an organization whose object would be to bind all members of the engineering profession in Canada more closely together and to imbue them with their responsibility towards society.

They enlisted the services of the late Rudyard Kipling, who developed an appropriate Ritual and the symbolic Iron Ring. The purpose was outlined by Rudyard Kipling in the following words:

"The Ritual of the Calling of an Engineer has been instituted with the simple end of directing the young engineer towards a consciousness of his[/her] profession and its significance, and indicating to the older engineer his[/her] responsibilities in receiving, welcoming and supporting the young engineers in their beginnings."

The Ritual has been copyrighted in Canada and the United States, and the Iron Ring has been registered. The Corporation of the Seven Wardens is entrusted with the responsibility of administering and maintaining the Ritual, which it does through a system of separate groups, called Camps, across Canada. There are presently 20 such Camps.

The Corporation of the Seven Wardens is not a "secret society." Its rules of governance, however, do not permit any publicity about its activities and they specify that Ceremonies are not to be held in the presence of the general public.

The original seven senior engineers who met in Montreal in 1922 were, as it happens, all past presidents of the Engineering Institute of Canada. There is, however, no direct connection between the Engineering Institute of Canada and the Corporation of the Seven Wardens.

The wearing of the Iron Ring, or the taking of the obligation, does not imply that an individual has gained legal acceptance or qualification as an engineer. This can only be granted by the provincial bodies so appointed and, as a result, it should also be mentioned that the Corporation of the Seven Wardens has no direct connection with any provincial association or order.

The obligation ceremonies for graduating students are held in cities where Camps are located, and for convenience, in some cases, on the university campus itself. Such ceremonies must not be misconstrued as being an extension of the engineering curriculum. The Iron Ring does not replace the diploma granted by the University of the School of Engineering nor is it an overt sign of having successfully passed the institution's examinations.

The purpose of the Corporation of the Seven Wardens and the Ritual is to provide an opportunity for men and women to obligate themselves to the standard of ethics and diligent practice required by those in our profession. This opportunity is available to any who wish to avail themselves to it, whether they be new graduates or senior engineers. The Ritual of the Calling of an Engineer is, of course, attended by all those who wish to be obligated, along with invited senior engineers and, when space permits, immediate family members. A complete explanation of the Ritual, its obligations and history is given to every man and woman before the ceremony so that they may decide in advance whether or not they wish to take part in the spirit intended. A few people, for one reason or another, have chosen to refrain from being obligated, and so cannot rightfully wear the Iron Ring. The Corporation of the Seven Wardens feels that this in no way detracts from their right to practise in the profession and further feels that the obligation should continue to be a matter of personal choice, taken only by those who wish to take part in the serious and sincere manner intended.

SOURCE: J.B. Carruthers, P.Eng., "The Ritual of the Calling of an Engineer," *Project Magazine: The National Magazine for Engineering Students* (April 1985): 19.

APPENDIX D

Professional Engineering Admission Requirements in the United States

GETTING THE LICENCE: KNOWING WHAT TO EXPECT CAN MAKE IT EASIER TO APPLY FOR THE PE LICENCE AND TAKE THE EXAMS

Applying for a professional engineer's licence is not a cut-and-dried procedure. Various combinations of educational and professional experience can fulfil the qualification criteria, and many applicants are unsure about how to document and evaluate their record to best advantage.

There are seven basic requirements, but these may vary slightly from state to state. It is therefore advisable to check with the state board for the latest rules. State board addresses can be obtained from the National Council of Engineering Examiners, Box 1686, Clemson, South Carolina 29633-1686; [Telephone:] (803) 654-6824. The seven requirements are concerned with:

- *Age:* The minimum age for most states is 25 for a full licence, and 21 for an engineer-in-training licence. In New York State, the minimum age to apply for a full licence is 18.
- *Citizenship:* Most states do not require U.S. citizenship. However, the boards recognize that inadequate proficiency in English can pose a risk to the safety and health of the public. Accordingly, they insist that a foreign applicant demonstrate proficiency in English.
- *Graduation:* The applicant must have a certificate of graduation from an accredited high school or the approved equivalent.
- *Degree:* An engineering degree is necessary from an engineering school approved by the Accreditation Board for Engineering and Technology (ABET), or the equivalent in approved practical engineering experience. Graduates of nonaccredited institutions must submit transcripts to the PE board of record. They must also have an additional number of years of board-approved experience. The board of registration will determine whether or not

an applicant's experience is equivalent to an engineering degree. Copies of "Accredited Programs Leading to Degrees in Engineering" are available from ABET, 345 E. 47th Street, New York, NY 10017.

- *Experience:* The applicant must supply documented evidence of sufficient qualifying engineering experience. If experience has been acquired overseas, the applicant must list the names and addresses of people who can attest to that experience.
- *Character:* References from licensed PEs personally acquainted with the applicant attesting to his or her moral character and integrity are required.
- *Examination:* The candidate is required to pass both parts of the written examination administered by the state board through the National Council of Engineering Examiners (NCEE). Waivers are seldom granted.

EXPERIENCE

Appropriate experience is the most important requirement for obtaining a licence. The NCEE defines qualifying experience as

> the legal minimum number of years of creative engineering work requiring the application of the engineering sciences to the investigation, planning, design, and construction of major engineering works. It is not merely the laying out of the details of design, nor the mere performance of engineering calculations, writing of specifications, or making tests. It is rather a combination of these things, plus the exercise of sound judgement, taking into account economic and social factors in arriving at decisions and giving advice to the client or employer, the soundness of which has been demonstrated in actual practice.

PROFESSIONAL EXPERIENCE

Although it is not always easy to determine whether or not specific activities are professional, some guidelines do exist. These include:

- design, specification, and/or supervision of the construction of major equipment; preparing plant layouts; performing economic balances;
- developing processes and pilot plants;
- performing research;
- preparing technical reports, manuals, and the like;
- taking charge of broader fields of engineering;
- consulting, appraising, evaluating, and working on patent laws;
- operating major plants; product testing; technical service; technical sales;
- teaching full time at the college level or at an ABET-accredited engineering school; editing and writing technical materials;

There are in addition some specific abilities that can be considered professional and that apply to all branches of engineering. These include:

- selecting technical procedures for use in problem solving and developing new approaches and methods when necessary;

- analyzing physical events in mathematical terms so as to determine the physical behaviour of materials and structures;
- converting designs into products and evaluating needs in terms of equipment and facilities;
- using the basic tools of engineering (mathematics, chemistry, physics, fluid mechanics, dynamics, thermodynamics, heat transfer, materials mechanics, and basic electricity) in the everyday performance of one's duties.

SUBPROFESSIONAL EXPERIENCE

This type of experience usually includes:

- constructing and installing major equipment, plants, piping, and pumping systems;
- working as a shift operator in manufacturing plants or pilot plants, or as a troubleshooter;
- drafting flowsheet layouts, instrument making, and servicing;
- selling standard equipment that does not involve the use and application of engineering knowledge;
- teaching as an assistant, without full responsibilities, in an accredited school or college, or teaching nonengineering subjects in an engineering college or elsewhere;
- completing correspondence school courses in engineering; taking in-house, non-credit courses sponsored by a company; continuing-education activities;
- performing routine analyses and computations, routine tests of equipment or apparatus, or routine tests of materials;
- working in nonengineering jobs in connection with engineering projects;
- finishing work projects before completing high school or reaching age 18; working in a co-operative college course or between terms of a college course;
- performing military service involving routine duties (no research, development, or design).

When the duties of the applicant involve both engineering and nonengineering assignments, the general practice is to grant partial credit. In describing his[/her] experience, the applicant should indicate the amount of time spent on the duties performed.

BORDERLINE EXPERIENCE

Included in an engineer's daily work are certain tasks that are not considered professional in themselves but which, when performed in combination with other tasks, may constitute professional experience. These borderline cases include:

- performing calculations concerned with heat transfer, fluid flow, fluid transport, drying, etc.;
- preparing flowsheets and logic diagrams;
- designing machinery components (tanks, pumps, dryers, etc.) and systems (noise, dust, and fume control);

- installing large-scale control, production, or environmental systems and specifying plan layouts, taking into consideration economics, from raw material to finished product.

DOCUMENTING EXPERIENCE

A state board's evaluation of an applicant's experience is conducted solely on the basis of his[/her] record and the testimony of references. One of the most common mistakes made in describing work experience is to resort to generalities. For example, an applicant might write, "My duties consisted of designing mechanical plants, including economics evaluations of the sites and original investment costs." This is merely a paraphrase of the definition of engineering in the statutes. Without a description of specific duties and responsibilities, the board cannot tell whether the applicant was a drafts[person], a designer, or print co-ordinator, or if he[/she] performed all or none of those functions. At the other extreme, an applicant may claim that he[/she] "had full responsibility for the following projects." Without amplification, this statement is meaningless, since it can apply to many upper-level administrators who have no knowledge of engineering.

APPROVED EXPERIENCE

For an applicant's experience to be considered satisfactory by the board of registration, it must demonstrate an increasing level of quality and responsibility, and must not have been obtained in violation of the state registration law. Ideally, it should have been gained under the direct supervision of a licensed PE. It should be in the branch of engineering in which the applicant claims proficiency, and it should not have been gained in jobs of short duration with various companies.

In addition to the major disciplines of chemical, civil/sanitary/structural, electrical, and mechanical engineering, examinations are now available in agricultural, ceramic, industrial, manufacturing, nuclear, petroleum, fire protection, mining/mineral, and metallurgical engineering. Examinations in the four major disciplines are offered twice a year; information regarding the others can be obtained from NCEE.

For sales experience to be creditable, it must be demonstrated that engineering principles and knowledge were actually employed.

RESPONSIBILITY

The basic criterion that the boards use to evaluate professional experience is that the applicant be "in responsible charge" of people or work. This implies a degree of competence and accountability sufficient for him[/her] to supervise or independently pursue engineering work. It is not enough for the applicant to be assigned work by a supervisor who is a licensed professional engineer. An engineer who is assigned work, but who requires a supervisor's continued attention, is not considered to be in responsible charge. The engineering graduate who started out at the drafting board must show that he[/she] has progressed to the point where he[/she] can independently apply basic principles to everyday work.

An applicant meeting the board's standard of responsibility makes decisions that could affect the health, safety, and welfare of the public. It is not sufficient to merely review the decisions made by subordinates. The applicant must be able to judge the qualifications of technical specialists and the validity and applicability of their recommendations, before those recommendations are incorporated into the work or design. Decisions made by an engineer in charge are generally at the project level or higher. These kinds of decisions include selecting the engineering alternative to be investigated and comparing those alternatives; selecting or developing design standards, methods, and materials; and developing and controlling operating and maintenance procedures.

It is not necessary for the applicant to defend decisions he[/she] has made, but only to demonstrate that he[/she] made them and possessed sufficient knowledge of the project to do so.

EVALUATING EXPERIENCE

A job title is not necessarily a good indicator of professional experience. An engineer engaged in flowsheet layout or flow diagrams may be doing high-grade work but, if the work is routine, he[/she] is not working at a professional level. Many engineers and drafts[people] become highly proficient in the use of handbooks, standard procedures, computers, and other tools for accelerating the design process, without ever attempting to understand the principles on which those tools are based.

Essentially, the board's evaluation of an applicant's experience is based on what he[/she] actually does. Each case is evaluated on its own merits, and the emphasis is placed on the total engineering effort involved in planning, organizing, scheduling, and controlling projects. It may be helpful to keep a journal of job assignments and to file away each completed project or assignment description chronologically for future reference.

EXAMINATIONS

The NCEE's uniform examinations are divided into two sections: Fundamentals of Engineering (Part A) and Principles and Practice of Engineering (Part B). Each part consists of two four-hour sessions. All states offer these sixteen-hour written examinations.

The Principles and Practice of Engineering examinations in all disciplines are now based on a matrix that lists clusters of tasks conducted by significant numbers of licensed engineers in each discipline, plotted against specialized technical areas in which those engineers work. The number of questions in each subject is intended to conform with the various grids in the matrix. Basing its test specifications on this matrix, NCEE has developed examinations that emphasize those areas in which licensed engineers are most active. NCEE also gives due weight to those areas in which beginning engineers need to be knowledgeable, based on the responses to a task-analysis survey of licensed engineers conducted by the council in 1981.

FUNDAMENTALS OF ENGINEERING

The morning section of this part of the examination consists of 140 multiple-choice questions graded by machine; there are five possible responses to each question, only one of which is correct. The questions are intended to establish the candidate's retention of knowledge and ability to apply that knowledge to fundamental problems. Abstruse concepts, and those that are rarely applied, are not included. The examination is offered in both SI and non-SI units. Part A is an open-book exam. Candidates may use textbooks, handbooks, bound reference books and materials, and battery-operated silent calculators. No writing tablets or unbound notes are permitted in the examination room. Sufficient paper is provided in the test books for scratchwork, and candidates are not permitted to exchange reference materials or aids during the examination.

The afternoon section of the exam is also graded by machine and consists of sets of multiple-choice problems for which there are five choices each. Of 50 questions, 15 are in engineering mechanics, 15 in mathematics, 10 in electrical circuits, and 10 in engineering economics. In addition, there are 20 questions in two other subjects, chosen by the applicant from among computer programming, electronics and electrical machinery, fluid mechanics, mechanics of materials, and thermodynamics/heat transfer.

Both the morning and the afternoon sections of the examination must be taken to receive a score. Scores are computed from the number of questions answered correctly; there is no penalty for incorrect answers. Each section carries equal weight, and scores are on a scale from zero to 100.

PRINCIPLES AND PRACTICE

This part of the examination consists of a four-hour morning and a four-hour afternoon section. It was developed by practising engineers and is graded by hand by professional engineers. Its purpose is to determine whether the candidate is prepared to take charge of engineering projects. Textbook questions, proofs, derivations, and abstruse problems are avoided. Like Part A, this is an open-book exam, and the same rules apply regarding materials that can be brought into the examination room.

There are four problems in each section, and only the most general principles of engineering are required to solve them. Of primary importance is the application of good judgement to the selection and evaluation of pertinent information, and the ability to make reasonable assumptions. Merely routine numerical solutions do not justify a passing grade. Careful definition of the problem and the method of its solution are as important as the numerical answers themselves. Partial credit is given when the correct principles have been applied, even if the solution is incomplete or the final answer incorrect. For example, an applicant will receive partial credit if he[/she] ran out of time but still managed to fill in the terms of the equation with the numerical values and to indicate the unit and dimension of the final answer.

In 1983, NCEE adopted new test specifications for Part B in mechanical engineering, based on the task analysis of licensed engineers. There are now approximately eight problems in mechanical design, one in management, six in energy systems, one in control systems, three in thermal and fluid processes, and one in engineering economics. The problem in engineering economics is common to all disciplines.

Local boards of registration do not make past exams available to applicants. However, typical exams and study aids are available from NCEE headquarters.

SOURCE: John D. Constance, PE (Consulting Engineer, Cliffside Park, New Jersey), *Mechanical Engineering* (December 1986): 67–69. Reprinted with permission of *Mechanical Engineering*.

APPENDIX E

THE INSTITUTE OF ELECTRICAL AND ELECTRONICS ENGINEERS (IEEE) CODE OF ETHICS

We, the members of the IEEE, in recognition of the importance of our technologies in affecting the quality of life throughout the world, and in accepting a personal obligation to our profession, its members and the communities we serve, do hereby commit ourselves to the highest ethical and professional conduct and agree:

1. to accept responsibility in making engineering decisions consistent with the safety, health, and welfare of the public, and to disclose promptly factors that might endanger the public or the environment;
2. to avoid real or perceived conflicts of interest whenever possible, and to disclose them to affected parties when they do exist;
3. to be honest and realistic in stating claims or estimates based on available data;
4. to reject bribery in all its forms;
5. to improve the understanding of technology, its appropriate application, and potential consequences;
6. to maintain and improve our technical competence and to undertake technological tasks for others only if qualified by training or experience, or after full disclosure of pertinent limitations;
7. to seek, accept, and offer honest criticism of technical work, to acknowledge and correct errors, and to credit properly the contributions of others;
8. to treat fairly all persons regardless of such factors as race, religion, gender, disability, age, or national origin;
9. to avoid injuring others, their property, reputation, or employment by false or malicious action;

10. to assist colleagues and co-workers in their professional development and to support them in following this code of ethics.

SOURCE: © 1998 IEEE. Reprinted with permission from The Institute of Electrical and Electronics Engineers.

AMERICAN SOCIETY OF MECHANICAL ENGINEERS (ASME) CODE OF ETHICS

THE FUNDAMENTAL PRINCIPLES

Engineers uphold and advance the integrity, honour, and dignity of the Engineering profession by:

I. using their knowledge and skill for the enhancement of human welfare;
II. being honest and impartial, and serving with fidelity the public, their employers, and clients, and
III. striving to increase the competence of the engineering profession.

THE FUNDAMENTAL CANONS

1. Engineers shall hold paramount the safety, health and welfare of the public in the performance of their professional duties.
2. Engineers shall perform services only in the areas of their competence.
3. Engineers shall continue their professional development throughout their careers and shall provide opportunities for the professional development of those engineers under their supervision.
4. Engineers shall act in professional matters for each employer or client as faithful agents or trustees, and shall avoid conflict of interest.
5. Engineers shall build their professional reputation on the merit of their services and shall not compete unfairly with others.
6. Engineers shall associate only with reputable persons or organizations.
7. Engineers shall issue public statements only in an objective and truthful manner.

SOURCE: Reproduced with permission of ASME.

AMERICAN SOCIETY OF CIVIL ENGINEERS (ASCE) CODE OF ETHICS*

FUNDAMENTAL PRINCIPLES**

Engineers uphold and advance the integrity, honor and dignity of the engineering profession by:

1. using their knowledge and skill for the enhancement of human welfare and the environment;
2. being honest and impartial and serving with fidelity the public, their employers and clients;
3. striving to increase the competence and prestige of the engineering profession; and
4. supporting the professional and technical societies of their disciplines.

FUNDAMENTAL CANONS

1. Engineers shall hold paramount the safety, health and welfare of the public and shall strive to comply with the principles of sustainable development in the performance of their professional duties.
2. Engineers shall perform services only in areas of their competence.
3. Engineers shall issue public statements only in an objective and truthful manner.
4. Engineers shall act in professional matters for each employer or client as faithful agents or trustees, and shall avoid conflicts of interest.
5. Engineers shall build their professional reputation on the merit of their services and shall not compete unfairly with others.
6. Engineers shall act in such a manner as to uphold and enhance the honor, integrity, and dignity of the engineering profession.
7. Engineers shall continue their professional development throughout their careers, and shall provide opportunities for the professional development of those engineers under their supervision.

GUIDELINES TO PRACTICE UNDER THE FUNDAMENTAL CANONS OF ETHICS

CANON 1.
Engineers shall hold paramount the safety, health and welfare of the public and shall strive to comply with the principles of sustainable development in the performance of their professional duties.

a. Engineers shall recognize that the lives, safety, health and welfare of the general public are dependent upon engineering judgments, decisions and practices incorporated into structures, machines, products, processes and devices.
b. Engineers shall approve or seal only those design documents, reviewed or prepared by them, which are determined to be safe for public health and welfare in conformity with accepted engineering standards.
c. Engineers whose professional judgment is overruled under circumstances where the safety, health and welfare of the public are endangered, or the principles of sustainable development ignored, shall inform their clients or employers of the possible consequences.
d. Engineers who have knowledge or reason to believe that another person or firm may be in violation of any of the provisions of Canon 1 shall present such information to the proper authority in writing and shall co-operate with the proper authority in furnishing such further information or assistance as may be required.

    e. Engineers should seek opportunities to be of constructive service in civic affairs and work for the advancement of the safety, health and well-being of their communities, and the protection of the environment through the practice of sustainable development.

    f. Engineers should be committed to improving the environment by adherence to the principles of sustainable development so as to enhance the quality of life of the general public.

CANON 2.

Engineers shall perform services only in areas of their competence.

    a. Engineers shall undertake to perform engineering assignments only when qualified by education or experience in the technical field of engineering involved.

    b. Engineers may accept an assignment requiring education or experience outside of their own fields of competence, provided their services are restricted to those phases of the project in which they are qualified. All other phases of such project shall be performed by qualified associates, consultants, or employees.

    c. Engineers shall not affix their signatures or seals to any engineering plan or document dealing with subject matter in which they lack competence by virtue of education or experience or to any such plan or document not reviewed or prepared under their supervisory control.

CANON 3.

Engineers shall issue public statements only in an objective and truthful manner.

    a. Engineers should endeavor to extend the public knowledge of engineering and sustainable development, and shall not participate in the dissemination of untrue, unfair or exaggerated statements regarding engineering.

    b. Engineers shall be objective and truthful in professional reports, statements, or testimony. They shall include all relevant and pertinent information in such reports, statements, or testimony.

    c. Engineers, when serving as expert witnesses, shall express an engineering opinion only when it is founded upon adequate knowledge of the facts, upon a background of technical competence, and upon honest conviction.

    d. Engineers shall issue no statements, criticisms, or arguments on engineering matters which are inspired or paid for by interested parties, unless they indicate on whose behalf the statements are made.

    e. Engineers shall be dignified and modest in explaining their work and merit, and will avoid any act tending to promote their own interests at the expense of the integrity, honor and dignity of the profession.

CANON 4.

Engineers shall act in professional matters for each employer or client as faithful agents or trustees, and shall avoid conflicts of interest.

    a. Engineers shall avoid all known or potential conflicts of interest with their employers or clients and shall promptly inform their employers or clients of any business association, interests, or circumstances which could influence their judgment or the quality of their services.

b. Engineers shall not accept compensation from more than one party for services on the same project, or for services pertaining to the same project, unless the circumstances are fully disclosed to and agreed to, by all interest parties.

c. Engineers shall not solicit or accept gratuities, directly or indirectly, from contractors, their agents, or other parties dealing with their clients or employers in connection with work for which they are responsible.

d. Engineers in public service as members, advisors, or employees of a governmental body or department shall not participate in considerations or actions with respect to services solicited or provided by them or their organization in private or public engineering practice.

e. Engineers shall advise their employers or clients when, as a result of their studies, they believe a project will not be successful.

f. Engineers shall not use confidential information coming to them in the course of their assignments as a means of making personal profit if such action is adverse to the interests of their clients, employers or the public.

g. Engineers shall not accept professional employment outside of their regular work or interest without the knowledge of their employers.

CANON 5.
Engineers shall build their professional reputation on the merit of their services and shall not compete unfairly with others.

a. Engineers shall not give, solicit or receive either directly or indirectly, any political contribution, gratuity, or unlawful consideration in order to secure work, exclusive of securing salaried positions through employment agencies.

b. Engineers should negotiate contracts for professional services fairly and on the basis of demonstrated competence and qualifications for the type of professional service required.

c. Engineers may request, propose or accept professional commissions on a contingent bases only under circumstances in which their professional judgments would not be compromised.

d. Engineers shall not falsify or permit misrepresentation of their academic or professional qualifications or experience.

e. Engineers shall give proper credit for engineering work to those to whom credit is due, and shall recognize the proprietary interests of others. Whenever possible, they shall name the person or persons who may be responsible for designs, inventions, writings or other accomplishments.

f. Engineers may advertise professional services in a way that does not contain misleading language or is in any other manner derogatory to the dignity of the profession. Examples of permissible advertising are as follows:

Professional cards in recognized, dignified publications, and listings in rosters or directories published by responsible organizations, provided that the cards or listings are consistent in size and content and are in a section of the publication regularly devoted to such professional cards.

Brochures which factually describe experience, facilities, personnel and capacity to render service, providing they are not misleading with respect to the engineer's participation in projects described.

Display advertising in recognized dignified business and professional publications, providing it is factual and is not misleading with respect to the engineer's extent of participation in projects described.

A statement of the engineers' names or the name of the firm and statement of the type of service posted on projects for which they render services.

Preparation or authorization of descriptive articles for the lay or technical press, which are factual and dignified. Such articles shall not imply anything more than direct participation in the project described.

Permission by engineers for their names to be used in commercial advertisements, such as may be published by contractors, material suppliers, etc., only by means of a modest, dignified notation acknowledging the engineers' participation in the project described. Such permission shall not include public endorsement of proprietary products.

g. Engineers shall not maliciously or falsely, directly or indirectly, injure the professional reputation, prospects, practice or employment of another engineer or indiscriminately criticize another's work.

h. Engineers shall not use equipment, supplies, laboratory or office facilities of their employers to carry on outside private practice without the consent of their employers.

CANON 6.

Engineers shall act in such a manner as to uphold and enhance the honor, integrity, and dignity of the engineering profession.

a. Engineers shall not knowingly act in a manner which will be derogatory to the honor, integrity, or dignity of the engineering profession or knowingly engage in business or professional practices of a fraudulent, dishonest or unethical nature.

CANON 7.

Engineers shall continue their professional development throughout their careers, and shall provide opportunities for the professional development of those engineers under their supervision.

a. Engineers should keep current in their specialty fields by engaging in professional practice, participating in continuing education courses, reading in the technical literature, and attending professional meetings and seminars.

b. Engineers should encourage their engineering employees to become registered at the earliest possible date.

c. Engineers should encourage engineering employees to attend and present papers at professional and technical society meetings.

d. Engineers shall uphold the principle of mutually satisfying relationships between employers and employees with respect to terms of employment including professional grade descriptions, salary ranges, and fringe benefits.

* As adopted September 25, 1976 and amended October 25, 1980, April 17, 1993, and November 10, 1996.
**The American Society of Civil Engineers adopted THE FUNDAMENTAL PRINCIPLES of the ABET Code of Ethics of Engineers as accepted by the Accreditation Board for the Engineering and Technology, Inc. (ABET). (By ASCE Board of Direction action April 12–14, 1975).

NATIONAL SOCIETY OF PROFESSIONAL ENGINEERS (NSPE) CODE OF ETHICS

PREAMBLE

Engineering is an important and learned profession. The members of the profession recognize that their work has a direct and vital impact on the quality of life for all people. Accordingly, the services provided by engineers require honesty, impartiality, fairness and equity, and must be dedicated to the protection of the public health, safety and welfare. In the practice of their profession, engineers must perform under a standard of professional behaviour which requires adherence to the highest principles of ethical conduct on behalf of the public, clients, employers and the profession.

I. FUNDAMENTAL CANONS

Engineers, in the fulfilment of their professional duties, shall:

1. Hold paramount the safety, health and welfare of the public in the performance of their professional duties.
2. Perform services only in areas of their competence.
3. Issue public statements only in an objective and truthful manner.
4. Act in professional matters for each employer or client as faithful agents or trustees.
5. Avoid deceptive acts in the solicitation of professional employment.

II. RULES OF PRACTICE

1. Engineers shall hold paramount the safety, health and welfare of the public in the performance of their professional duties.
   a. Engineers shall at all times recognize that their primary obligation is to protect the safety, health, property and welfare of the public. If their professional judgement is overruled under circumstances where the safety, health, property or welfare of the public are endangered, they shall notify their employer or client and such other authority as may be appropriate.

b. Engineers shall approve only those engineering documents which are safe for public health, property and welfare in conformity with accepted standards.

c. Engineers shall not reveal facts, data or information obtained in a professional capacity without the prior consent of the client or employer except as authorized or required by law or this Code.

d. Engineers shall not permit the use of their name or firm name nor associate in business ventures with any person or firm which they have reason to believe is engaging in fraudulent or dishonest business or professional practices.

e. Engineers having knowledge of any alleged violation of this Code shall co-operate with the proper authorities in furnishing such information or assistance as may be required.

2. Engineers shall perform services only in the areas of their competence.

   a. Engineers shall undertake assignments only when qualified by education or experience in the specific fields involved.

   b. Engineers shall not affix their signatures to any plans or documents dealing with subject matter in which they lack competence, nor to any plan or document not prepared under their direction and control.

   c. Engineers may accept assignments and assume responsibility for co-ordination of an entire project and sign and seal the engineering documents for the entire project, provided that each technical segment is signed and sealed only by the qualified engineers who prepared the segment.

3. Engineers shall issue public statements only in an objective and truthful manner.

   a. Engineers shall be objective and truthful in professional reports, statements or testimony. They shall include all relevant and pertinent information in such reports, statements or testimony.

   b. Engineers may express publicly a professional opinion on technical subjects only when that opinion is founded upon adequate knowledge of the facts and competence in the subject matter.

   c. Engineers shall issue no statements, criticisms or arguments on technical matters which are inspired or paid for by interested parties, unless they have prefaced their comments by explicitly identifying the interested parties on whose behalf they are speaking, and by revealing the existence of any interest the engineers may have in the matters.

4. Engineers shall act in professional matters for each employer or client as faithful agents or trustees.

   a. Engineers shall disclose all known or potential conflicts of interest to their employers or clients by promptly informing them of any business association, interest, or other circumstances which could influence or appear to influence their judgement or the quality of their services.

   b. Engineers shall not accept compensation, financial or otherwise, from more than one party for services on the same project, or for services pertaining to the same project, unless the circumstances are fully disclosed to, and agreed to by, all interested parties.

   c. Engineers shall not solicit or accept financial or other valuable consideration, directly or indirectly, from contractors, their agents, or other parties in connection with work for employers or clients for which they are responsible.

d. Engineers in public service as members, advisors or employees of a governmental body or department shall not participate in decisions with respect to professional services solicited or provided by them or their organizations in private or public engineering practice.

e. Engineers shall not solicit or accept a professional contract from a governmental body on which a principal or officer of their organization serves as a member.

5. Engineers shall avoid deceptive acts in the solicitation of professional employment.

a. Engineers shall not falsify or permit misrepresentation of their, or their associates', academic or professional qualifications. They shall not misrepresent or exaggerate their degree of responsibility in or for the subject matter of prior assignments. Brochures or other presentations incident to the solicitation of employment shall not misrepresent pertinent facts concerning employers, employees, associates, joint ventures or past accomplishments with the intent and purpose of enhancing their qualifications and their work.

b. Engineers shall not offer, give, solicit or receive, either directly or indirectly, any political contribution in an amount intended to influence the award of a contract by public authority, or which may be reasonably construed by the public of having the effect or intent to influence the award of a contract. They shall not offer any gift or other valuable consideration in order to secure work. They shall not pay a commission, percentage or brokerage fee in order to secure work except to a bona fide employee or bona fide established commercial or marketing agencies retained by them.

III. PROFESSIONAL OBLIGATIONS

1. Engineers shall be guided in all their professional relations by the highest standards of integrity.

a. Engineers shall admit and accept their own errors when proven wrong and refrain from distorting or altering the facts in an attempt to justify their decisions.

b. Engineers shall advise their clients or employers when they believe a project will not be successful.

c. Engineers shall not accept outside employment to the detriment of their regular work or interest. Before accepting any outside employment, they will notify their employers.

d. Engineers shall not attempt to attract an engineer from another employer by false or misleading pretenses.

e. Engineers shall not actively participate in strikes, picket lines, or other collective coercive action.

f. Engineers shall avoid any act tending to promote their own interest at the expense of the dignity and integrity of the profession.

2. Engineers shall at all times strive to serve the public interest.

    a. Engineers shall seek opportunities to be of constructive service in civic affairs and work for the advancement of the safety, health and well-being of their community.

    b. Engineers shall not complete, sign, or seal plans and/or specifications that are not of a design safe to the public health and welfare and in conformity with accepted engineering standards. If the client or employer insists on such unprofessional conduct, they shall notify the proper authorities and withdraw from further service on the project.

    c. Engineers shall endeavour to extend public knowledge and appreciation of engineering and its achievements and to protect the engineering profession from misrepresentation and misunderstanding.

3. Engineers shall avoid all conduct or practice which is likely to discredit the profession or deceive the public.

    a. Engineers shall avoid the use of statements containing a material misrepresentation of fact or omitting a material fact necessary to keep statements from being misleading or intended or likely to create an unjustified expectation; statements containing prediction of future success; statements containing an opinion as to the quality of the Engineers' services; or statements intended or likely to attract clients by the use of showmanship, puffery, or self-laudation, including the use of slogans, jingles, or sensational language or format.

    b. Consistent with the foregoing, Engineers may advertise for recruitment of personnel.

    c. Consistent with the foregoing, Engineers may prepare articles for the lay or technical press, but such articles shall not imply credit to the author for work performed by others.

4. Engineers shall not disclose confidential information concerning the business affairs or technical processes of any present or former client or employer without his[/her] consent.

    a. Engineers in the employ of others shall not without the consent of all interest parties enter promotional efforts or negotiations for work or make arrangements for other employment as a principal or to practise in connection with a specific project for which the Engineer has gained particular and specialized knowledge.

    b. Engineers shall not, without the consent of all interested parties, participate in or represent an adversary interest in connection with a specific project or proceeding in which the Engineer has gained particular specialized knowledge on behalf of a former client or employer.

5. Engineers shall not be influenced in their professional duties by conflicting interests.

    a. Engineers shall not accept financial or other considerations, including free engineering designs, from material or equipment suppliers for specifying their product.

    b. Engineers shall not accept commissions or allowances, directly or indirectly, from contractors or other parties dealing with clients or employers of the Engineers in connection with work for which the Engineer is responsible.

6. Engineers shall uphold the principle of appropriate and adequate compensation for those engaged in engineering work.

   a. Engineers shall not accept remuneration from either an employee or employment agency for giving employment.

   b. Engineers, when employing other engineers, shall offer a salary according to professional qualifications.

7. Engineers shall not attempt to obtain employment or advancement or professional engagements by untruthfully criticizing other engineers, or by other improper or questionable methods.

   a. Engineers shall not request, propose, or accept a professional commission on a contingent basis under circumstances in which their professional judgement may be compromised.

   b. Engineers in salaried positions shall accept part-time engineer work only to the extent consistent with policies of the employer and in accordance with ethical consideration.

   c. Engineers shall not use equipment, supplies, laboratory, or office facilities of an employer to carry on outside private practice without consent.

8. Engineers shall not attempt to injure, maliciously or falsely, directly or indirectly, the professional reputation, prospects, practice or employment of other engineers, nor untruthfully criticize other engineers' work. Engineers who believe others are guilty of unethical or illegal practice shall present such information to the proper authority for action.

   a. Engineers in private practice shall not review the work of another engineer for the same client, except with the knowledge of such engineer, or unless the connection of such engineer with the work has been terminated.

   b. Engineers in governmental, industrial or educational employ are entitled to review and evaluate the work of other engineers when so required by their employment duties.

   c. Engineers in sales or industrial employ are entitled to make engineering comparisons of represented products with products of other suppliers.

9. Engineers shall accept responsibility for their professional activities; provided, however, that Engineers may seek indemnification for professional services arising out of their practice for other than gross negligence, where the Engineer's interests cannot otherwise be protected.

   a. Engineers shall conform with state registration laws in the practice of engineering.

   b. Engineers shall not use associations with a nonengineer, a corporation, or partnership, as a "cloak" for unethical acts, but must accept personal responsibility for all professional acts.

10. Engineers shall give due credit for engineering work to those to whom credit is due, and will recognize the proprietary interests of others.

   a. Engineers shall, whenever possible, name the person or persons who may be individually responsible for designs, inventions, writings, or other accomplishments.

   b. Engineers using designs supplied by a client recognize that the designs remain the property of the client and may not be duplicated by the Engineer for others without express permission.

   c. Engineers, before undertaking work for others in connection with which the Engineer may make improvements, plans, designs, inventions, or other records which may justify copyrights or patents, should enter into a positive agreement regarding ownership.

    d. Engineers' designs, data, records, and notes referring exclusively to an employer's work are the employer's property.

11. Engineers shall co-operate in extending the effectiveness of the profession by interchanging information and experience with other engineers and students, and will endeavour to provide opportunity for the professional development and advancement of engineers under their supervisions.

    a. Engineers shall encourage engineering employees' efforts to improve their education.

    b. Engineers shall encourage engineering employees to attend and present papers at professional and technical society meetings.

    c. Engineers shall urge engineering employees to become registered at the earliest possible date.

    d. Engineers shall assign a professional engineer duties of a nature to utilize full training and experience, insofar as possible, and delegate lesser functions to subprofessionals or to technicians.

    e. Engineers shall provide a prospective engineering employee with complete information on working conditions and proposed status of employment, and after employment will keep employees informed of any changes.

"By order of the United States District Court for the District of Columbia, former Section 11(0) of the NSPE Code of Ethics prohibiting competitive bidding, and all policy statements, opinions, rulings or other guidelines interpreting its scope, have been rescinded as unlawfully interfering with the legal rights of engineers, protected under the antitrust laws, to provide price information to prospective clients; accordingly, nothing contained in the NSPE Code of Ethics, policy statements, opinions, rulings or other guidelines prohibits the submission of price quotations or competitive bids for engineering services at any time or in any amount."

STATEMENTS OF NSPE EXECUTIVE COMMITTEE

In order to correct misunderstandings which have been indicated in some instances since the Issuance of the Supreme Court decision and the entry of the Final Judgement, it is noted that in its decision of April 25, 1978, the Supreme Court of the United States declared: "The Sherman Act does not require competitive bidding."

It is further noted that as made clear in the Supreme Court decision:

1. Engineers and firms may individually refuse to bid for engineering services.
2. Clients are not required to seek bids for engineering services.
3. Federal, state, and local laws governing procedures to procure engineering services are not affected, and remain in full force and effect.
4. State societies and local chapters are free to actively and aggressively seek legislation for professional selection and negotiation procedures by public agencies.
5. State registration board rules of professional conduct, including rules prohibiting competitive bidding for engineering services, are not affected and

remain in full force and effect. State registration boards with authority to adopt rules of professional conduct may adopt rules governing procedures to obtain engineering services.

6. As noted by the Supreme Court, "nothing in the judgement prevents NSPE and its members from attempting to influence governmental action. . . ."

NOTES

In regard to the question of application of the Code to corporations vis-à-vis real persons, business forms or type should not negate nor influence conformance of individuals to the Code. The Code deals with professional services, which services must be performed by real persons. Real persons in turn establish and implement policies within business structures. The Code is clearly written to apply to the Engineer and it is incumbent on a member of NSPE to endeavour to live up to its provisions. This applies to all pertinent sections of the Code.

SOURCE: NSPE Publication No. 1102, as revised January 1990. Reproduced with permission of NSPE.

APPENDIX F

NSPE Guidelines to Professional Employment for Engineers and Scientists

SUPPORTED AND ENDORSED BY:

American Association of Cost Engineers
American Association of Engineering Societies
American Council of Independent Laboratories
American Institute of Chemical Engineers
American Institute of Chemists
American Institute of Plant Engineers
American Microscopical Society
American Nuclear Society
American Society of Agricultural Engineers
American Society of Civil Engineers
American Society of Mechanical Engineers
American Society of Naval Engineers
Board of Certified Safety Professionals
Engineering Society of Detroit
Institute of Electrical and Electronics Engineers
Institute of Industrial Engineers
Institute of Transportation Engineers
Instrument Society of America
The Minerals, Metals & Materials Society
Mycological Society of America
National Council of Teachers of Mathematics
National Institute of Ceramic Engineers
National Society of Professional Engineers
Policy Studies Organization
Sigma Xi, The Scientific Research Society
Society for the Advancement of Material and Process Engineering
Society for Economic Botany

Society for Experimental Mechanics
Society of Fire Protection Engineers
Society of Packaging Professionals
Society of Women Engineers
System Safety Society
U.S. Metric Association

FOREWORD

The following Guidelines were developed for use by employers in evaluating their own practices, by professional employees in evaluating their own responsibilities and those of their employers, and by new graduates and other employment seekers in evaluating their prospective employment posture. They are intended to promote a satisfactory employer–employee working relationship.

Because of variations in individual circumstances and organization practices, it is inappropriate to consider evaluations on the basis of any single policy or benefit or on the basis of certain policies or benefits. Rather, attention should be focussed on evaluating the entire employment "package," including compensation (salary and other benefits) and such factors as opportunities for advancement, participation in profits, job location, and local cost of living.

These Guidelines should continue to be viewed as general goals rather than a set of minimum standards. Where practices do not fully measure up to the spirit of these Guidelines, it is recommended that employers initiate action for improvement; employers and employees should discuss situations and work together to minimize personnel problems, reduce misunderstandings, and generate greater mutual respect. Taking a constructive and flexible approach is essential to deal with individual circumstances and varying organization practices.

The Guidelines reflect the combined experience and judgement of many employers and professional employees. This Third Edition includes changes and additions incorporated to improve clarity and to reflect current practice trends and legal implications.

Questions regarding interpretation of these Guidelines should be referred to the headquarters office of any of the endorsing societies.

OBJECTIVES

The endorsing societies, with their avowed purpose to serve the public and the professions they represent, recognize that professional employees and employers must establish a climate conducive to the proper discharge of their mutual responsibilities and obligations. Prerequisites for establishing such a climate include:

1. Developing a sound relationship between the professional employee and the employer, based on ethical practices, co-operation, mutual respect, and fair treatment.
2. Recognizing employers' and employees' responsibility to safeguard the public's health, safety, and welfare.

3. Encouraging employee professionalism and creativity to support the employer's objectives.
4. Providing employees the opportunity for professional growth, based on employee initiative and employer support.
5. Recognizing that discrimination based on age, race, religion, political affiliation, or sex must not enter the professional employee–employer relationship. Employers and employees should jointly accept the concepts reflected in the "Equal Employment Opportunity" regulations.
6. Recognizing that local conditions may result in differences in the interpretation of, and deviations from, details of these Guidelines. Such differences should be discussed to gain a mutual understanding that meets the spirit of these Guidelines.

I. RECRUITMENT

Hiring should be based on a professional's competence and ability to meet specific job requirements. Employee qualifications and employment opportunities should be represented in a factual and forthright manner. An employer's employment offer and a prospective employee's acceptance of the offer should be in writing. Agreements between employers or between an employer and a professional employee that limit the opportunity for professional employees to seek other employment or, subsequent to separation, establish independent enterprises, are contrary to the spirit of these Guidelines.

PROFESSIONAL EMPLOYEE

1. Prospective employees (applicants) should attend interviews and accept reimbursement only for those job opportunities in which they have a sincere interest. Applicants should charge the prospective employer(s) for no more than the expenses incurred for the interview(s).
2. Applicants should carefully evaluate past, present, and future confidentiality obligations regarding trade secrets and proprietary information connected with potential employment. They should not seek or accept employment on the basis of using or divulging any trade secrets or proprietary information. All applicants should be aware of their legal rights and obligations in this regard.
3. Having accepted an employment offer, applicants are ethically obligated to honour the commitment unless they are formally released after giving adequate notice of intent.
4. Applicants should not use a current employer's funds or time to seek new employment unless approved by the current employer.

EMPLOYER

1. Employers should make clear their policy on paying expenses incurred by the applicant for attending an arranged interview prior to the interview. Potential employers should reimburse all legitimate expenses incurred by an applicant when the employer requests the interview.

2. Applicants should be interviewed by the prospective employer and, if possible, by the prospective supervisor, so the applicant will understand clearly the technical and business nature of the job opportunity.
3. Ethically, prospective employers should be responsible for all representations regarding the conditions of employment.
4. Employment applications should be kept confidential. Prospective employers should seek the expressed consent from applicants before contacting a current employer.
5. Employers should minimize hiring during periods of major reductions of personnel.
6. Hiring professional employees should be planned, wherever possible, to provide satisfying careers.
7. An employer's written offer of employment should state all relevant terms, including salary, relocation assistance, expected type and duration of employment, and patent obligations. Prospective employees should be informed of any documents requiring signature.
8. Having accepted an applicant, an employer who finds it necessary to rescind an offer of employment should make equitable compensation to the applicant for any resulting monetary loss.

II. EMPLOYMENT

Terms of employment should be in accordance with applicable laws and be consistent with generally accepted ethical and professional practices. These terms should be based on mutual respect between employer and employee.

PROFESSIONAL EMPLOYEE

1. Professional employees should accept only those assignments for which they are qualified; should diligently, competently, and honestly complete assignments; and should contribute creative, resourceful ideas to the employer while making a positive contribution toward establishing a stimulating work atmosphere and maintaining a safe working environment.
2. Professional employees should have due regard for the health, safety, and welfare of the public and fellow employees in all work for which they assume responsibility. When the technical adequacy of a process or product is unsatisfactory, professional employees should withold approval of the plans and should state the reasons for such action. If an employee's professional judgement is ignored or overruled under circumstances where public safety, health, property or welfare is endangered, the employee should first formally notify the employer and then, if necessary, notify such other authority as may be appropriate.
3. Professional employees should sign or seal only plans or specifications they prepared, or personally reviewed and satisfactorily checked, or those prepared by employees under their direct supervision.
4. Professional employees are responsible for the effective use of time in the employer's interest and the proper care of the employer's facilities.
5. Professional employees should avoid conflicts of interest with their employers, and should immediately disclose any actual or potential conflicts.

6. Professional employees should co-operate fully with their employers in obtaining patent protection for inventions.
7. Professional employees should not divulge proprietary information.
8. Professional employees should not accept any payments or gifts of significant value, directly or indirectly, from parties dealing with a client or employer.
9. Professional employees should act in a manner consistent with their profession's code of ethics.

EMPLOYER

1. Employers should keep professional employees informed of the organization's objectives, policies, and programs.
2. Employers should provide professional employees with salary and other benefits commensurate with a professional's contribution, taking into account the employee's abilities, professional status, responsibilities, education, experience, and the potential value of the work to be performed.
3. Employers should establish a salary policy that takes into account current salary surveys for professional employees. The salary established should be commensurate with those for other professional and nonprofessional employees within the organization. The salary structure should be reviewed periodically with respect to the current economy.
4. Each individual position should be properly classified in the overall salary structure. The evaluation of each position should consider such factors as skills required for acceptable performance, the original thinking required for solving problems involved, and the accountability for actions and their consequences.
5. Duties, levels of responsibility, and the relationship of positions within the organization should be defined.
6. Employers should restrict use of titles denoting professional engineering or scientific status to those employees qualified by graduation from an appropriate baccalaureate program, or by professional licensure. [Professional licence is essential in Canada.] Appropriate titles and career patterns not denoting professional status should be developed for other categories of employees such as those holding associate degrees in engineering technology.
7. Economics advancement should be based on merit. Provision should be made for accelerated promotion or extra compensation for superior performance or special accomplishments, including generation of significant proprietary information, patents, or inventions. Compensation should be evaluated at least annually.
8. Employers should encourage [a] continuing dialogue with professional employees emphasizing the relationship between current activities and potential future activities in support of organizational goals. This may be accomplished through regular performance evaluations. Professional employees should be informed when their performance is unsatisfactory and should be advised of steps required for improvement. This information should be documented and a copy should be provided to the employee.
9. Employers should consider an equivalent ladder for compensation and advancement of professional employees whose aptitudes and interests are technical rather than managerial.

10. It is inappropriate for professional employees to use a time clock to record arrival and departure.

11. If the work demanded of professional employees regularly exceeds the normal working hours for extended periods, the employer should provide extra compensation for this continuing extra effort according to the employer's clearly stated policy.

12. Employers should also provide such benefits as pensions, life insurance, health insurance (including coverage of catastrophic illness and long-term disability), sick leave, vacations, holidays, and savings or profit-sharing plans consistent with current industrial practices. To the extent such benefits are not provided, equivalent additional compensation should be provided.

13. Employers should provide a pension plan for employees who meet minimum participation standards. Based on a full career, the minimum employer-sponsored pension benefit at retirement should be no less than 50 percent of the average best five years' salary. Employer-sponsored pension plans should provide for early participation and vesting, full portability, and survivor benefits. Consideration should be given to periodic increases in pension benefits relating to increases in the cost of living. Pension benefits should not be integrated with Social Security. The fund that supports the plan should not be terminated until all obligations to vested employees and retirees have been met. Tax-sheltered savings plans should be available to provide incentives for individual investment for retirement.

14. Employers should provide support staff and physical facilities that promote the maximum personal effectiveness, health, and safety of their professional employees.

15. Employers should not require professional employees to accept responsibility for work not performed or supervised by those employees.

16. Employers should have established policies for reviewing all items that involve public safety, health, property, and welfare that are brought to their attention by a professional employee. The results of this review should be reported to the employee in a timely manner and opportunity should be given for further input by the employee. Employers should not penalize employees for invoking these policies.

17. Employers should defend any suits and indemnify claims against present or former individual professional employees in connection with their authorized activities on behalf of the employer.

18. There should be no employer policy that requires or forbids a professional employee to join a labour organization as a condition of continued employment.

19. Employers should clearly identify proprietary information and should release employees' inventions and other information that is not useful to the employer.

20. Employers should not discriminate on the basis of national origin, ethnicity, age, race, religion, political affiliation, or sex, with regard to compensation, job assignment, promotion, or other matters.

21. In the event of transfer, employers should allow adequate time for transferring employees to settle personal matters before moving. All normal moving costs of transfers should be paid by the employer, including household moving expenses, realtor fees, travel expenses to the new location to search for housing, and reasonable living expenses for the families until perma-

nent housing is found. Unusual moving expense reimbursement should be settled in a discussion between the employee and employer.

III. PROFESSIONAL DEVELOPMENT

Both professional employees and their employers have responsibilities for professional development — the employee to establish goals and take the initiative to reach them and the employer to provide a supportive environment and appropriately challenging job assignments.

PROFESSIONAL EMPLOYEE

1. Professional employees should maintain technical competence through continuing education programs and by broadening experience.
2. Professional employees should belong to, and participate in, the activities of appropriate professional societies in order to obtain additional knowledge and experience. Such participation should include preparing professional and technical papers for publication and presentation.
3. Professional employees should achieve appropriate registration and/or certification as soon as they are eligible.
4. Professional employees should participate in public service activities, including civic and political activities of a technical and nontechnical nature. If such participation interferes with the timely execution of work, employees should seek the agreement of their employers.

EMPLOYER

1. Employers should encourage their employees to maintain technical competence and broaden experience, for example, through appropriate work assignments of a rotational nature, and support of continuing education by self-improvement, courses in-house and at institutions of higher learning, and meetings and seminars on appropriate subjects. They should also encourage and support employees' membership and participation in professional society activities.
2. Employers should consider compensated leaves of absence for professional studies that will improve competence and knowledge.
3. Consistent with employer objectives, employees should be given every opportunity to publish work promptly and to present findings at technical society meetings.
4. Employers should encourage and assist professional employees to achieve registration and/or certification in their respective fields.

IV. TERMINATION

Adequate notice of termination of employment should be given by the employee or employer as appropriate.

PROFESSIONAL EMPLOYEE

1. When professional employees decide to terminate employment, they should provide sufficient notice to enable the employer to maintain a continuity of function. When termination is initiated by the employee, no severance pay is due.
2. Upon termination, professional employees should maintain all proprietary information as confidential.

EMPLOYER

1. Employers should inform employees in a personal interview of the specific reasons for termination.
2. Additional notice of termination, or compensation in lieu thereof, should be provided by employers in consideration of responsibilities and length of service. Employees should receive notice or equivalent compensation equal to one month, plus at least one week per year of service. In the event that the employer elects notice in place of severance compensation, then the employer should allow the employee reasonable time and facilities to seek new employment.
3. Employers should make every effort to relocate terminated professional employees either within their own organizations or elsewhere. Provision should be made to continue major employee protection plans for a reasonable period following termination, and to reinstate them fully in the event of subsequent reemployment.
4. If employers seek to encourage employees to retire, employers should do so without using coercion, and solely by means of offering an adequate financial incentive.
5. Employers should seek agreement with employees on the amount of compensation to be paid for the employees' assistance in obtaining patent protection or in patent litigation on behalf of the employer.

SOURCE: National Society of Professional Engineers, *Guidelines to Professional Employment for Engineers and Scientists*, 3rd ed. (Alexandria, VA, 31 October 1989). Reproduced with permission of NSPE.

Index